Forensic Science Laboratory Benchmarking

Forensic Science Laboratory Benchmarking: The FORESIGHT Manual takes a step-by-step instructional approach to utilizing FORESIGHT data, detailing how labs can participate in the process to improve efficiencies. The FORESIGHT Project—a business benchmarking process for forensic service providers—was created in 2008 to collect and report data while offering improvement to processes through analysis, comparisons, and best practice evaluations. The program has grown to include more than 200 participating forensic laboratories worldwide.

FORESIGHT offers the capability for labs to improve core functions, provide and benefit from metrics, and thus, improve the labs capabilities and functioning for the public good, while maintaining their often limited, fixed budgets. Due to ever-increasing caseloads, forensic laboratories are constantly plagued by backlogged casework—cases submitted to the laboratory but not yet worked. This leads to inefficiencies, delays, and unhappy agencies expecting timely results. Unfortunately, even if a lab's slates were wiped clean and the backlog were erased, many of the inefficient processes—that created the backlog—would still be in place. Eventually, and inevitably, the lab would develop a new backlog.

Unique coverage and features:

- Presents critical and proven cutting-edge measures to utilize FORESIGHT data improve laboratory testing, operational efficiency, and policies without added additional costs.
- Synthesizes the data input from more than 200 labs and a decade's worth of analytics to illustrate process improvements and the advantages of participating.
- Outlines how to develop data-driven responses to solve current and future problems.

Forensic Science Laboratory Benchmarking will be of interest to quality assurance specialists, economists, supervisors in the parent agencies of the labs, managers at all levels of any of the hundreds of public laboratories around the world, and anyone concerned about the effectiveness and efficiency of laboratory testing. As an operational guide, the book provides a helpful roadmap to help public science agencies and forensic labs analyze how they operate, improve on what works, and change what doesn't to better meet their mission and serve their community's goals.

Max M. Houck, PhD, FRSC is an international expert in the forensic sciences. Dr. Houck has nearly 30 years of expertise in casework, research, teaching, and writing. His casework includes the Branch Davidian Investigation, the September 11 attacks on the Pentagon, the D.B. Cooper case, and the West Memphis Three case, among hundreds of others. His committee work includes the White House, the National Academies of Science, the Royal Society, and Interpol. Dr. Houck is one of the most published professionals in his field. He is a founding Editor of the journal *Forensic Science Policy and Management* and has also co-authored numerous books, including the best-selling textbooks, *Fundamentals of Forensic Science*, *Science of Crime Scenes*, and *Successful Expert Testimony*. Dr. Houck is a Fellow of the Royal Society of Chemistry.

Paul J. Speaker, PhD, is a Finance Professor in the John Chambers College of Business & Economics at West Virginia University. He earned his MS Economics and PhD from Purdue University and a BA in Economics from LaSalle College. Dr. Speaker also holds the position of Chief Executive Officer of Forensic Science Management Consultants LLC, a firm which specializes in the business of forensics using the forensics of business. He serves as a member of NIJ's Forensic Laboratory Needs Technology Working Group. Dr. Speaker is the Principal Investigator for Project FORESIGHT, a business-guided analysis of Forensic Science Laboratories. His research activity is concentrated in economic modeling, including the role of public sector and not-for-profit institutions. These research projects include numerous studies in the business of forensics, including estimation of the return on investment from forensic laboratory analyses.

Forensic Science Laboratory Benchmarking

The FORESIGHT Manual

Max M. Houck and Paul J. Speaker

CRC Press
Taylor & Francis Group
Boca Raton London New York

CRC Press is an imprint of the
Taylor & Francis Group, an **informa** business

Designed cover image: Gorodenkoff / Shutterstock

First edition published 2024
by CRC Press
2385 NW Executive Center Drive, Suite 320, Boca Raton FL 33431

and by CRC Press
4 Park Square, Milton Park, Abingdon, Oxon, OX14 4RN

CRC Press is an imprint of Taylor & Francis Group, LLC

Library of Congress Cataloging-in-Publication Data
Names: Houck, Max M., author.
Title: Forensic science laboratory benchmarking: the foresight manual /
Max M. Houck, Ph.D., FRSC and Paul J. Speaker, Ph.D.
Description: First edition. | Boca Raton, FL: CRC Press, 2024. |
Includes bibliographical references and index.
Identifiers: LCCN 2023044565 (print) | LCCN 2023044566 (ebook) |
ISBN 9780367251581 (hardback) | ISBN 9781032691909 (paperback) |
ISBN 9780429318528 (ebook)
Subjects: LCSH: Crime laboratories—Management. |
Benchmarking (Management) | Forensic sciences—Laboratory manuals.
Classification: LCC HV8073 .H765 2024 (print) |
LCC HV8073 (ebook) | DDC 363.25—dc23/eng/20240112
LC record available at https://lccn.loc.gov/2023044565
LC ebook record available at https://lccn.loc.gov/2023044566

ISBN: 9780367251581 (hbk)
ISBN: 9781032691909 (pbk)
ISBN: 9780429318528 (ebk)

DOI: 10.4324/9780429318528

Typeset in Minion Pro
by codeMantra

Contents

Preface

This book is based on dozens of peer-reviewed articles and many workshops about FORESIGHT. It is a unique project that has literally gathered more data about how forensic service providers operate than has ever existed before. Additionally, it has provided a data set that actually shows some key economic concepts that aren't normally seen due to market forces. Hundreds of forensic agencies use FORESIGHT to improve their operations; FORESIGHT is the *de facto* standard process for understanding the state of the global forensic industry.

Many of the core peer-reviewed articles appeared in the journal *Forensic Science Policy & Management* (Disclosure: One of the authors, Houck, was a co-editor of that journal). Thanks to the graciousness of Taylor & Francis, the publisher of that now-defunct journal, the articles relating to FORESIGHT published in *Forensic Science Policy & Management* appear in this book. Rather than re-writing and re-hashing all of the content from those many papers, this book presents a less formal introduction to each topic covered in the articles and then provides the actual articles for those who want to go deeper into the literature. This approach makes the book like a workshop you hold in your hand, rather than an academic textbook you need to slog through before the final exam.

Project FORESIGHT has been one of the most fascinating, exciting, and difficult research arcs of our careers. We hope this book, this manual, helps you engage more fully with the process and brings the insights you need to understand and improve how your forensic organization works.

Acknowledgments

We'd like to thank, in no particular order, the following:

- The National Institute of Justice, for the original awards that made FORESIGHT possible.
- The forensic directors and managers who helped us refine and define the foundations that FORESIGHT is built on.
- The WVU business faculty who helped us put FORESIGHT on a solid intellectual footing.
- Jay Siegel, who co-founded *Forensic Science Policy & Management* and saw the need for our community to not only talk about forensic science but what we do with that science.
- Taylor & Francis, for returning the publication rights so that many of the original FORESIGHT articles could be included in this book.
- The hundreds of forensic service providers across the world that participate in FORESIGHT and strive to make themselves better in the name of justice.

Why You Should Read This Book

1

You want your forensic organization to do better. Resources are limited, case submissions are growing, and you are under pressure. Trying random or fad management approaches will not work. You need to understand your forensic organization as a *production system*: Your group uses money, instrumentation, and expertise to produce reports and testimony. They do this for the police, the court system, and the public.

- Unless you understand what your organization does and how to measure it, you will always be guessing about what might improve things.
- You were educated as a scientist, not as a business manager but the jobs are not that different: Data, analysis, results, interpretation.
- Change the variables from alleles, molecules, and minutiae to costs, full-time equivalents (FTEs, that is, employees), and ratios and you can work out how you function, where the problems might be, and come up with ways to address them.

This book will take you from guessing about what to do to making evidence-based strategies and methods to make your forensic organization cheaper, faster, and more resilient.

This Is <u>Not</u> an Academic Book

It is based on research, however, about 12 years of it. In 2008, the authors and faculty from West Virginia University's College of Business (now the John Chambers College of Business), along with guidance from forensic leadership, created a formal approach to define forensic operations, measure them, and analyze them to keep what works, change what doesn't, and improve their operations. That approach, called Project FORESIGHT, is now the *de facto* standard for the forensic industry.

- "Industry?" you may ask. Yes, industry. An industry is a distinct group of related productive or profit-making enterprises that produce a particular kind of goods or services based on their primary business activities.
- Laboratories, crime scene units, police services, medical examiners, death investigators, and other forensic service providers all work in the same technical domain, producing reports and testimony as outputs. Forensic organizations have all of the characteristics of a publicly traded company with one exception: Profit.
- Therefore, **a forensic organization, generally called a *forensic service provider or FSP*, is part of the forensic industry**.
- By the way, **participation in FORESIGHT is free**; more on how to join later.

DOI: 10.4324/9780429318528-1

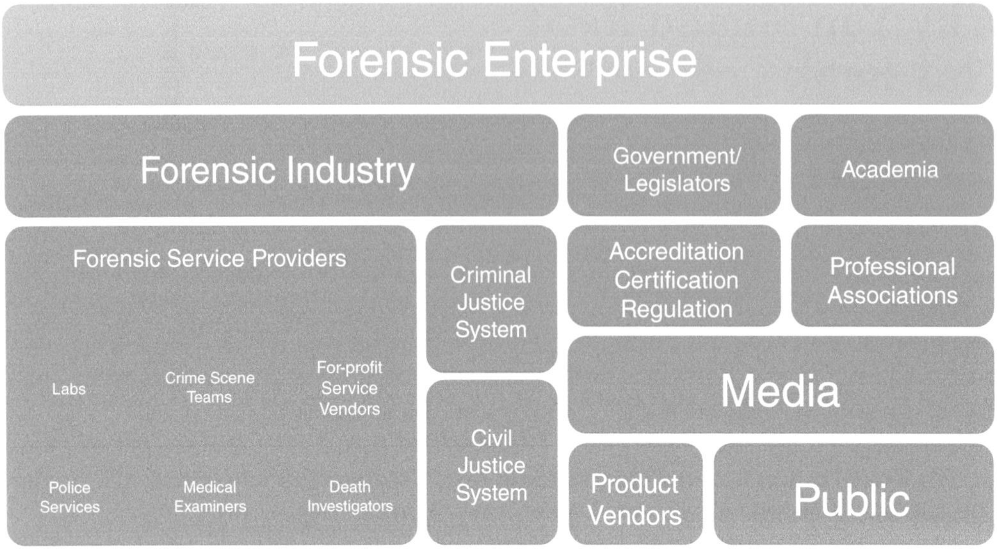

Figure 1.1 The forensic enterprise (From Tontarski et al., 2012) An **enterprise** is a defined group of all of the organizations, their qualities, and inter-relationships in a given sector. An enterprise includes all the people, information, technology, and operations or processes involved. Think of a large network with looser and tighter connections bound by an idea or topic, like "transistors," "health care," or "forensic."

What is a "Forensic Service Provider"?

The governmental units (and the vast majority are governmental) that provide forensic services have many forms and structures.

- They may be under a policing agency, a prosecutorial office, a medical examiner, or a more general unit of government.
- Not all forensic service providers are strictly laboratories and potentially thousands of them exist across the roughly 18,000 jurisdictions in the U.S. These include city, county, state, and federal branches of governments.
- **In this book, all forensic organizations that provide some type of forensic service, regardless of their type, will be called *forensic service providers (FSPs)* to keep it simple.**
- The number and variety of jurisdictions mean that a universal language is needed for different FSPs to talk to each other about their business processes and solutions; **FORESIGHT provides that common business language**.

This Is a How-to Manual

This book walks through the three-dozen or so academic papers related to Project FORESIGHT and provides clear, practical, and relevant explanations and examples to help you apply this system to your FSP.

- Each paper or group of papers on a topic will be explained. The heading **Why This Matters** will give you the "so what" points to make sure that the relevance to your FSP is clear.
- Examples using data from two made-up FSPs, City Lab and State Lab, "show the work" and give you pointers on how to apply FORESIGHT to your FSP.
- The academic papers are included so that you can "go deeper" and dig into the details should you choose.

Why This Matters

Hundreds of forensic FSPs across the world use FORESIGHT to make improvements, gain resources, and create efficiencies. For example,

- A state forensic FSP was losing employees to a federal forensic FSP within commuting distance; this cost the state FSP time and money finding and training new employees. The state laboratory director used FORESIGHT to make an analysis of the differences in pay and convinced his legislature to approve a 20% pay increase, for all employees. Think the employees stuck around? You bet.
- A county FSP that also served a large city was found to be over-testing on postmortem toxicology even though their costs were below average. When it suggested that they could reduce their testing intensity to save money, the laboratory countered that they were over-testing because of their constituency: An aging population served by multiple nationally ranked hospitals. Without a unified health records system, the doctors and specialists would often prescribe medications that unknowingly affected other medications. The laboratory found that a large number of deaths due to heart attacks were actually due to misaligned medications. They were over-testing **on purpose** to better serve their citizens. What an amazing way to support budgets and future requests!

By using this book along with your FORESIGHT data, you will be able to know more about how your FSP works, where the problems may be, and how to develop data-driven responses to solve current and future problems.

Go Deeper

Perrow, C., 1986. *Complex organizations. A critical essay* (3rd ed.). Glenview, IL: Scott-Forcs.

Tontarski Jr, R.E., Houck, M.M., Grose, W.P. and Gialamas, D.M., 2012. Alternative models promote self-regulation of the forensic enterprise. *Forensic Science Policy & Management: An International Journal*, 3(3), pp. 139–150.

Article: Survey of Forensic Service Providers

1.1

RANDALL A. CHILDS[1], TOM S. WITT[1], AND KANYBEK NURTEGIN[2]

[1]*West Virginia University, Bureau of Business and Economic Research, PO Box 6025, Morgantown, WV 26506-6025, USA*

[2]*Florida Atlantic University, Economics, 5353 Parkside Drive, Jupiter, FL 33458, USA*

Summary results from the Survey of Forensic Identification Professionals (sent to IAI members in the United States.) include:

- 79.4% of units serve local city or county jurisdictions
- More than two-thirds of units have ten or fewer employees
- Over three-fourths of units conduct latent print identification
- Units serving city (or city-equivalent) jurisdictions spend 88% of their budget on personnel, leaving only 7.2% ($40,600) for equipment and supplies
- Units serving city jurisdictions have budgets of $77,000 per FTE compared to $150,000 for units serving county jurisdictions
- Units serving city jurisdictions received 214 cases per FTE compared to 226 for units serving county jurisdictions
- Units serving states report backlogs of four months.

Summary results for the Forensic Awareness Survey (sentto every police chief and sheriff in the United States) include:

- Average employment of departments is 120
- 60% of departments report having one or more employees that work directly on forensic services
- 85% of departments report that they conduct crime scene investigations
- Departments with forensic services have an average of 8.6 employees that work directly on forensic services.

[Supplementary materials are available for this article. Go to the publisher's online edition of Forensic Science and Policy Management to view the free supplemental file(s).]

DOI: 10.4324/9780429318528-2

These results provide the first look at forensic service providers outside of the Census of Publicly Funded Crime Labs from the Bureau of Justice Statistics.[2] As such, the research indicates the larger scope of forensic services undertaken in the United States.

Introduction

Mention the word "forensics" and many of us think of the popular television crime-drama programs where crimes occur and are solved, including the "forensics," all in one hour. Unfortunately, the reality is that forensic analysis takes much longer and, with recent improvements in forensic science, the demand for these services has increased to the point that significant backlogs exist in many disciplines.

The issue of backlogs at the nation's publicly funded forensic laboratories has received considerable attention, especially with regards to DNA analysis. There have been numerous surveys of forensic laboratories over the years, including two recent studies[3] from the Bureau of Justice Statistics, the Census of Publicly Funded Crime Laboratories, 2002 and the Census of Publicly Funded Crime Laboratories, 2005, that have quantified and benchmarked the forensic analysis that takes place within the nations publicly funded forensic laboratories. In addition, research studies have specifically targeted DNA analysis.

While the existing research has provided much-needed information on forensic analysis that takes place within the forensic laboratories, there has been a lack of research that quantifies and benchmarks other forensic services. According to the International Association for Identification's (IAI) 180 Day Study Final Report, two-thirds of fingerprint identification occurs outside the traditional forensic laboratory setting. The study also suggests that backlogs are prevalent in many of the identification units. The current study, funded through the West Virginia University Forensic Science Initiative by a grant from the National Institute of Justice and conducted in collaboration with the IAI, examines forensic identification that is conducted outside of forensic laboratories. Data on forensic identification was gathered through two surveys. The first survey, a detailed survey about the organization's jurisdiction, employment, functions performed, caseload, backlog, budget, and outsourcing, was sent to IAI members in the United States The second survey, a brief survey, was sent to every police chief and sheriff in the United States An overview and summary of the results for each survey is provided below.

Forensic Identification Surveys

Although studies examining forensic science within forensic laboratories exist, little is known about other forensic services, which are often performed outside the traditional forensic laboratory setting. In an effort to quantify and benchmark this forensic identification activity, two surveys were conducted, both sponsored by a grant (#2001-RC-CX-K003) from the National Institute of Justice. Copies of the two surveys, Survey of Forensic Identification Professionals and Forensic Awareness Survey, are available on this journal's Web site.

The Survey of Forensic Identification Professionals, was conducted in collaboration with the International Association for Identification (IAI), one of the largest forensic professional organizations. IAI's membership includes members from throughout the United States and around the world. Use of the IAI member mailing list, for active United States members, provided an ideal sample of forensic identification professionals, not only geographically, but also across many forensic disciplines, including: crime scene, latent print, photography, tenprint, bloodstain pattern, digital evidence, footwear, tire track, firearms, tool mark, questioned documents, forensic art, polygraph, and odontology.

The Forensic Awareness Survey was distributed to every police chief and sheriff throughout the United States. The Forensic Awareness Survey was designed as a brief survey that would provide a snapshot of these local departments regarding their employment and role in forensic analysis.

Survey of Forensic Identification Professionals

Using active United States resident IAI members, the Survey of Forensic Identification Professionals was distributed to 5,353 forensic identification professionals. Surveys from 815 individuals (15.2 percent) were returned. In order to complement the 2002 and 2005 Census of Publicly Funded Crime Labs, members were asked if they worked "at a laboratory that is a member of the American Society of Crime Laboratory Directors (ASCLD)." Surveys from these members were not included in the analysis because the goal of the project was to quantify and benchmark forensic identification being conducted outside of the forensic laboratories (and thus not included in the 2002 and 2005 Censuses). In addition, respondents were excluded from the final results if they were retired or if more than one survey was completed for their organization. The final completed surveys used in this analysis represent 300 organizations that perform some level of forensic service.

Survey respondents were asked to provide information regarding their organization's jurisdiction, employment, functions performed, caseload, backlog, budget, outsourcing, and more. Several methods were utilized to increase the survey response rate, including:

- A cover letter from the IAI president was mailed with each initial survey;
- Researchers attended meetings of forensic service professionals and made presentations on the goals and benefits of the survey;
- Articles were published in the IAI newsletter and a prominent forensic magazine;
- A follow-up postcard was mailed reminding members to complete the survey;
- Members returning a completed survey were included in two drawings for an iPod nano;
- Reminder emails were sent to members with a valid e-mail address;
- Surveys could be returned via fax or mailed back using an enclosed postage-paid envelope (or scanned and returned via e-mail); and
- Multiple copies of the survey were provided by either a mailed hard copy and/or an e-mailed Adobe PDF file.

Table 1 Jurisdiction served, Survey of Forensic Identification Professionals, IAI members

Jurisdiction Served	Percent of Organizations
City, Borough, Village, or Town	50.9%
County	28.5
State	10.0
Federal/National	3.8
Other (Private, International, etc.)	6.9

Table 2 Total full-time and part-time employment, Survey of Forensic Identification Professionals, IAI members

Employment	Percent of Organizations	Percent of Total Employment
1–10 Employees	68.3%	11.5%
11–50 Employees	27.1	23.9
51–100 Employees	2.8	12.9
Greater Than 100 Employees	1.8	51.7

Table 3 Sworn versus civilian units, Survey of Forensic Identification Professionals, IAI members

	Percent of Organizations
100% Sworn Employees	25.5%
80%–99% Sworn Employees	7.8
80%–99% Civilian Employees	9.8
100% Civilian Employees	42.4

Results

The following tables summarize the results of the survey. Over one-half of all respondents serve local city (or city-equivalent) jurisdictions (Table 1). Nearly 80 percent of the forensic identification units serve sub-state areas (city or county).

The forensic identificat in units are generally small units with 68.3 percent of organizations consisting of ten or fewer employees (Table 2). While more than 90 percent of the forensic identification units have fifty or fewer employees, they account for just over one-third (35.4 percent) of total employment. Over 50 percent of the total employment comes from organizations with more than 100 employees.

The forensic identification units may include both sworn and civilian employees, but these make up less than one-third of all units (Table 3). Forensic identification units are much more likely to be completely staffed by civilian employees (42.4 percent of all units) or completely staffed by sworn employees (25.5 percent of all units).

Forensic identification units perform numerous forensic functions. Crime scene investigation and latent print identification/AFIS are most common with 78.5 percent of all units conducting crime scene investigation and 76.5 percent conducting latent print identification (Table 4). Less than one in five units conduct firearms/tool mark examinations, as staffing

Table 4 Forensic functions performed, Survey of Forensic Identification Professionals, IAI members

Forensic Function	Percent of Organizations
Crime Scene	78.5%
Latent Print	76.5
Photography	72.0
Tenprint	66.6
Bloodstain	40.6
Digital Evidence	34.1
Footwear	31.4
Firearms	18.1
Forensic Art	12.6
Questioned Documents	12.3
Polygraph	7.2
Odontology	1.4

of these positions has become challenging with many of the existing analysts nearing retirement and with a shortage of qualified firearms analysts available.

Budgets of forensic identification units vary widely. Not surprisingly, forensic service units that serve larger jurisdictions are likely to require, and indeed do have, larger average total budgets. What is potentially alarming is that the forensic service units that serve the city jurisdictions (accounting for more than one-half of all units) spend nearly 88 percent of their budget on personnel (wages, salaries, and benefits), while units at the county and state

Respondents that provided detailed budget, number of cases, and full-time equivalent (FTE) employment are included. Averages for units serving state jurisdictions are not included due to small sample size.

level spend 49.6 percent and 42.6 percent, respectively. This leaves forensic service units at the city-level with little money left over for purchasing equipment and supplies ($40,600 on average). This is compared to equipment and supplies budgets of $333,400 for county-level forensic service units and $1,349,000 for state-level forensic service units.

While the forensic service units that serve cities have little money for equipment and supplies compared to units with county or state jurisdictions, the units with city jurisdiction have nearly as many cases per full-time employee (FTE) and almost double the number of cases per $1,000 of budget (Table 6). The budget per FTE of units serving city jurisdictions falls well short of budget per FTE of units serving county jurisdictions. Units serving city jurisdictions have budgeted $77,233 per FTE compared to $149,765 for units serving county jurisdictions. On the other hand, units serving county jurisdictions averaged 3,508 cases, more than double that of units serving city jurisdictions that averaged 1,542 cases.

The following figures plot the individual units using cases per FTE on the vertical axis and budget per FTE of the unit on the horizontal axis. Figure 1 provides all units with respect to the average cases per FTE and budget per FTE. As the figure shows, relatively few units with very large cases per FTE or budgets per FTE can influence the averages. This is most evident with regard to the budget per FTE.

Table 5 Budget by jurisdiction served, Survey of Forensic Identification Professionals, IAI members

		Percent of Total Budget		
Jurisdiction Served	Average Total Budget	Personnel (including Fringe)	Equipment And Supplies	Other
City, Village, or Town	$561,000	87.9%	7.2%	4.9%
County	$1,741,000	49.6	19.1	31.3
State	$8,436,000	42.6	16.0	41.4

Table 6 Cases and budget per FTE, Survey of Forensic Identification Professionals, IAI members

Average Unit	City, Village, or Town	County
Number of Cases	1,542	3,508
Cases per FTE	214	226
Budget per FTE	$77,233	$149,765
Cases per $1,000 Budget	2.8	1.5

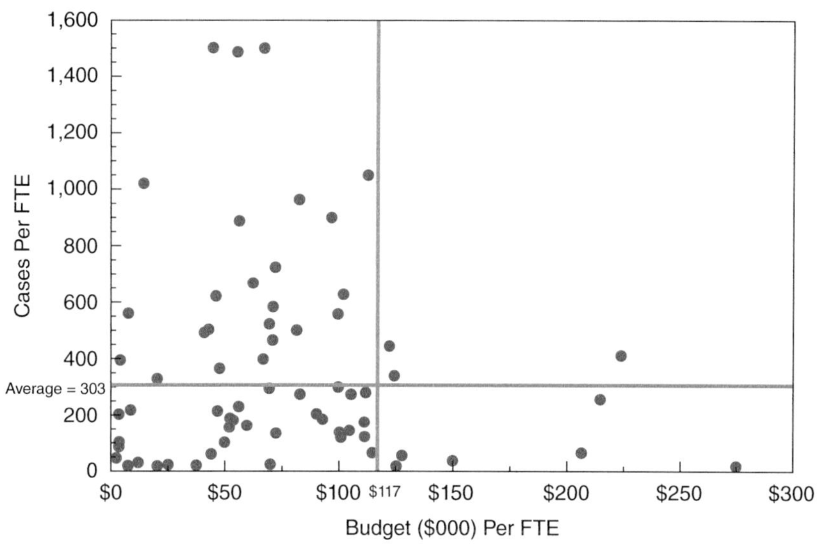

Figure 1 Cases and budget per FTE, Survey of Forensic Identification Professionals, IAI Members.

While economies of scale may suggest that the larger units would have an efficiency advantage over the smaller units, as measured by cases and budget per FTE, the above figures do not support this suggestion. A closer examination of the individual units may provide the answer. For example, are the larger units analyzing the more difficult cases with more forensic requests per case or with forensic requests that take more time to complete?

Table 7 provides information regarding the backlog of cases that exist in units by type of jurisdiction they serve. A case was to be considered backlogged if it remained

Table 7 Backlogged cases, Survey of Forensic Identification Professionals, IAI members

Jurisdiction Served	Percent of Respondents That Provided # of Cases	Percent of Respondents with # of Cases That Provided # of Backlog	Backlog as a Percent of Cases Received in 2006	Backlog in Weeks
City, Village, or Town	75.7%	29.5%	12.3%	6.4
County	73.5	31.1	7.3	3.8
State	65.5	36.8	31.2	16.2

unreported for a period of 30 days or more. The number of backlogged cases was reported as of December 31, 2006.

Approximately three-fourths of units that serve local city or county jurisdictions provided the number of cases received by their unit in 2006, compared to two-thirds of units that serve a state jurisdiction. Those units providing information on the number of cases and the number of backlogged cases were used to compute the average backlogs presented in the table. Most notably, forensic service units that provide services for a state jurisdiction average significantly longer backlogs (31.2 percent of annual cases received). The backlog in these units is equivalent to four months in terms of annual cases received. Units serving local jurisdictions (city or county) also had significant numbers of backlogged cases, albeit much less than units serving entire states. Units serving a city reported an average backlog of 6.4 weeks and units serving counties reported backlogs of nearly 1 month. Given the nature of many of the smaller units and their budget concerns, it is not surprising to see the longer backlogs at the state-level units since many of the local-level units rely on the state-level units for analysis they are not equipped to handle.

Additionally, respondents were asked about outsourcing and quality assurance. Fifty percent of units reported outsourcing, with tenprint, latent print, and firearms/tool mark examinations being the most frequent requests outsourced. Nearly 15 percent of units also outsource peer review for verification. Many units value accreditation with 13 percent already accredited and another one-third are currently considering accreditation. Two-thirds of units have a system to verify results and one-third having LIMS or other system.

Figure 2, plotting the same units by case and budget per FTE, shows the mean (average) and median (middle value) for each measure. Using the median cases per FTE and median budget per FTE provides a more meaningful way of looking at these individual units. Thus the remaining figures are shown with the median values only.

The earlier tables provided a look into the differences of units based on the jurisdiction they serve. While this is useful from a descriptive standpoint, it may be useful from a policy view to examine the units based on a different measure because some units with city jurisdiction may actually be serving a larger area than some of the units that are serving a county jurisdiction. The following figures take a look at units of varying employment size, plotting the cases per FTE against the budget per FTE as in Figure 1 and 2.

Figure 3 shows the units with 1–5 employees (plotted as black triangles). These very small units account for the vast majority of units with low cases and budget per FTE. These units are also likely to have above the median number of cases per FTE, regardless of budget per FTE, while they are not likely to have a high (relative to the median) budget per FTE with a low number of cases per FTE.

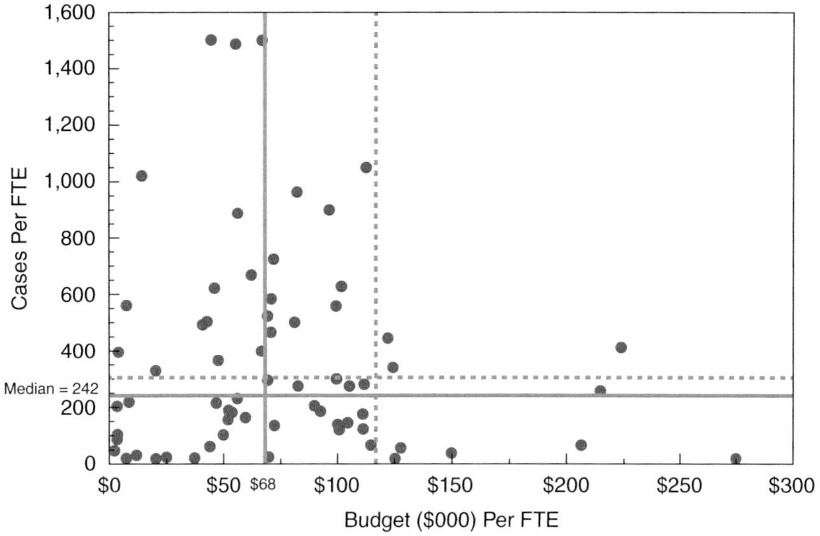

Figure 2 Cases and budget per FTE — mean vs. median, Survey of Forensic Identification Professionals, IAI members.

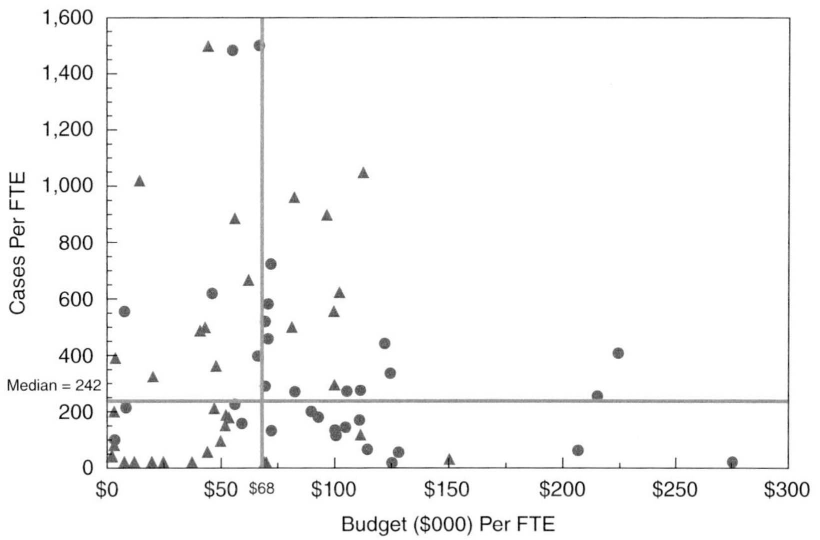

Figure 3 Cases and budget per FTE — 1–5 employees, Survey of Forensic Identification Professionals, IAI members.

Figure 4 shows the units with 6–10 employees (plotted as solid black triangles). These units tend to have above the median budget per FTE, while they may be above or below the median cases per FTE. Units with 11–20 employees (plotted as solid black triangles in Figure 5), on the other hand, are just as likely to be in all four quadrants: low cases-low budget; low cases-high budget; high cases-low budget; high cases-high budget. Finally, Figure 6 shows the units with more than 20 employees (plotted as black triangles). These units, almost exclusively, have above the median budget per FTE and are just above or below the median number of cases per FTE.

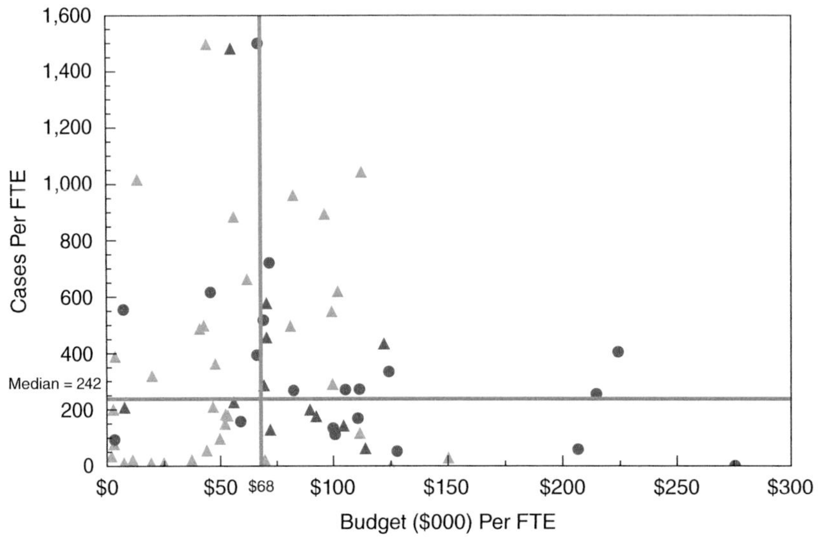

Figure 4 Cases and budget per FTE—6– 10 employees, Survey of Forensic Identification Professionals, IAI members.

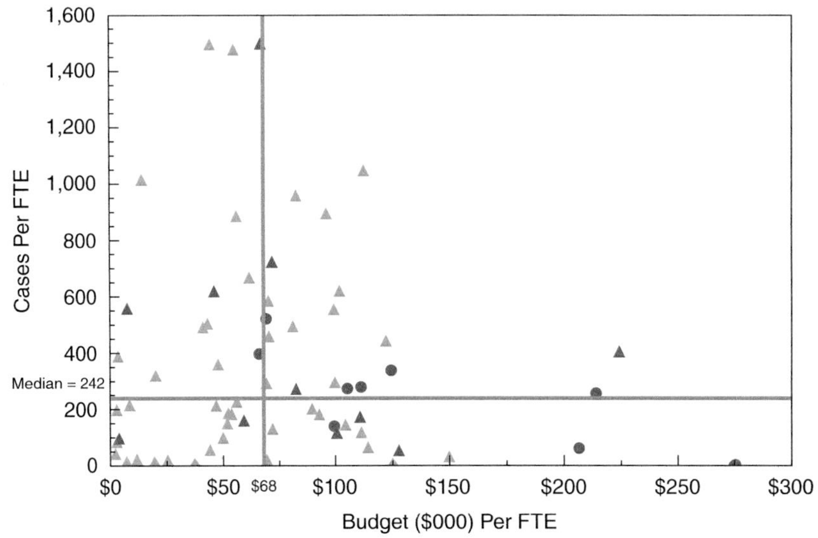

Figure 5 Cases and budget per FTE — 11–20 employees, Survey of Forensic Identification Professionals, IAI members.

Forensic Awareness Survey

The Forensic Awareness Survey was mailed to every United States police chief and sheriff. In all, 15,918 surveys were mailed. Surveys were completed for 1,219 departments. The Forensic Awareness Survey was intended to identify how many of the local city police and county sheriff departments had personnel to work on forensic services and what types of forensic functions they performed.

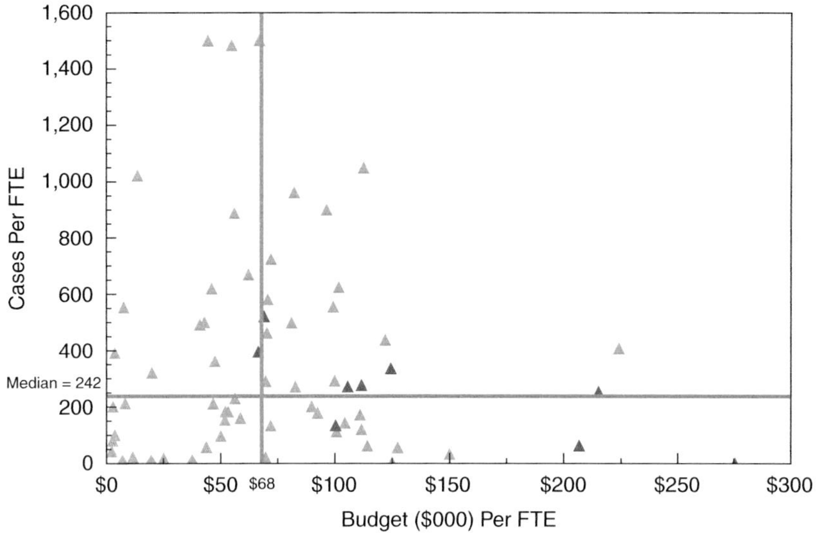

Figure 6 Cases and budget per FTE — 21+ employees, Survey of Forensic Identification Professionals, IAI members.

This survey was sent to every department, regardless of whether or not they had an employee who was a member of IAI. For this reason, these results stand on their own and must be used with caution when comparing to the limited sample of units covered by the detailed IAI survey. Survey respondents were asked to provide information regarding their organization's jurisdiction, total employment, FTE employment, forensic employment, and forensic functions performed. Due to the number of departments included in the survey, the survey instrument was mailed only once and no reminder post cards or e-mails were sent to the departments. In order to increase the response rate, the survey could be returned without postage and a version of the survey was provided online. Only 16.4 percent of respondents chose to fill out the online version of the survey. The remaining 83.6 percent returned the survey via the United States Postal Service.

Results

The following tables summarize the results of the survey. The city police and sheriff departments averaged 120 full-time and part-time employees (Table 8). These departments had, on average, 4.3 percent of their employees, who worked directly on forensic services. Only 59 percent of departments reported having one or more employees work directly on forensic services. These departments had an average of 8.6 employees, who worked directly on forensic services (median = 7.5).

Survey respondents were also asked what forensic functions they performed. These results are provided in Table 9. While less than 60 percent of departments reported having at least one employee, who worked directly on forensic services, nearly 85 percent of departments listed crime scene investigation as a forensic function they performed. This indicates that many of these departments do not consider crime scene work as forensic services.

Table 8 Selected characteristics of city police and sheriff departments, Forensic Awareness Survey

City Police and Sheriff Departments	
Average Number of Employees	120
Average Number of FTE Employees	101
Percent of Employees that Work Directly on Forensics	4.3%

Table 9 Forensic functions performed by city police and sheriff departments Forensic Awareness Survey

Forensic Function	Percent of City Police and Sheriff Departments
Crime Scene	84.7%
Photography	44.9
Footwear	34.0
Latent Print	33.9
Digital Evidence	25.3
Bloodstain	22.6
Tenprint	21.3
Firearms	18.3
Polygraph	17.6
Questioned Documents	11.7
Forensic Art	6.5
Odontology	1.0

The other forensic functions, possibly more generally accepted as "forensic services," all fall below the 60 percent threshold. While several of the forensic functions provided from the Forensic Awareness Survey are similar to those reported from IAI members, others are quite different. Most notable is latent print identification where IAI members reported 76.5 percent of units performed this function compared with 33.9 percent for this survey.

Conclusion

While previous government-funded studies have focused primarily on forensic science that takes place in the laboratory setting, recent studies have suggested that a significant amount of forensic analysis takes place in law enforcement agencies across the United States. This report summarizes the findings from two surveys aimed at these "non-laboratory" forensic service providers.

The first survey, Survey of Forensic Identification Professionals, was sent to members of the International Association for Identification and asked detailed questions about the organization's jurisdiction, employment, functions performed, caseload, backlog, budget, and outsourcing. The second survey, Forensic Awareness Survey, was sent to every police chief and sheriff in the United States. The second survey asked only for the organization's total employment, employees working on forensics, and the forensic functions performed by the organization.

The results from these surveys suggest that forensic services are widely performed outside of the traditional laboratory setting. In fact, 85 percent of these local (city and county) jurisdictions conduct crime scene investigations with 60 percent of the departments having one or more employees that work directly on forensic services.

The results also suggest a significant difference in funding across organizations. Units serving city (or city-equivalent) jurisdictions spend 88 percent of their budget on personnel, leaving only 7.2 percent ($40,600) for equipment and supplies, while units serving county jurisdictions spend less than 50 percent of their budgets on personnel. The results also indicate a difference in funding per full-time equivalent employee (FTE) with units serving city jurisdictions reporting budgets of $77,000 per FTE compared to $150,000 per FTE for units serving county jurisdictions.

These results provide the first look at forensic service providers outside of the Census of Publicly Funded Crime Labs from the Bureau of Justice Statistics. As such the research indicates the larger scope of forensic services undertaken in the United States and provides a snapshot of these service providers that can be used as a benchmark for future research.

Endnotes

1 The National Institute of Justice's report, "Status and Needs of the Forensic Science Service Providers: A Report to Congress," suggests that two-thirds of fingerprint identification is performed outside of forensic laboratories. A copy of this report is available at http://www.ncjrs.gov/pdffiles1/nij/213420.pdf.

2 The *Census of Publicly Funded Forensic Crime Laboratories, 2005* is available online at http://www.ojp.usdoj.gov/bjs/pub/pdf/cpffc105.pdf.

3 *Census of Publicly Funded Forensic Crime Laboratories, 2002*, February 2005, is available online at <http://www.ojp.usdoj.gov/bjs/pub/pdf/cpffc102.pdf> *Census of Publicly Funded Forensic Crime Laboratories, 2005*, July 2008, is available online at <http://www.ojp.usdoj.gov/bjs/pub/pdf/cpffc105.pdf>

Article: Alternative Models Promote Self-Regulation of the Forensic Enterprise[1]

1.2

TONTARSKI, R.E. JR.[2], GROSE, W.P.[3], HOUCK, M.M.[4], AND GIALAMAS, D.M.[5]

"Every problem interacts with other problems and is therefore part of a set of interrelated problems, a system of problems...I choose to call such a system a mess." (Ackoff, 1974)

Introduction

There is little fundamental disagreement about the areas of forensic science needing improvement and standardization when forensic services are considered globally (NAS, 2009). Studies conducted periodically over the last twenty-five years by government-funded working groups, committees, and organizations, have identified similar concerns about shortcomings in forensic science, from fundamental education to consistent standardization in analysis methods to continuing need for research. The National Academies of Science National Research Council's 2009 report, *Strengthening Forensic Science in the United States: A Path Forward* (NAS, 2009), is only the latest study to echo these needs and enjoys the greatest current visibility.

As the NAS Report reminds us, the forensic science has no overarching structure or unifying entity able to provide direction, guidance, or oversight to the profession. The solution proposed by the NAS Report is the creation of a federally operated regulatory body, the National Institute of Forensic Science:

> "To promote the development of forensic science into a mature field of multidisciplinary research and practice, founded on the systematic collection and analysis of relevant data, Congress should establish and appropriate funds for an independent federal entity, the National Institute of Forensic Science (NIFS). NIFS should have a full-time administrator and an advisory board with expertise in research and education, the forensic science disciplines, physical and life sciences, forensic pathology, engineering, information technology, measurements and standards, testing and evaluation, law, national security, and public policy."

The NAS Report then outlines the areas the NIFS should focus on:

- "establishing and enforcing best practices for forensic science professionals and laboratories;
- establishing standards for the mandatory accreditation of forensic science laboratories and the mandatory certification of forensic scientists and medical examiners/

DOI: 10.4324/9780429318528-3

forensic pathologists—and identifying the entity/entities that will develop and implement accreditation and certification;
- promoting scholarly, competitive peer-reviewed research and technical development in the forensic science disciplines and forensic medicine;
- developing a strategy to improve forensic science research and educational programs, including forensic pathology;
- establishing a strategy, based on accurate data on the forensic science community, for the efficient allocation of available funds to give strong support to forensic methodologies and practices in addition to DNA analysis;
- funding state and local forensic science agencies, independent research projects, and educational programs as recommended in this report, with conditions that aim to advance the credibility and reliability of the forensic science disciplines;
- overseeing education standards and the accreditation of forensic science programs in colleges and universities;
- developing programs to improve understanding of the forensic science disciplines and their limitations within legal systems;
- and assessing the development and introduction of new technologies in forensic investigations, including a comparison of new technologies with former ones." NAS, 2009)

Two legislative initiatives have been since introduced in Congress. The first initiative by Senator Leahy[6] and the second by Senator Rockefeller[7]. Each proposes creating a federal entity with far reaching scope and authorities to regulate forensic science and drive research and education requirements. There are high costs associated with the creation of such entities with no funding authorized. Neither proposal has gotten significant support from Congress.

Federally legislated regulatory models are not the only viable model to fill this gap. Many laudable organizations originated to improve distinct aspects of forensic science, including accreditation, certification, sharing professional information and research results, and educational curricula. All have contributed significantly to advancing the profession. A lot of the work is voluntary, because most of the organizations have little to no professional staff, or their staff function is limited to operations that focus on the organization. These aggregate organizations represent the main professional aspects of the forensic enterprise and their activities need to shape the forensic profession. What is lacking is a strategic architecture and processes to coordinate and synchronize these aspects, and more, of the forensic science profession. Without such a strategic approach, professional progress will continue to lag behind expectations and need.

A National Forensic Science Board (NFSB) could coordinate forensic programs and activities nationwide. Forensic sciences across the U.S. already have many of the resources in place to lead and significantly advance all aspects the profession. What are lacking are consistent coordinated efforts. The creation of an NFSB would integrate and synchronize the on-going efforts leveraging the subject matter expertise that has been developed by each of the recognized forensic professional organizations. The NFSB would ensure that all the stakeholders, including legal community, victim advocates, wrongfully accused, community needs, laboratory management and individual examiners, have a defined place where they can go to be heard. With non-regulatory oversight, a collaborative and holistic approach can be developed to gain an accurate understanding of capabilities, resource gaps, and professional requirements.

Defining the Forensic Enterprise

The jurisdictional patchwork quilt of the US criminal justice system creates enormous inefficiencies in the provision of forensic services. Over 18,000 federal, state, and local law enforcement agencies exist in the US with control focussed at the local level; by comparison, the UK, which is one quarter the size of the US, has 43 police departments (Walker and Katz, 2002). These myriad roles and responsibilities create a fragmented, decentralized system. The downstream effect of this is the hodgepodge delivery of forensic services, some of which overlap significantly either in services, geography, or both. In a 2008 US Bureau of Justice Statistics report, 254 public forensic laboratories were recognized (Durose, 2008); moreover, a 2009 survey identified an additional 300 forensic service providers (Childs and Witt, 2009). As an example, consider the Dallas-Fort Worth metroplex, encompassing 12 counties with a 2010 population of nearly 6.4 million people. The City of Fort Worth has a police department forensic laboratory; the surrounding county (Tarrant) has a medical examiner's office that also has a full-service forensic laboratory. Likewise, the City of Dallas police department offers forensic services as does the Dallas county medical examiner's office. In between theses jurisdictions, the city of Arlington, TX police department also has forensic service provision for its citizens. Overlaid on all of this is the statewide Texas Department of Public Safety's Forensic Laboratories, in addition to numerous private forensic laboratories. The various law enforcement agencies (city, county, state) throughout the metroplex have numerous opportunities for forensic service provision, some of which are contractual (both public-public and public-private). Further recognizing that federal agencies may have forensic service provision in the area (like a Drug Enforcement Administration laboratory in the Dallas area) who may offer assistance to local agencies pro bono, the forensic landscape becomes complicated, to say the least.

This patchwork jurisdictional situation creates an environment of "extreme disaggregation" which retards or even precludes a more effective and coordinated approach:

> The forensic science enterprise [is] marked by multiple types of practitioners with different levels of education and training and different professional cultures and standards for performance and a reliance on apprentice-type training and a guild- like structure of disciplines, which work against the goal of a single forensic science profession...The fragmented nature of the enterprise raises the worrisome prospect that the quality of evidence presented in court, and its interpretation, can vary unpredictably according to jurisdiction." (NAS, 2009, pages 16-17).

As troubling--and accurate--as this statement is, it begs the question, "Has the forensic enterprise been defined?" The fragmented nature of forensic science provision, as noted above, has also hindered a more global view of the scope of the profession, how its components articulate, and the role of each in the wider pursuit of its endeavors. An enterprise can be defined as the totality of business entities, the externally visible properties of those

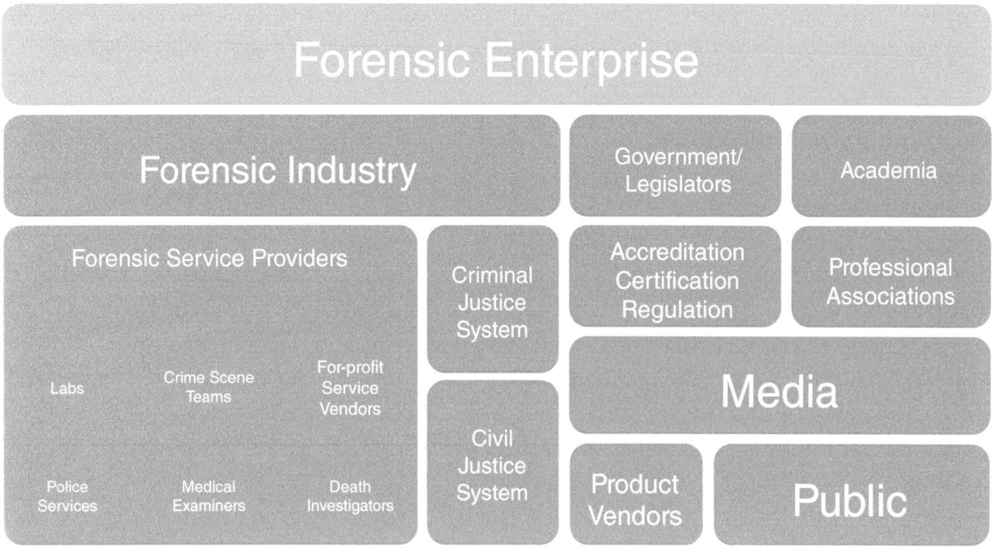

Figure 1 The definition of an enterprise is the business entities, the externally visible properties of those components, and the relationships between them; this includes people, information, technology, and business (operations or process).

components, and the relationships between them; this includes people, information, technology, and business operations or process (Giachetti, 2010). The term may seem broad and encompassing; this is intentional as it allows it to be used widely in many circumstances. Defining an enterprise is an important first step in addressing the goals of effectiveness, efficiency, agility, and durability while providing a conceptual architecture that permits current action ("as is") and future strategic planning ("to be") (Figure 1); this step is part of the formal establishment of a concept of operations, or CONOPS.

For the current purposes, an industry constitutes the aggregate of manufacturing or technically productive entities in a particular field, often named after its principal product: The forensic industry, for example. This definition would include all forensic service providers (a term coined by Childs, et. al (2009)), any organization or agency that provides one or more forensic service to a customer or client. This term is inclusive of laboratories, crime scene units, police service units (fingerprinting or so-called "ID sections"), medical examiners, death investigators, and any private provisioner (typically, for profit businesses) of a forensic service. Forensic service providers are thus the "manufacturing centers" of the forensic industry.

In this scheme, forensic professionals function as knowledge workers. "Knowledge worker" is a term coined by Peter Drucker in 1959 to describe the then-rising group of workers whose jobs required extensive education, the application of theoretical and analytical knowledge, and continuous learning (Drucker, 1995). Knowledge work, as defined by Drucker, "is not experience-based as all manual work has always been. It is learning-based." (1995, page 227). Knowledge work has high entry costs:

> Knowledge work and most of services work, in their work characteristics, are nontraditional. Displaced industrial workers thus cannot simply move into knowledge work or services work

> the way displaced farmers and displaced domestic workers moved into industrial work. At the very least they have to make a major change in their basic attitudes, values, and beliefs. (1995, page 227)

Specialization, not generalization, is what makes knowledge useful and the more specialized knowledge is, the more useful it becomes. Increased specialization does not imply that the knowledge will become more "applied," however, as many knowledge workers with highly specialized knowledge conduct very basic research, as with high-energy particle physics. With specialization comes two concomitant features of knowledge workers: They operate in teams and they have to have access to an organization. Teams balance out the necessary specialization for knowledge to be applied properly and the organization provides the basic continuity that allows the knowledge workers specialization to be converted into performance. For forensic scientists, the collection of investigators, laboratory colleagues, and the larger organization constitutes one or more participatory teams and their parent agency provides the explicit organization. Note that the teams and organizations are not necessarily limited to the worker's employer and may extend across jurisdictional, functional, or professional boundaries. Forensic knowledge workers, then, take raw materials (evidence) and convert them into products and information (reports and testimony) while meeting specific industry standards (quality and accreditation) and production levels (efficiency) within resource constraints (effectiveness).

The remainder of the forensic industry includes the criminal and civil justice systems which, along with law enforcement agencies, are direct consumers of forensic services and products. Investigators and attorneys can also function as part of the team or organization within which forensic knowledge workers operate.

The other elements in the forensic enterprise, such as the media or academia, play solid roles in defining much of how the enterprise is structured, operates, and evolves. One could consider the elements offered in Figure 1 as the totality of stakeholders in the forensic enterprise, moving from internal on the left-hand side to more external on the right. Other elements could be included or the offered elements parsed differently; in fact, many elements will have multiple functions or identities depending on the situation.

The boundaries created by the aforementioned resource constraints are part of another key aspect in defining the forensic enterprise. Money, time, and personnel are all resources in achieving goals and, of necessity, are limited. Choices have to be made on how best to effectively allocate those resources to meet one or more goals. The concept of "trade space," the set of program and system parameters, attributes, and characteristics required to satisfy performance standards, is integral to any operation. Within any level of an enterprise, inclusive, decision makers define and refine the developing system by making tradeoffs with regard to cost, schedule, risk, and performance, all of which fall within the systems trade space. Work on defining and analyzing the forensic trade space at the laboratory level has begun (Houck, Riley et al., 2009; Speaker, 2009; Speaker, 2009) and can be eventually extended to the industry level.

The final concept of importance in the current work is that of policy, which is a plan or course of action, as of a government, political party, or business, intended to influence and determine decisions, actions, and other matters. The interaction of policy and science, in particular, can be difficult, in that both systems use information to come to conclusions but with different perspectives; policy is overtly political while science is (or consistently attempts to be) objectively data-driven. A useful matrix for understanding this relationship can be found in Stine, 2009 and is presented below with forensic examples

	Policy ***influencing*** science and technology	Science and technology ***informing*** policy
Science	**Policy for Science** "Should the US federal government tie research funding to accredited forensic educational programs?"	**Science for Policy** "Should microfluidics be used for forensic DNA casework?"
Technology	**Policy for Technology** "Should the emerging field of familial searching be supported and regulated?"	**Technology for Policy** "Should policy actions be taken on AFIS interoperability?"

Figure 2 The relationship between science and policy as outlined by Stine, 2009 with forensic examples offered by the authors.

(Figure 2). Science and technology can inform policy while policy can influence science and technology.

Science and policy applied to solve the challenge of the provision of forensic services is problematic, in that a purely rational solution cannot be found because "the problem" lacks clear definition and each stakeholder has differing perspective of what "the problem" is and how to "solve it," ignoring that multiple solutions probably exist (Conklin, 2005). Social policy solutions sought via purely scientific or rational means will not succeed because of the fundamental nature of the problems:

> "Policy problems cannot be definitively described. Moreover, in a pluralistic society there is nothing like the indisputable public good; there is no objective definition of equity; policies that respond to social problems cannot be meaningfully correct or false; and it makes no sense to talk about 'optimal solutions' to these problems...Even worse, there are no solutions in the sense of definitive answers." (Rittel and Webber, 1984)

The issue of policy to govern, oversee, or regulate forensic science thus may constitute a "wicked problem" which is a problem that is difficult to solve because of incomplete, contradictory, and changing requirements (Rittel and Webber, 1984; Conklin, 2005). Wicked problems share the following traits:

1. Until a solution is devised, the problem is not understood.
2. Wicked problems have no stopping rule.
3. Solutions to wicked problems are "more" or "less", not "right" or "wrong".
4. Every wicked problem is essentially novel and unique.
5. Solutions to wicked problems are rarely if ever systemic.
6. No alternative solutions present themselves.
7. The solution depends on how the problem is framed and vice-versa,
8. Each stakeholders approaches the problem from a different viewpoint.
9. Problem constraints and resources to solve the problem change.
10. The problem is never solved definitively.

The remainder of this article will address Item 6 in the above list, alternative solutions to the oversight of the forensic enterprise. The alternative models that will be discussed include self-regulation, governmental intra-agency (using the NIST Office of Weights and Measures as an example), governmental independent (such as the National Transportation Safety Board), and non-profit independent (like the National Fire Protection Association), with examples.

Self-regulation

Calls for an industry or an enterprise often start with "self regulation," as can be gleaned from any newspaper article. A self-regulatory organization is one that has some level of authority over an industry or a profession, either alone or in addition to some other type of regulation, such as governmental, like licensing. Such an organization does not get its authority or mandate from government; rather it derives its control from the constituents of the relevant group it oversees. Furthermore, self-regulating organizations will provide resources to constituents and stakeholders, hold meetings, promulgate policies (such as professional ethical standards), and provide public information to those interested in the industry.

The American Medical Association is an example of a typical professional self-regulating organization and one that works over one or more layers of regional and federal government. Another is the National Association of Realtors, which establishes and oversees the rules and policies for the multiple listing services (the data bases of property that is available for purchase or lease). Although organizations, such as the American Academy of Forensic Sciences, provide many resources to its members and stakeholders, it has no authority over the practice of forensic science as a profession or the livelihood of its members to conduct their work. Likewise, The American Society of Crime Laboratory Directors and the American Society of Crime Laboratory Directors / Laboratory Accreditation Board (ASCLD/LAB) play a significant self-regulation role in forensic science. ASCLD was formed in 1973 and is a not-for-profit organization. It was created to share best practices among forensic science laboratory leadership. All its members and leadership are volunteers, most of whom have a full-time management position in a crime laboratory. ASCLD/LAB, derived from ASCLD efforts to improve the quality of all aspects of forensic laboratories by establishing crime laboratory accreditation standards, became a separate organization in 1992. As of December 2012, 390 laboratories are ASCLD/LAB accredited. While both organizations play leadership roles in the forensic community their impact and authority is limited. Authority comes from the span of control exercised by the individual crime laboratory director and his or her desire to improve operations. Both organizations historically have limited the positions they have taken on other forensic enterprise issues.

There are also a number of other forensic enterprise leaders in critical areas such as peer-based certification, pockets of forensic research, and educational initiatives. However, there is no one voice for the forensic enterprise able to listen to all stakeholder needs and integrate recommended courses of action.

Governmental Inter-Agency: NIST Office of Weights and Measures

Inter-governmental organizations are composed of sovereign members who willingly collaborate to achieve their goals; each member remains authoritative in their own field or domain but works with the other governmental agencies to address issues of importance to all involved.

The Office of Weights and Measures at the National Institute of Standards and Technology (NIST) is a relevant example of such a governing body. The Office, established under President Washington, eventually became the National Bureau of Standards on July 1, 1901, which was reorganized into the National Institute of Standards and Technology in 1988. NIST's Office of Weights and Measures (OWM) promotes uniformity in U.S. weights and measures laws, regulations, and standards. Buyers and sellers alike use these to have fair and equitable trade around the world, promoting trade, easing transactions, and supporting consumer confidence.

The OWM achieves its goals not through direct legislative oversight but rather through the promulgation of laws, regulations, and methods of practice in partnership with international, state, and local officials and representatives of business, industry, consumer groups, and federal agencies. OWM works with the International Organization of Legal Metrology (OIML) and the National Conference on Weights and Measures (NCWM) to achieve its goals internationally. Through the Physical Measurement Laboratories at NIST, the OWM ensures traceability of state weights and measures standards in the international metric system (SI), creates procedures and methods for metrology tests and inspections, and trains scientists, metrologists, and officials.

The OWM is thus a "first among equals" body, providing technical guidance and authority under the aegis of the US government; it does not, however, have direct regulatory authority but rather provides standards through a collaborative process of interagency partners who de facto represent their industry.

Governmental Independent: National Transportation Safety Board (NTSB)

Another existing government model of oversight to consider that is independent, highly respected, and successful is the National Transportation Safety Board (NTSB). The National Transportation Safety Board (NTSB) was originally established in 1926 as part of the Civil Aeronautics Board, and was initially tasked with determining the cause of aircraft accidents. Its role expanded to the safety of all transportation modes when it was put in the newly formed Department of Transportation (DOT) in 1967. The NTSB is a non-regulatory, non-legislative board composed of five members (technical staff provide discipline expertise). The NTSB as an independent entity provides recommendations to promote safe transportation and investigates and publicly reports findings of transportation incidents; in these activities, the NTSB provides transportation industry oversight.

The NTSB is offered as an example of a successful government model of efficiency and objectivity for three main reasons. First, it is an independent entity. The initial approach was to make the NTSB an independent entity within the Department of Transportation (DOT), for administrative purposes since it was thought to be too small to stand alone. The real or perceived political influences within the DOT eventually began to impugn the perception of impartiality of the NTSB, resulting in a call for total independence. In 1974, it was removed from Department of Transportation, allowing it to be free from the politically charged arena filled with lobbyists that is inherent within a regulatory organization such as the DOT (49 U.S.C. §1131). All findings of the NTSB are reviewed and approved in a public forum. Stakeholders are invited and their perspectives are welcomed and considered before the final recommendations are made; concerns can be brought to the Board. The NTSB does not need to wait for an invitation to investigate: By statute, it is automatically the lead investigative agency. This eliminates confusion when a transportation crisis warrants an investigation, and is critical component in the success of any investigation, allowing nearly immediate access to the site being investigated. Additionally, witnesses tend to cooperate since their interviews cannot be used against them, and the Federal Code prohibits use of findings and recommendations in lawsuits. Having this legal authority is critical, and is a benefit of being an independent governmental agency.

Second, the NTSB only provides recommendations. By statute, the NTSB has no authority to regulate, fund, or be directly involved in the operation of any mode of transportation. Initially, some called for it to do all of those things so that it would have the "hammer" necessary to be an effective tool in bringing about change in the powerful transportation industry. Yet, it ended up that the very lack of that hammer is what has made it to be such an effective and internationally respected body. It can appeal to Congress and the public when its recommendations are not heeded and each of its recommendations must be addressed within 90 days by the entity which is the object of its recommendation. From a practical perspective, lobbyists ignore the NTSB and go to the regulatory agencies within the DOT, such as the Federal Aviation Administration (FAA), to push their position. This again verifies the value of an independent board and further strengthens the

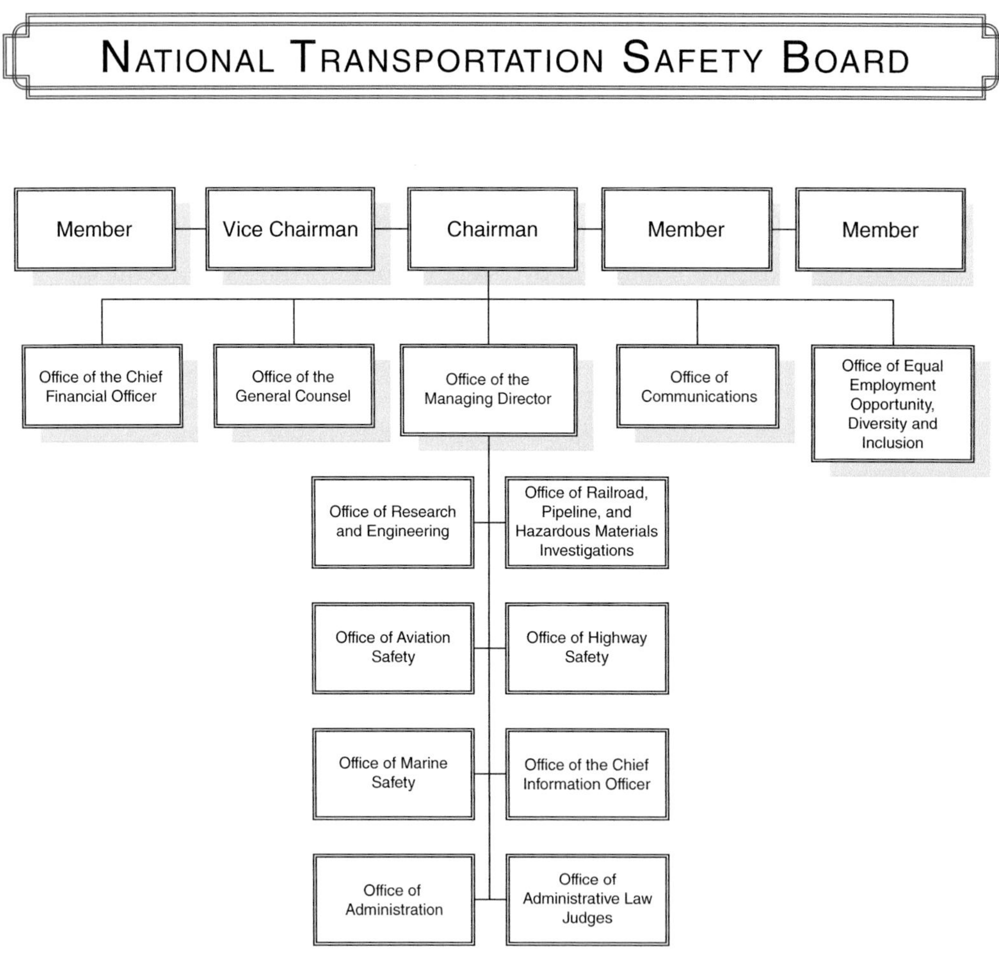

Figure 3 The Structure of the NTSB Board.

NTSB recommendations as being unbiased and based on an evaluation of the evidence, not swayed to fit into a particular industry's pocket. The NTSB's focused objective--with limited oversight--provides a lean and effective structure that does not permit straying from its primary mission. Although a transportation accident can have far reaching implications, when the facts surrounding the cause can be objectively collected and evaluated there is a better chance that the truth about the cause can be determined leading to a safer transportation environment.

Third, the NTSB is comprised of experts. The five Board Members are appointed by the President with the consent of Congress, with at least 3 having substantial expertise in safety and no more than 3 from any particular political party are allowed at a time to reduce political influence. The actual investigative and research staff is hired, not appointed, with their technical expertise as the primary selection criteria; it is their work which is provided to the Board for consideration. The staff members are career experts, not affected by the rotation of the Board, yet benefitting from the fresh perspective and insight that "new eyes" bring to any process. The Board Members serve a five year term, with one appointed each year. This reduces the politicization of the Board since the terms of the members extend beyond any one Presidential term. Areas of safety to which Board Members have a personal interest can also be studied, which has become another side benefit of the regular rotation of the members.

Non-profit Independent: NFPA

In contrast to a governmental model, the private sector can provide the needed leadership and oversight for forensic science without legislated or enforcement authority. The National Fire Protection Association (NFPA) has shown that a private organization that effectively leverages it enterprise partners can set national and international standards and meet industry and public safety requirements. It is a model that can also be applied to forensic science.

The NFPA is an international nonprofit formed in 1896 whose mission is "To reduce the worldwide burden of fire and other hazards on the quality of life whose mission by providing and advocating consensus codes and standards, research, training and education." The requirement for standards, research, training and education are the same essential needs set out in the NAS report for forensic science.

The NFPA origins were fiscally driven (Grant, 1997)[8]. Property insurance companies were interested in improving fire safety to reduce insurance risk and loss. Their initial focus was the importance of sprinkler systems to fire safety and the lack of standards for their installation including pipe sizes and sprinkler head placement. The inconsistent installation of sprinklers increased fire safety risks for manufacturers and insurance companies. Meetings began in 1895 among insurance companies' leaders and sprinkler systems manufacturers. Those meetings were the genesis of an association to administer the sprinkler standards. In late 1896, the articles of incorporation for the National Fire Protection Association were adopted and the first NFPA meeting was held in 1897.

Initially membership was limited to insurance groups who began to adopt formally NFPA standards around 1900. In 1904, industry members and individuals were permitted to join the association. Quickly NFPA had recognized that to be effective they needed to include industry members and scientific testing organizations as full partners and involve them in the development of standards. NFPA turned to Underwriters Laboratory to support the development of the codes and standards they were developing. The membership continued to expand to include other testing laboratories such as Factory Mutual Laboratories, and industry manufacturers including Grinnell Fire Protection, a major sprinkler manufacturer. By 1903 NFPA membership had expanded to include members from London, Sydney and Russia. In 1905, a NY City Fire Department Battalion Chief joined the NFPA, as did the Ohio State Fire Marshal. Soon other organizations including the International Association of Fire Chiefs, American Society of Mechanical Engineers joined, as did individuals in public sector jobs (e.g. fire department officers and building inspection officials).

Electricity was coming into its own in the late 1800s but electricity also presented significant fire safety hazards. By 1896, the US had five standards for the use of electrical equipment. The "Joint Conference of Electrical and Allied Interests" met to develop international standards. The Conference reviewed the five US codes as well as codes from Germany and England. In 1897, through an international consensus process, the "National Code" was published by the National Board of Fire Underwriters. In 1911, the organization that was administering the National Code was dissolved and NFPA took on the responsibility for administering the National Electrical Code. This unified the standard setting

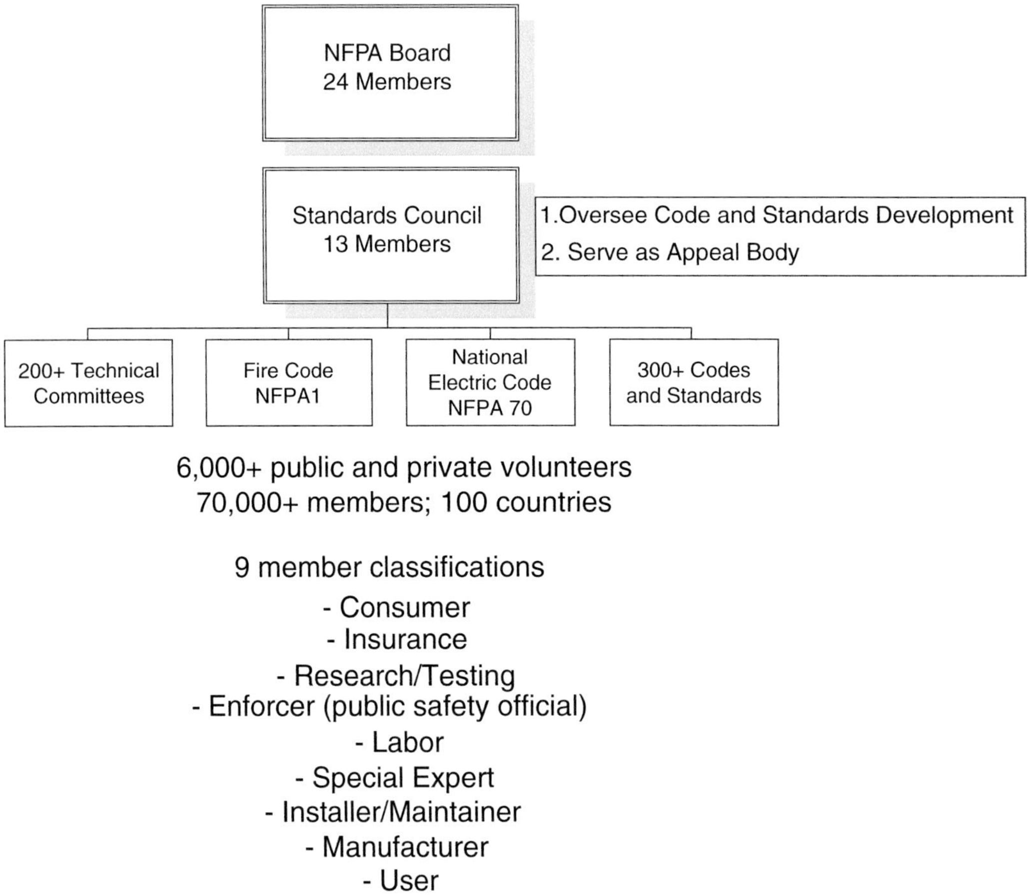

Figure 4 The Structure of the NFPA Board.

and gave NFPA oversight of two major safety issues, fire and electricity. NFPA's scope has expanded to engage in virtually all aspects of building safety. They have published over 300 codes and standards that are focused on minimizing hazards. Without legislated or regulatory authority, NFPA has become, arguably, the authoritative source on fire safety in the US and internationally. NFPA currently has more than 70,000 members from approximately 100 countries. Its membership represents industry, consumers and public safety entities (Figure 4). The Standards Council is appointed by the NFPA Board of Directors to administer the standards development process. The Council's responsibilities include oversight of the Technical Committees including Technical Correlating Committees to integrate codes and standards where needed.[9] NFPA standards cover a broad range of topics from cooking fires and coffee makers to the function of fire doors and the handling of fireworks.

It can be postulated that a measure of NFPA's success comes from the inclusive nature of their "enterprise." NFPA embraced all key stakeholders to carry out its standard development process public sector interests, industry users, equipment and product manufacturers, enforcers, researchers, and consumers. NFPA activities address all the major aspects of fire and building safety and influence components that are key building blocks to current and future building and fire safety in much the same leadership roles the NAS Report suggests forensic science.

- Professional development includes seminars and the publication of authoritative texts and monographs on building and fire safety.
- Certification programs for fire inspectors and other fire professionals are administered by NFPA.
- Initiatives on current safety issues are developed as issues emerge and currently include home sprinkler systems, wildfire awareness and safety, and preparing to address first responder hazards associated with electric vehicles.[10]
- The Fire and Analysis Research Division conducts studies on contemporaneous fire safety issues and incidents. This is supplemented by on-scene investigations of significant safety events by the Fire Investigations Division.[11] They have addressed issues that range from night club fires[12] to dormitory fire safety.[13]
- In 1982, the NFPA sponsored an independent Fire Protection Research Foundation. The Research Foundation sets a national fire research agenda that supports the creation and improvement of codes and standards. Each research project is directed by a research panel drawn from a range of stakeholders.[14]
- NFPA is also active as an advocate promoting initiatives and legislation to enhance public safety.

These NFPA processes address six of the nine roles the NAS report outlines for a forensic science leadership entity listed in the introduction to this paper: best practices, certification, strategic research approach, public and legislative advocacy, and assessing new technologies. The NFPA has taken an inclusive approach. Their wide-ranging approach has created perceived authority in fire and building safety issues.

This perceived authority translates into effective enforcement of critical issues because there is a "cost" for not complying with NFPA standards. NFPA standards are virtually universally accepted and often enacted in codes and legislation locally. As a result, cost takes the form of public scrutiny, liability, and local legislative interest when there is a fire and loss of life and property. Ultimately, this translates into attendant penalties and rewards to businesses and manufacturers. That includes businesses who properly comply with safety standards and those who manufacturer materials and items that comply with their respective standards.

The NFPA model illustrates the need to be inclusive and involve all stakeholders in whatever form forensic enterprise leadership may take. The buy-in that comes from the holistic approach and range of perspectives results in standards that can be met, are well understood, and as a result become enforceable. The perceived authority and its associated strengths are what can be the basis for providing clear direction, consistency of purpose, continuity and sustainability to the forensic enterprise.

A Discussion: The National Forensic Science Board

The NFSB model proposes to integrate and synchronize the on-going efforts leveraging the subject matter expertise that has been developed by each of the recognized forensic professional organizations. It would also ensure that all the stakeholders that have labored so diligently to identify needs and propose solutions are heard, including the legal community, victim advocates, wrongfully accused, community needs, laboratory management, and individual examiners. With non-regulatory oversight, a collaborative, holistic approach can be developed to gain an accurate understanding of capabilities, resource gaps, and profession requirements and to address the profession's needs.

Any investigative findings or reviews of forensic enterprise issues are only beneficial if transparent and parties participate in the reviews free of reprisal or retribution. The medical and the aviation communities both have mechanisms that the forensic enterprise could consider. Shield laws in the medical community allow medical professionals the opportunity to openly discuss system or individual failures to improve future outcomes. They are controversial in some aspects because they can be seen as limiting an individual's rights and protections against malpractice. The aviation community has a system that allows persons in that community to identify safety issues that need to be addressed without the fear of government penalty (Helmreich, 2000).

Some key concepts for how such a Board might be organized and operate include:

- Serving as an independent advisory body for the Executive, Legislative, and Judicial Branches of the government.
- The majority of the Board composed of forensic scientists, with other advisory members coming from other sciences and stakeholder communities.
    - Overlapping terms with limits; political restrictions (such as party or branch affiliation)
- A full-time professional staff for consistency and continuity of business process.
- The legal authority and independence to act without political or industry influence.
- Providing consistent leadership and oversight in the furtherance of the forensic sciences by issuing recommendations on topics that include but are not limited to:
    - Proper analytical procedures and processes
    - Appropriate formal education and training
    - Initial and ongoing training requirements for forensic practitioners
    - Appropriate training for crime lab managers

  - Methods for competency
  - Findings prohibited from lawsuits

- Acting as an investigative body for and issuing recommendations on acute forensic science issues including the:

  - Receiving, investigating, and addressing stakeholder concerns and complaints
  - Evaluation of laboratory failures
  - Appropriate use and limitations of forensic analyses or procedures
  - Current research needs for forensic disciplines

- Providing consistent direction and support to forensic science discipline groups that include:

  - Recommending certification and accreditation requirements
  - Defining purpose, responsibilities, and expectations of practitioners
  - Coordinating defining and prioritizing the research needs Reviewing and promulgating work products as authoritative forensic guidance

- Full transparency of process, products, goals, and outcomes

  - Public forum for communication, suggestions, complaints, and review

There are several significant challenges to creating and operating an entity like a NFSB, all of which could be overcome. Financial resources are not the primary hurdle; the costs to start and sustain such a Board with professional staff are anticipated to be modest. Three major challenges to launching this concept are:

1 Territoriality: Forensic science has demonstrated a tremendous capacity for terrioriality. Until the profession can move beyond immediate, operationally-discrete functions (with their perceived political threats, concerns, and challenges) and trust in a larger process that will encompasses, recognizes, and addresses our collective concerns, significant collective progress will continue to be stymied and limited to individual initiative and its necessarily limited scope of success. Overcoming the universal desire of the practitioners and stakeholders to be in control
2 Inertia: Any undertaking of this magnitude and scope requires managing expectations and building relationships that will transcend the individual needs of any one forensic organization.
3 Fear: Any proposal to accomplish a model like what is outlined will have naysayers; no plan will be quite right for all parties and a "Goldilocks" outcome may or may not be suitable. Leadership will be required to stem inevitable territoriality, to lead people to act on their better natures, and to provide a consistent vision of goals and outcomes. Moving forward in spite of the fear of criticism and failure will be key.

These are political and cultural changes, not financial or mechanical ones. They are inherently more difficult to address, for many of the reasons that have been elucidated earlier in this work; the greatest may be the splintered, multi-jurisdictional nature of forensic science provision in the U.S. Nevertheless, territoriality, inertia, and fear are just obstacles (or challenges, depending) and could be overcome.

Discussion

Forensic enterprise evolution is hampered by the multiplicity of voices and organizations that speak for forensic science. As discussed, excellent efforts that have been made to standardize forensic science education create discipline specific standards or guidelines, coordinate research, and establish accreditation and certification. Those singularly-focused efforts have contributed to the lack of strategic forensic enterprise leadership. Each effort speaks with a single voice with no effective mechanism for "hearing" public stakeholders or integrating their program products in a way that maximizes their effectiveness for the forensic enterprise. The forensic enterprise lacks the three main components necessary for effective establishment and operation of a system:

A concept of operations, an enterprise architecture, and a governance structure. The enterprise architecture is the fundamental organization of a complex program and has been initially described in this paper. A governance structure covers the management principles and decision making in the system (Giachetti, 2010); part of the oversight for the governance structure—the oversight of enterprise--has also been described in this paper, albeit in expanded discussion of relevant analogous models. A CONOPS provides a holistic framework for establishing and evaluating forensic service providers. On its own, a CONOPS can provide these benefits:

- Sets context for measuring system efficiency and effectiveness
- Provides a logic trail of capability
- Creates a basis for communication between stakeholders (who now have a common vocabulary)
- Illustrates how the system should function
- Sets the "as is" and bridges the "to be" states
- Helps scope the situation and proposed solutions (Frittman and Edson, 2010).

The importance of a CONOPS, essentially an operational covering philosophy, cannot be overstated. Forensic science standards enforcement will be legislatively unenforceable, in a very practical sense. The patchwork nature of law enforcement and forensic services cannot be directed by any one federal entity or legislative act; state's rights alone would preclude this. Resources are not effectively administered from the federal level to provide for state and local needs. Resourcing is an unrelenting challenge and forensic science will not compete favorably with the need for new highways or patrol cars. Any governance structure that seeks to establish a formidable bureaucracy has little chance fiscally. The change in the forensic community has to come from a shared and validated sense of continuity for operations, structure, and governance.

The NAS Report focuses on the need to put more science in forensic science, both in terms of education and research. It is important for forensic enterprise leadership to ensure science is protected as policy is developed and evolves. While there may be no "optimal solutions" to policy problems, there are optimal forensic analysis procedures. It is important to understand the limitations of those procedures in the context of what investigative or judicial answer is being sought; that understanding has to inform decisions made about the use and application of the science. For example, scientific advances may result in only one definitive method for performing a certain forensic examination while fiscal constraints may prevent the use of that scientific technique. What is the correct policy concerning the effected forensic analysis? More discussion is warranted; policy and method collide and must be mitigated before change can be effected.

The form the forensic enterprise takes is likely to include both a public and private sector aspect; only then will the entire range of interests be represented. Melding the two will also help ensure that public safety, judicial requirements, and practitioners are heard. The solution must also include viable "listening" mechanisms that reassure not only the traditional forensic stakeholders, but also the public, victims, and offenders.

The public, investigative, and legal communities have justifiably high expectations for forensic examinations. A defendant's freedom and justice for victims are critical and emotionally charged issues. Unfortunately, forensic scientists have greater challenges associated with getting the correct scientific answer than almost any other science. Virtually all evidence submitted for examinations is compromised in some way--our work is not done on pristine samples. The,ultimate crime scene source is always an unknown with multiple possible results. These are limitations that many forensic stakeholders do not consider, yet are important factors in any analysis standard and must be "heard" and understood by stakeholders as policies are developed.

Because of stakeholder diversity and perspectives, and level of understanding, listening mechanisms will likely need to have to take several forms to maximize understanding and dialog. Once there is an understanding, the enterprise must be able to prioritize and respond by taking positions on the directions forensic science needs to take to address concerns that are raised. The forensic enterprise should consider aspects of both the NTSB and NFPA models as it wrestles with the evolution of its leadership.

The forensic community needs to decide its fate actively, rather than respond to sniping, attacks, and political pressures; however, this includes considering and possibly embracing stakeholders' and critics' perspectives. Failure to coordinate and synchronize the forensic profession's collaborative efforts to do so will result in continued misunderstandings among stakeholders and critics, resulting in an increasing loss of faith in forensic science and, ultimately, the criminal justice system.

Endnotes

1 The opinions or assertions contained herein are the private views of the author and are not to be construed as official or as reflecting the views of the Department of the Army, the Department of Defense or the Los Angeles County Sheriff's Department, the District of Columbia, and are those solely of the authors.
2 Chief, Forensic Analysis Division, U.S. Army Criminal Investigation Laboratory, Atlanta, GA
3 Assistant Director, Scientific Services Bureau, Los Angeles County Sheriff's Department
4 Director, Department of Forensic Sciences, Consolidated Forensic Laboratory, Washington, D.C.
5 Director, Scientific Services Bureau, Los Angeles County Sheriff's Department
6 http://www.leahy.senate.gov/download/?id=5ce82fe7-cb94-408c-ab4a-4904b03c9b91
7 http://www.opencongress.org/bill/112-s3378/text
8 Downloaded at www.nfpa.org/displaycontent.asp?categoryid=500.
9 www.nfpa.org/displayContent.asp?categoryID=495
10 www.nfpa.org/displayContent.asp?categoryID=2293
11 For example - Oklahoma City Bombing 1995, NFPA Documentation of Response, *National Fire Protection Journal, 1996:1, p. 50.*

12 Handling and Planning for Fires in Overcrowded Clubs, *National Fire Protection Journal, 2009:6, p. 32.*
13 NFPA Programs Promote College Dormitory Fire Safety, *National Fire Protection Journal, 2002:1, p. 38.*
14 Research Foundation Policies for the Conduct of Research Programs, adopted August 12, 2005.

References

Ackoff, R. (1974). Redesigning the Future. Hoboken, John Wiley and Sons.
Childs, R. and T. Witt (2009). "Survey of Forensic Science Service Providers." Forensic Science Policy and Management 1(1): 49-56.
Conklin, J. (2005). Dialogue Mapping: Building Shared Understanding of Wicked Problems. Hoboken, John Wiley and Sons.
Drucker, P. F. (1995). Managing in a time of great change. New York, Truman Talley Books.
Durose, M. (2008). Census of Publically Funded Crime Laboratories. Washington, DC, Bureau of Justice Statistics.
Frittman, J. and R. Edson (2010). Illustrating the Concept of Operations (CONOPs) Continuum and its Relationship to the Acquisition Lifecycle. Excerpt from the Proceedings of the 7th Annual Acquisition Research Symposium. Monterey, CA, Naval Postgraduate School: 49.
Giachetti, R. E. (2010). Design of Enterprise Systems, Theory, Architecture, and Methods. Boca Raton, FL, CRC Press.
Grant, C. C. (1997). "The Birth of the NFPA." National Fire Protection Journal 1: 58.
Helmreich, R. L. (2000). "On error management: Lessons from aviation." British Medical Journal 320: 781-785.
Houck, M. M., R. Riley, et al. (2009). "FORESIGHT: A business approach to improving forensic science services." Forensic Science Policy and Management 1: 85-95.
NAS (2009). Strengthening Forensic Science in the United States: A Path Forward. Washington, DC, National Academies Press.
Rittel, H. and M. Webber (1984). Dilemmas in a general theory of planning. Develoments in Design Methodology. N. Cross. Hoboken, John Wiley and Sons.
Speaker, P. (2009). "The decomposition of return on investment for forensic laboratories." Forensic Science Policy and Management 1: 96-102.
Speaker, P. (2009). "Key Performance Indicators and Managerial Analysis for Forensic Laboratories." Forensic Science Policy and Management 1(1): in press.
Stine, D. (2009). Science and Technology Policymaking: A Primer. Washington, DC, American Association for the Advancement of Science.
Walker, S. and C. Katz (2002). The Police in America. New York, McGraw-Hill.

How Much Good Are You Doing? 2

Forensic service providers face the classic economic problem: With limited resources, like people, time, and money, how do they meet an increasing demand for their services while maintaining high-quality standards? In addressing this existential problem, questions like the following come up:

- Are the limited resources being put to best use?
- Is performance efficient, that is, "the most bang for the buck"?
- Are there options that will improve your services, either in cost, time, or in quality?
- Are safeguards in place to meet the necessary quality of analysis? Do they work?
- Is investment in equipment, training, and personnel enough to improve performance?
- Is the public getting a good return on its investment (ROI) in the FSP? Is the FSP being a good steward of the public's funding and providing quality and timely services?

Peter Drucker, a pioneer in organizational management, once said that every institution, public or private, needs to be analyzed and evaluated against the beliefs and promises of those it serves, that is, the public. When an institution is central to the beliefs and promises of a society, its performance represents the society; in other words, certain institutions are important to a society that the people in it see themselves in the performance of that institution. **Forensic service providers are a central institution in the life of a society and are a core part of the rule of law.**

Why This Matters

Unlike for-profit businesses, governmental FSPs can't use profit as a standard indicator of performance; other measures must be used to indicate how well they are doing, that is, how much good are they providing to the public with the resources they've been given. This is a truth about governmental and nonprofit entities, as Anheier notes, and they face persistent uncertainties about not only their revenue but also about performance. Forensic services are a public good, like clean water and firefighters, and must be responsible and transparent in how they operate. **At some point, your FSP will be called to account for its performance, to demonstrate how much good you are doing. How will you respond?**

 DOI: 10.4324/9780429318528-4

Project FORESIGHT

Project FORESIGHT was conceived at a meeting of international forensic managers and business school faculty at West Virginia University's College of Business (now the John Chambers College of Business). Using an award from the U.S. National Institute of Justice (NIJ), a group of faculty, forensic experts, and FSP directors adapted a European pilot study that attempted to measure the performance of four EU FSPs. FORESIGHT is a measurement system to identify, adapt, and refine standardized metrics to FSP managers to measure, assess, and assess performance for improving efficiencies, quality, and service.

FORESIGHT has been operating since 2009 and **is free for FSPs to participate.** Hundreds of FSPs around the world participate in FORESIGHT and benefit from the process. **FORESIGHT is the largest project of this type and has more data on how FSPs operate than has ever existed.**

- It uses standardized definitions for services the FSP provides, called **investigative areas.**
- Definitions for metrics, like **case**, **item**, **sample**, and **test**, are also provided.
- FORESIGHT uses **ratios** to make robust comparisons of performance within and between FSPs.
- Data is entered via the **Laboratory Reporting and Analysis Tool**, or **LabRAT**, which gives immediate results about your FSP's metrics.

The goal of FORESIGHT is to measure what matters, share what works, and improve what does not. It does this by generating a measured description of your FSP and allows you to compare its performance to *any other FSP in Project FORESIGHT.* FORESIGHT is, therefore, a benchmarking process.

What Is Benchmarking?

Benchmarking is comparing your operational processes and performance metrics to others in your industry so you can learn better practices from other FSPs. And do think of them as "better" practices: No process is perfect and anything can be improved. Benchmarking involves:

- Using dimensions like people, time, and cost.
- Identifying the best performing firms in your industry or in another industry where similar processes exist.
- Comparing your results to theirs.
- Learning how well you and the others perform in regard to a process.
- And using the results to analyze and evaluate why you or they are successful.

This is a key point: **The numbers are not the answer; the numbers *lead you* to the answer.** Finding out that your FSP's drug analysis cost per case is $62.50 and another's is $45.21 is just the start of the evaluation. Why is the other FSP's drug analysis cost per case $17.29 less than yours? Are they more efficient? Do they have automation? Or are they cutting corners? The $17.29 is not the answer but it leads you to the analysis as to **why the costs**

are different. The analysis and evaluation of benchmarking create flags and indicators for where to investigate further. The reason for the differences in measurements will be found by looking deeper into processes, materials, and methods.

Benchmarking is used to measure performance using one or more specific indicators, like

- cost per unit of measure
- productivity per unit of measure
- cycle time per unit of measure or defects per unit of measure.

Ratios work well because they are easy to calculate (X divided by Y); they are robust and resistant to outliers. Ratios are also easy to compare.

Why Worry about Benchmarking?

Measuring is good for understanding performance. Using units, counts, scales, and categories helps us to understand what is getting done and when it gets done. For example, if your FSP measured cases submitted in a year, it might look like the graph in Figure 2.1.

It would be interesting to know how submissions this year compared to last year, right? That would help you see if demand is going up or down, as shown in Figure 2.2.

Demand seems to be holding steady, so that's good. But what about production, that is, reports issued? Comparing this year to last year results in the graph depicted in Figure 2.3.

And production seems to be the same year to year. All good, yes? Not so fast. What is dangerous about this comparison? **Comparisons need to be meaningful**. These charts answer the questions, "Is demand changing?" and "Is production changing?" but they do

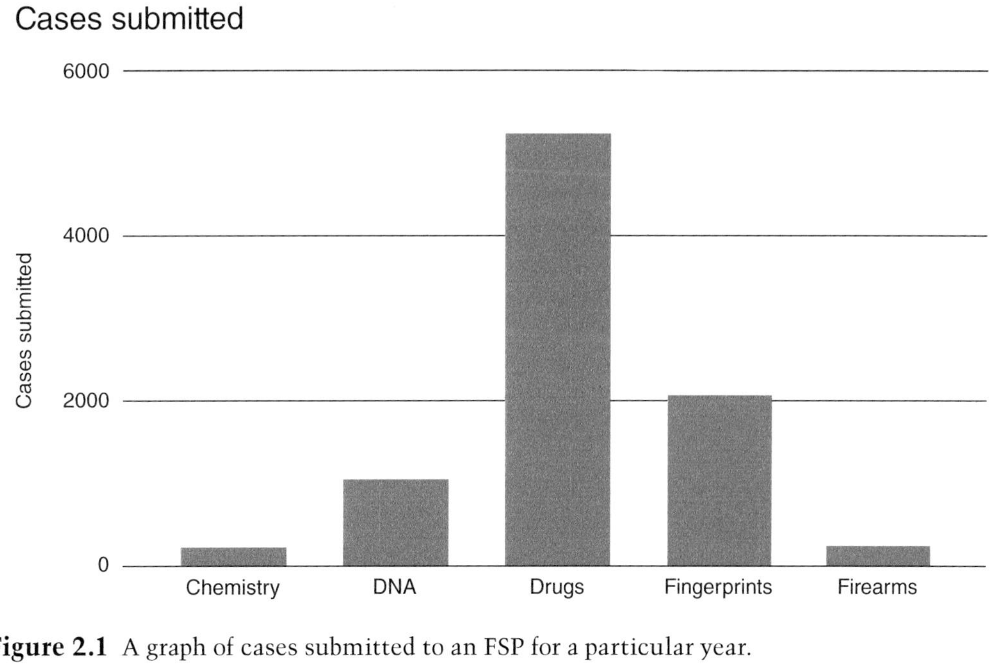

Figure 2.1 A graph of cases submitted to an FSP for a particular year.

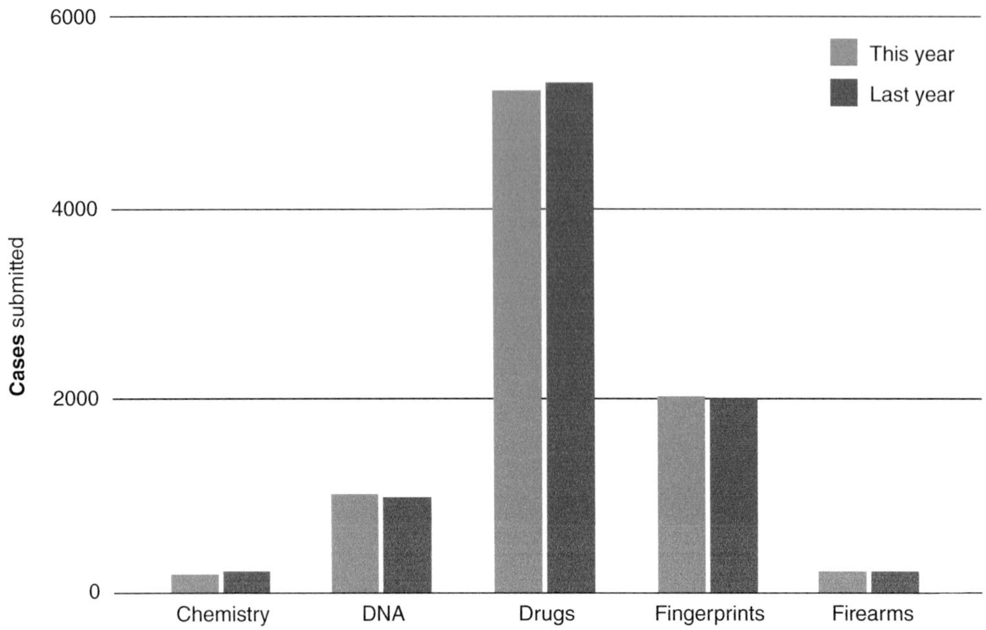

Figure 2.2 A graph of <u>cases</u> submitted for this year (blue) and last year (red).

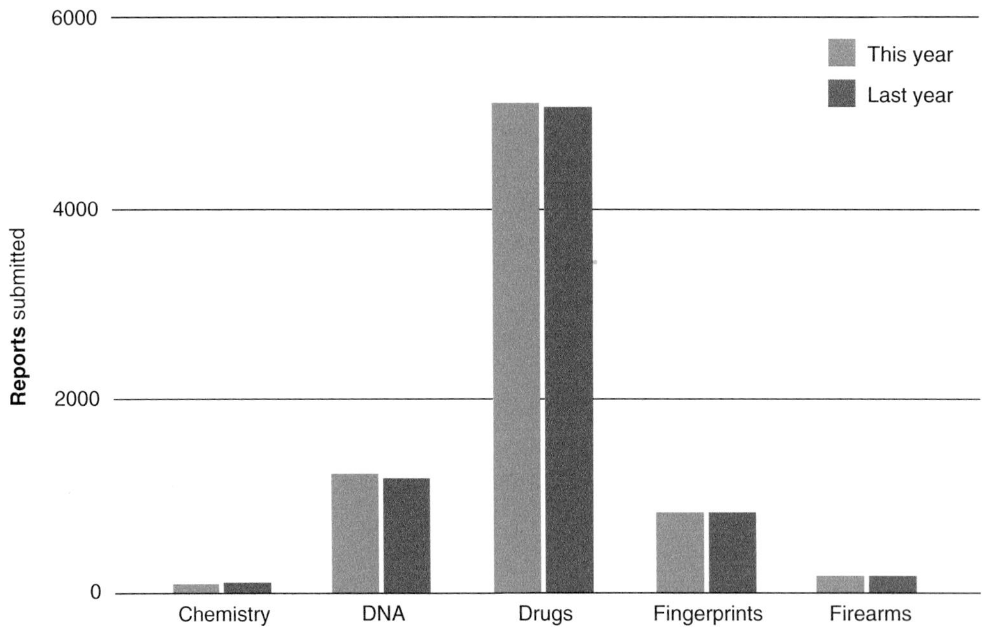

Figure 2.3 A graph of <u>reports</u> issued for this year (blue) and last year (red).

not answer the question, "Is production keeping up with demand?". If you compared submissions this year to reports this year, you would see results as shown in Figure 2.4.

And now the problems are flagged. What's happening in the fingerprints section? Why are they lagging in getting reports out? Also, with an even smaller caseload, the chemistry section is struggling, as well. You can measure the differences between submissions (input) and reports (output) but the key message here is **the numbers are not**

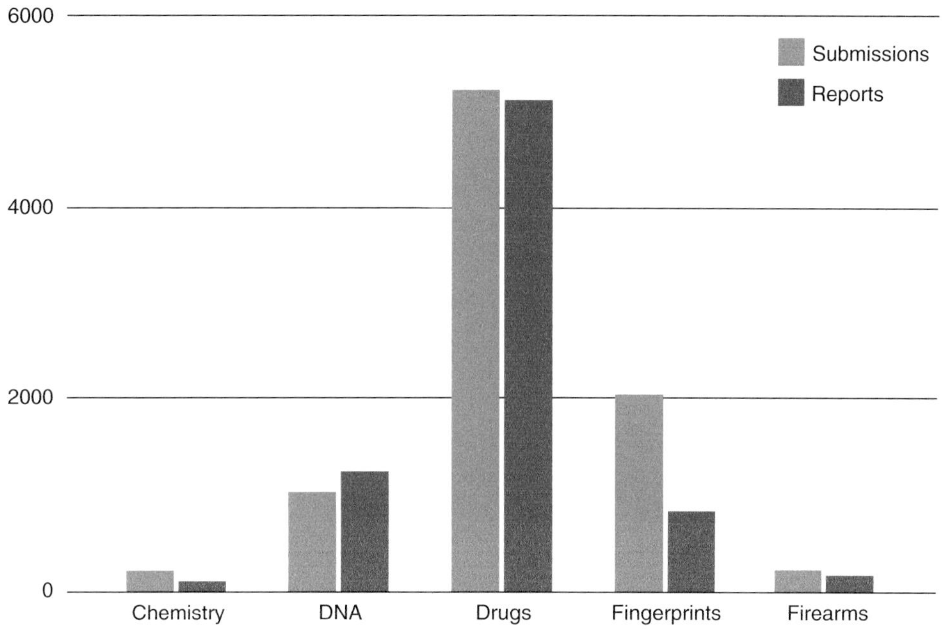

Figure 2.4 A graph of cases submitted (blue) and reports issued (red) for this year. Note the discrepancies in the Chemistry and Fingerprints Units.

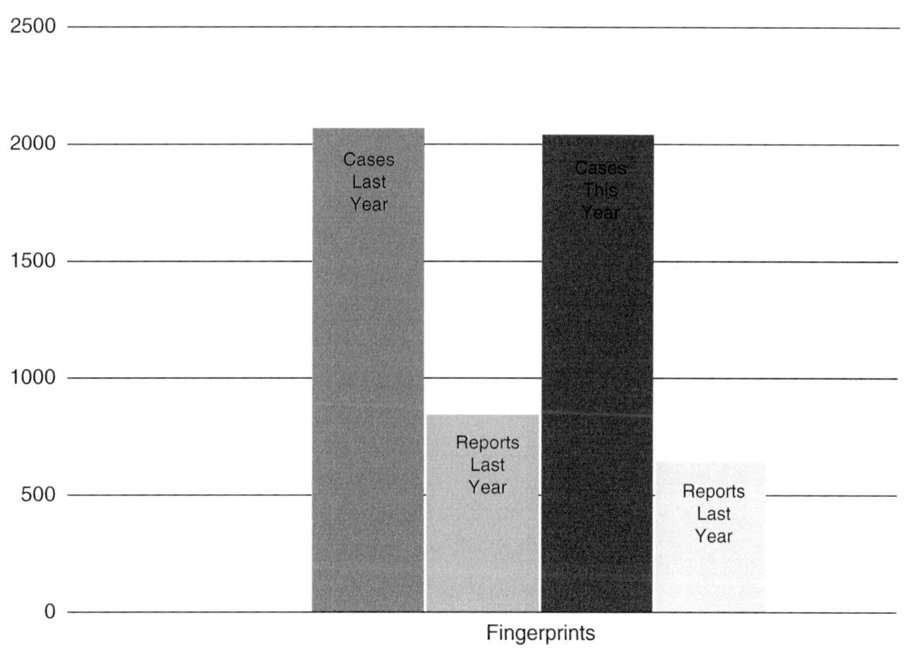

Figure 2.5 A graph of cases submitted last year (darker blue) and this year (darker green) alongside reports issued last year (lighter blue) and this year (lighter green) for the Fingerprint Unit.

the answer, they show you where the problems are. Any two things can be compared but that's not typically useful. In a quote misattributed to Drucker, Ridgway actually said: "What gets measured gets managed — even when it's pointless to measure and manage it, and even if it harms the purpose of the organization to do so". Salience or relevance is key: Numbers in a vacuum are dangerous. In benchmarking and in FORESIGHT, context is king.

Now that you have identified the problem areas, what do you want to know next? Is it getting better or worse would be good to know. Therefore, you look at fingerprints over this year and last year, as depicted in Figure 2.5.

Whatever is going on in fingerprints is getting worse: The reports to case ratio went from 0.41 last year to 0.32 this year. Whatever they are, serious issues are preventing the fingerprint staff from issuing reports. Having flagged the situation, now is the time to go down and talk with the staff to see what's going on.

"My Best Time Yet!"

Comparisons within FSPs over time are good; they keep you on track. Assuming your internal definitions are standardized, within-FSP comparisons are useful for spotting trends, evaluating any changes in your processes, quality, and efficiency. But internal comparisons will only get you so far because you are only comparing yourself against yourself.

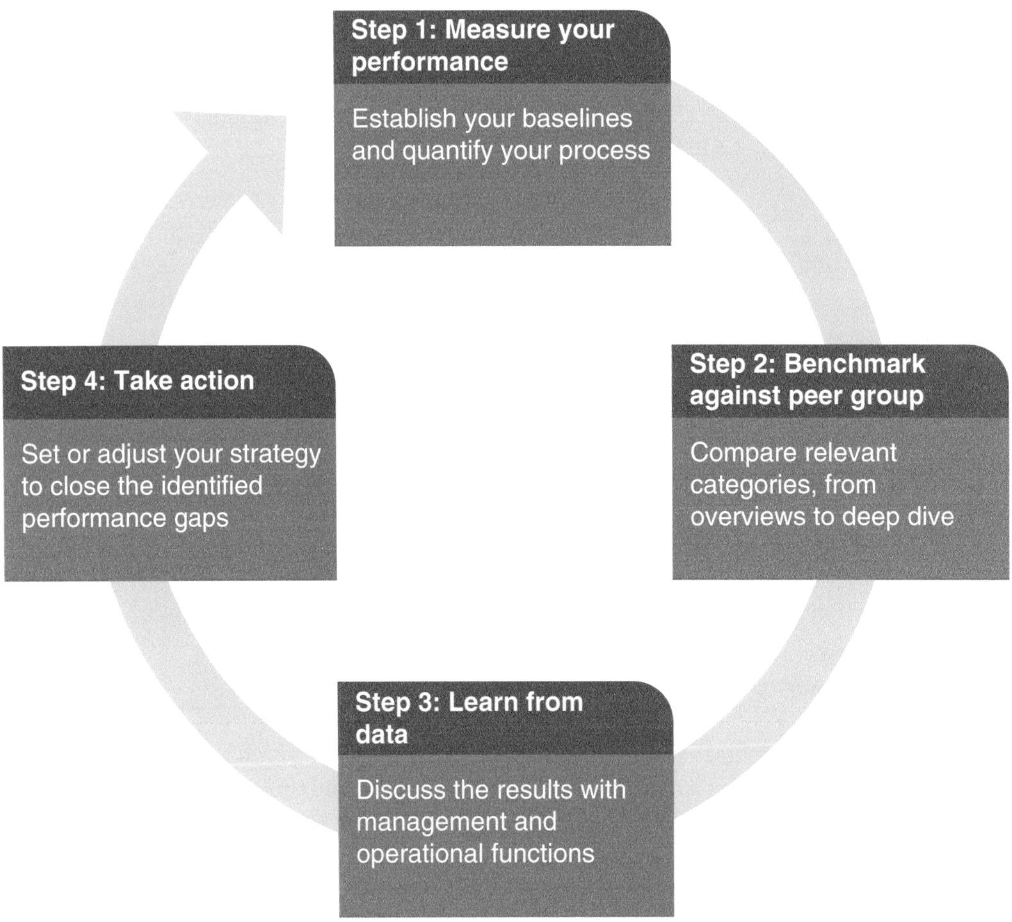

Figure 2.6 The Four Steps of Process Improvement.

It's like running a 15-minute mile and saying: "My best time yet!" That may be but that time could be greatly improved (the men's record is 3:43 and the women's is 4:12). Only looking inwards limits innovation and your strategy can become insular.

How good is your track compared with other FSPs doing the same work? Comparing between FSPs allows comparisons across jurisdictions, methods, and processes. It provides a platform for FSPs to discuss methods, approaches, and quality. Remember that $17.29 difference in drug case cost? Now you can discuss it with that other FSP using the same language. The process, like any quality or improvement effort, is circular.

Over time, inter-FSP comparisons provide for trend analysis and forecasting and they give the industry better information about itself and allow for comparison to other industries, like public health or clinical laboratories. Ultimately, they can inform policy and decision-makers about the status and needs of FSPs.

Five Common Pitfalls in Benchmarking

If you are new to benchmarking and FORESIGHT, you need to avoid the common traps that can keep you from being successful at it.

1 Skipping the Definition of Clear Objectives

Often, organizations neglect to define clear and unambiguous objectives and they jump straight into evaluating performance. Without a clear purpose, benchmark analysis fails to meet expectations. You need to decide what you really want to know and based on that, decide on the measurements that are critical for your goals. The good news is **FORESIGHT has done this for you by establishing definitions and metrics that are useful across the forensic industry.**

2 Choosing the Wrong Benchmarks

Choosing benchmarks is a very crucial step of your analysis. Too often organizations measure "coulda, not shoulda"; just because you can use a metric does not mean you should. The handy or traditional metrics may not be helpful. **Again, FORESIGHT uses flexible performance indicators that allow you to compare a variety of the standardized metrics.**

3 Benchmarking too Many Things

"The more, the merrier" is a phrase that does not apply for benchmarking. Avoid choosing performance indicators that "would be nice to know": What do you *really need to know*. Going after "nice to know" instead of "need to know" could hinder your analysis and will distract your attention from the true critical factors. Therefore, choose your Key Performance Indicators (KPIs) wisely based on your benchmarking analysis and stated strategic or operational goals. **FORESIGHT has standard performance indicators and you can choose which ones are most relevant to your FSP.**

4 Ending the Benchmark Analysis with a Descriptive Report

Even though it may seem that the work is done once you have written a report which shows all the variances between your benchmarks and your goals, it is just the beginning. The aim of benchmarking is not to produce a report, but to take action in order to achieve the stated goals. Taking action is hard but necessary and planning and implementing are among the biggest challenges. **FORESIGHT provides annual updates for your ongoing comparison and can provide analyses tailored to your FSP's needs.**

5 Assuming that Numbers and Performance Stay Stable for Long Periods

Benchmarking is a process, not an event. Benchmarking is a fundamental shift in how you see your operations: You are changing the way you do business, for the better. You should always have in mind that numbers and performance may be changing. Pick a timeframe (weekly, monthly, quarterly, etc.) for review and analysis that would support you in identifying trends in time and react to them. **This is why FORESIGHT is updated annually.**

Finally, prioritize metrics that reflect value to customers, rather than simple volume or efficiency. For example, Rolls Royce changed its metrics from unit sales to a new metric, called Power-by-the-Hour. An airplane engine and maintenance service was offered on a fixed-cost-per-flying-hour basis. This new metric aligned the interests of Rolls Royce, who wanted sales, and the customers, who only paid for engines that performed well. A forensic example: An FSP had been striving to achieve a target 30-day turnaround time for case reports to be issued. Sometimes, they made it, sometimes not. But it was a metric they established against which they were being rated. The agency was "paperless" and could see how long it took clients to read their reports. On average, it took a minimum of **45 days** for agents to open the reports. Therefore, the laboratory changed their target to 45 days, easily meeting it and not upsetting their clients.

Go Deeper

Anheier, H., 2005. *Nonprofit organizations*. London: Routledge.
Department of Justice. 2019. Needs Assessment of Forensic Laboratories and Medical Examiner/Coroner Offices: A Report to Congress. U.S. Department of Justice, Washington, D.C.
Drucker, P., 2017. *Concept of the corporation*. Routledge.
Houck, M.M., Riley, R.A., Speaker, P.J. and Witt, T.S., 2009. FORESIGHT: a business approach to improving forensic science services. *Forensic Science Policy and Management*, 1(2), pp.85–95.
Nussle, J., Orszag, P.R. and Ayotte, K. eds., 2015. *Moneyball for government*. New York: Disruption Books.
Ridgway, V.F., 1956. Dysfunctional consequences of performance measurements. *Administrative Science Quarterly*, 1(2), pp.240–247.
Spendolini, M.J., 1992. The benchmarking process. *Compensation & Benefits Review*, 24(5), pp.21–29.

Article: FORESIGHT A Business Approach to Improving Forensic Science Services

2.1

MAX M. HOUCK[1], RICHARD A. RILEY[2], PAUL J. SPEAKER[3], AND TOM S. WITT[3]

[1]*West Virginia University, Forensic Science Initiative, 1600 University Ave, Morgantown, WV 26506-6217, USA*

[2]*West Virginia University, Division of Accounting, Morgantown, WV 26506, USA*

[3]*West Virginia University, College of Business and Economics, P.O. Box 6025, Morgantown, WV 26506-6025, USA*

> "It is not for me to change you. The question is, how can I be of service to you without diminishing your degrees of freedom?"
>
> —*Buckminster Fuller*

Preparation to become a laboratory manager is largely neglected in the education of scientists (Forde, 2005). The assumption is that management skills will come through on-the-job experience, if at all necessary. Scientists who manage laboratories see themselves primarily as scientists and then as managers; consequently, they tend to devalue the managerial aspects of their jobs (Geles et al., 2000), particularly what are considered the "softer" skills, such as time and project management (Forde, 2005).

As the Chronicle of Higher Education put it,

> The world of science offers a perplexing career track: Scientists spend nearly a decade learning how to do great, clean experiments, interpret data accurately, and think creatively and independently. Then they land a professorship and are faced with the responsibility of overseeing their own laboratory. All of a sudden they are thrust into a new type of job for which they've never been trained: management. And like any business, a laboratory can flourish or flounder by the quality of that management. (Beckman, 2003)

Scientists rarely receive managerial education or training; the expectation on the scientists' part is that being highly educated in the sciences is enough of a prerequisite for management. However, as a professor of strategy at UCLA said, "If you know how to design a great motorcycle engine, I can teach you all you need to know about strategy in a few days. If you have a PhD in strategy, years of labor are unlikely to give you the ability to design great new motorcycle engines" (as quoted in Mintzberg, 2005, p. 13). Scientists thus move from a "science-only" environment to one where the motivations are different, money matters, and the education and training are very different (Picker et al., 2005).

 DOI: 10.4324/9780429318528-5

Forensic laboratory managers are no different, but the stakes may be much higher given the importance of quality science to the criminal justice system. The need for training and support in forensic laboratory management has been recognized for many years (NIST, 1999), but little has been done to transition the tools of business to the forensic laboratory environment. The FORESIGHT Project is a volunteer project to identify, adapt, field test, and refine standardized definitions and metrics to forensic laboratory managers to measure, assess, and benchmark performance for improving efficiencies, quality, and service. Providing a uniform basis for evaluation is the first step toward training and support—without a consensus of what specific words, terms, and processes mean, no meaningful conversations can take place. An agency's strategic value is a function of benefits as well as costs, and managers frequently will focus on the wrong measures (Campbell, Whitehead, & Finkelstein, 2009).

Motivation for the Project

A study in Europe called QUADRUPOL (2003) conducted an in-depth analysis of four forensic laboratories in the European Union, namely Sweden, the Netherlands, Poland, and Finland. At the 2006 International Forensic Business and Economics Colloquium, sponsored by the West Virginia University Forensic Science Initiative, it was proposed that a similar study would benefit North American forensic laboratories.

Benchmarking is improving performance by recognizing, understanding, and integrating best—or at least better—practices from either inside the organization or from outside entities (Camp, 1995). To move forward on a benchmarking project, a standard of comparison must be established. Currently in forensic science, no such standard, as it relates to business practices, exists.

The FORESIGHT Project created the standard methods to collect management data, and these provide the basis for broad, yet deep, comparisons between forensic laboratories using robust measures. The FORESIGHT Project is a business-guided self-evaluation of forensic science laboratories across North America. The project involved standardizing definitions for metrics to evaluate work processes, and linking financial information to work tasks, and functions. Laboratory managers can use these functions to assess resource allocations, efficiencies, and value of services—the mission is to measure, preserve what works, and change what does not. While the Census of Publicly Funded Crime Laboratories (Durose, 2008) and other surveys (Childs & Witt, 2009) approach the forensic industry broadly, FORESIGHT uses cross-laboratory data comparisons to identify and examine processes, strate gies, resources, and allocations at a detailed level. The goal was not simply more information; most forensic laboratories have laboratory information management systems (LIMS; Durose, 2008) and, therefore, the assumption is that information already may be useful to some degree. From discussions with the participant laboratories, it is the authors' opinion that LIMS systems in forensic science are designed more for routine record-keeping tasks, like chain of custody, than truly managing information. Some participants knew the data were in their systems but could not extract it because of software limitations. More information by itself solves nothing. The goals here are human goals, after all, to improve the efficiency of the science offered to the criminal justice system from a business perspective, while also maintaining quality. In FORESIGHT, the intent was to distinguish between "know that" (data or information) and "know how" (putting "know that" to use; Ryle, 1949), and facilitating the latter through standardization, collaboration, and analysis while retaining innovation to solve complex problems (Brown & Duguid, 2002).

A project of this magnitude for forensic laboratories has not been carried out anywhere. Previous studies have been limited in scope or are simply dated (Law Enforcement Assistance Administration, 1968; Benson & Stacy, 1970; Rosenthal & Travnicek, 1974). National and international interest has increased participation in the FORESIGHT Project from nine to fourteen laboratories; it is hoped that international cooperation will improve forensic laboratory performance and increase the quality and efficiency of their services to their respective justice systems. As will be discussed later, any public forensic science laboratory can participate in the FORESIGHT Project. The participating laboratory managers see the value of this type of project and how it can aid them in managing their laboratories' resources, communicating achievements and needs up and down the hierarchy, supporting and justifying decisions, and laying the groundwork for improvement processes. Ultimately, the more laboratories that participate, the more statistically significant and representative the data become and, therefore, the greater benefit to laboratories who participate. The cost of the central project is free to public forensic laboratories, the costs initially being borne by a cooperative agreement with the National Institute of Justice.

Building a Performance Evaluation Model: Methodology

FORESIGHT is based on the plan-do-check-act cycle of Shewhart (1939) and Deming (1986), an iterative four-step process to solve problems and improve processes.

Plan: Establish objectives processes necessary to achieve the stated output.
Do: Implement the processes.
Check: Measure the new processes and compare them against previous and expected results evaluating any variances.
Act: Analyze any differences or variances to determine cause.

A single pass through this cycle rarely solves the problem, and the cycle continues until the expected output is achieved. FORESIGHT is predicated on facilitating the plan, check, and act steps of the cycle, leaving the "do" step to the laboratories to implement the improvements in their own laboratories. Interestingly, the cycle is based upon the scientific method (Shewhart, 1939) and so should be readily sensible to scientific laboratory managers. The FORESIGHT Project managers also collaborate to identify and share best or "better" practices through periodic meetings. The West Virginia University (WVU) faculty helps guide the participants with experience, resources, and input. In this method, the FORESIGHT Project managers collect the data; the WVU faculty analyzes it; and the participants—as a collective—discuss and vet processes, innovations, and solutions.

The FORESIGHT Project laboratories participate on a volunteer basis. The participating laboratories represent local, regional, state, and federal agencies. The authors from the WVU College of Business and Economics provide assistance, guidance, and analysis to the laboratories in the design of business process definitions. Standardization of terminology is critical to successful comparison and benchmarking. Definitions were kept as similar as possible to the QUADRUPOL study to promote collaboration; Appendix A contains the glossary for the FORESIGHT Project definitions, and Appendix B contains the investigative area definitions for the project. Most of the FORESIGHT Project managers came to the initial meeting with a clear, if parochial, idea of what professional terms, such as "case," "item of evidence," and "test," meant. The discussion about these words lasted three days over two meetings. Clearly, the meanings were not uniform across all laboratories. For example, one of the participant laboratories had been criticized for having a longer turnaround time (that is, cycle time; De Feo & Barnard, 2005) than a laboratory with fewer employees and a smaller budget; the implication being that the larger, better-resourced laboratory should be able to perform proportionately to its resources. The larger laboratory defined turnaround time as the time from when the first item of evidence was submitted in a case until the date the case report was issued. However, the smaller laboratory measured turnaround time as the time from when the last item of evidence was submitted until the date the case report was issued; the rationale was that the casework could not be completed until all the evidence was available for analysis. This obviously made a huge difference in the reported metrics, but no one had thought to ask if the terms were defined differently—and the assumption was that all laboratories use the same definitions. Extensive discussions among the participants revealed that differences in human resource and management structure made some terms unrelated or irrelevant, while others had to be redefined for use in North America (in italics in the Tables).

There is a seemingly endless array of potential metrics that might be devised from the data in any crime laboratory. The FORESIGHT Project created but has not completely evaluated numerous metrics requested by participants or as viewed by other standards in the accounting and finance literature. Not all of these measures were used initially but may

Correlation between pay and mobility	Analysis of backlog
Correlation between pay and training (per FTE)	Ratio: Staff FTEs/Manager FTEs
Correlation between pay and experience	R&D: Hours as a percent of tota
Turnaround Times	R&D: Dollars as a percent of total expenditure
Casework versus non-casework	Training funding/FTE
FTE/Case	Expenditure by area/case
Correlation between T&E and productivity	Expenditure by area/item
Correlation between experience and productivity	Expenditure by area/exam (test)
Case output/FTE (by investigative area)	Gap analysis/best practices
Floor space/FTE	Cost per case by investigative area
Floor space/Case	Training of clients & customers/total hours
FTEs by job type (e.g., analysts) per population	Correlation between training and mobility
Percent of casework by investigative area	Casehours/total hours
Percent of items by investigative area	Training/demographics

be useful for specific evaluative purposes germane to laboratories' needs. These measures include but are not limited to:

It was decided that ratios proved to be the most robust, easily compared metrics (Speaker, 2009). The ratios used are collected through the Laboratory Reporting and Analysis Tool (LabRAT). LabRAT is a condensed, active, datacollection tool that allows for easy entry of information most forensic laboratory directors should have or be able to get readily (Appendix C). With LabRAT, examples are offered to assist in the tallying of items, cases, and tests (Appendix D). The data collected from LabRAT is collated and compared to produce analyses and reports; an executive summary on the 2008 FORESIGHT study is available online at www.be.wvu.edu/forensic. Any public forensic science laboratory can participate in FORESIGHT by completing a LabRAT form; using the standard investigative areas, definitions, and tallies; and submitting it to the corresponding author (MMH).

The goal of the FORESIGHT project is improvement, not punishment—inherent in the goal of quality is the promise of redemption (Deming, 1986). Laboratories not in the upper quartile were self-aware of their results and performance relative to the other laboratories. The laboratories in the upper quartile then represented a "better practice" contact to discuss what processes may have led to that performance.

Conclusions

The FORESIGHT Project stresses that the metrics themselves are not the answers, but are pointers or flags to the processes that are the answers to questions of improved performance. The job ofmanagement is "the messy stuff—the intractable problems, the complicated connections" (Mintzberg, 2005, p. 13), and information should feed judgment, not command it (Brown & Duguid, 2002); regrettably, that is often not the case. Critics may be quick to latch onto one or another number from this type of study and start laying blame. Poor laboratory management or obvious mistakes of science are easy enough to identify. The forensic profession will not be able to distinguish the good from the better, however, unless some standardized measures are in place as a platform for the conversations about processes, methods, and solutions.

Acknowledgements

The authors gratefully appreciate the financial support of Senator Robert Byrd (D-WV) and the National Institute of Justice, without which this project would not have been possible. The authors are also grateful to the participating forensic laboratory managers, directors, supervisors, and employees whose hours of devotion to the FORESIGHT Project and contributions of quality work product vastly improved the research outcomes. We also deeply appreciate the support of the laboratories' parent organizations that freed up the time of the participants on the project.

References

Beckman, M. 2003. The scientist as manager. *The Chronicle of Higher Education*, January 28. http://chronicle.com/jobs/news/2003/01/2003012801c.htm

Benson, W., & J. Stacy. 1970. Systems analysis of criminalistics operations. Washington, DC: Law Enforcement Assistance Administration.

Brown, J., & P. Duguid. 2002. *The social life of information*. Boston: Harvard Business School Press.

Camp, R. 1995. *Business process benchmarking*. Milwaukee, WI: Irwin Professional Publishing/ASQC Quality Press.

Campbell, A., J. Whitehead, & S. Finkelstein. 2009. Why good leaders make bad decisions. *Harvard Bus. Rev.* (February): 60–66.

Childs, R., & T. Witt. 2009. Survey of forensic science service providers. *Forensic Sci. Policy Manage.* 1(1): in press.

De Feo, J., & W. Barnard. 2005. *JURAN Institute's six sigma breakthrough and beyond—Quality performance breakthrough methods*. New York: McGraw-Hill Professional.
Deming, W. E. 1986. *Out of the crisis*. Boston: MIT Center for Advanced Engineering Study.
Durose, M. 2008. Census of publicly funded forensic crime laboratories, 2005. Washington, DC: U.S. Dept. Justice, Bureau of Justice Statistics.
Forde, A. 2005. Training scientists as managers. *Science Career Magazine*. July 15. http://sciencecareers.sciencemag.org/career magazine
Geles, C., G. Lindecker, M. Month, & C. Roche. 2000. *Managing science*. New York: John Wiley & Sons.
Law Enforcement Assistance Administration. 1968. Crime laboratories—Three study reports. Washington, DC: Office of Law Enforcement Assistance.
Mintzberg, H. 2005. *Managers not MBAs*. San Francisco: Berrett-Koehler.
National Institute of Justice. 2004. *Status and needs of forensic science service providers: A report to Congress*. Rockville, MD: NCJRS.
National Institute of Standards and Technology (NIST). 1999. Forensic sciences: Review of status and needs. Rockville, MD: NIST.
Picker, S., C. Vehring, L. Hahn, J. Lecker, & M. Vala. 2005. Why are scientists not managers? *J. Bus. Chem.* 2(1): 1–3.
QUADRUPOL. 2003. European Network of Forensic Science Institutes.
Rosenthal, P., & D.A. Travnicek. 1974. Analysis of criminalistics laboratory effectiveness in criminal justice systems. Buffalo, NY: CALSPAN.
Ryle, G. 1949. *The concept of mind*. London: Hutchinson.
Shewhart, W. 1939. *Statistical method from the viewpoint of quality control*. New York: Dover.
Speaker, P. 2009. Key performance indicators and managerial analysis for forensic laboratories. *Forensic Sci. Policy Manage.* 1(1): 32–42.

Appendix A

Terminology Definitions for FORESIGHT Glossary

Assistant/Analyst An individual carrying out general casework examinations or analytical tests under the instruction of a reporting scientist or reporting analyst and who is able to provide information to assist with the interpretation of the tests.

Backlog Open cases that are older than 30 days as measured at the end of the year.

Case-Institute Case A request from a crime lab "customer" that includes forensic investigations in one or more investigative areas.

Case-Area Case A request for examination in one forensic investigation area. An area case is a subset of an institute case.

Casework All laboratory activities involved in examination of cases.

Casework Time Total FTEs for the operational personnel in the investigation area (in hours) subtracted by the hours of R & D, E & T, and support and service given to external partners.

Crime Perceived violation of the law that initiates a case investigation.

Direct Salary Total salary paid to employees, including overtime compensations, vacation salary, bonuses, etc.

Examinations (Exams) The word QUADRUPOL used for "test"; see both "test" and "sample" in this glossary for the changes adopted by FORESIGHT.

Facility Expense Sum of rents, cleaning and garbage collection, security, energy, water, communication, ICT infrastructure, and facility maintenance.

Floor Area Total of all floor area including office, laboratory and other.

Full-Time Equivalent (FTE) The work input of a full-time employee working for one full year.

Full-Time Researcher A forensic scientist whose primary responsibility is research and who is not taking part in casework.

Investigation Area Area limited by item type and methods as they are listed in the benchmarking model.

Investment Expense Sum of purchases of equipment, etc., with a lifetime longer than three years and a cost above $1,000 (alternatively capital expenses).

Item A single object for examination submitted to the laboratory. Note: **One item may be investigated and counted in several investigation areas.**

Laboratory Area Floor area used for forensic investigation, including sample and consumable storage rooms.

Non-Reporting Manager An individual whose primary responsibilities are in managing and administering a laboratory or a unit thereof, and who is not taking part in casework.

Office Area Floor area of offices (square feet).

Operational Personnel Personnel in operational units providing casework, research and development (R & D), education and training (E & T), and external support services. Non-reporting unit heads are included.

Other Area Floor area of space not belonging to laboratories or offices, i.e., corridors, lunch corners, meeting rooms, etc. (square feet).

Overhead Time Total FTEs in hours in the investigation area subtracted by the total hours of casework, R & D, E & T, and support and service given to external partners.

Personnel Expense Sum of direct salaries, social expenses (employer contribution to FICA, Medicare, workers comp, and unemployment comp), retirement (employer contribution only towards pensions, 401(k) plans, etc.), personnel development and training (internal or external delivery, including travel), and occupational health service expenses (employer contribution only).

Report A formal statement of the results of an investigation, or of any matter on which definite information is required, made by some person or body instructed or required to do so.

Reporting Analyst An analyst responsible in non-complicated cases (e.g., simple drugs analysis) for performing the examination of the items submitted, interpreting the analysis results, writing the analysis report and, if necessary, providing factual evidence for the court.

Reporting Scientist The forensic scientist responsible in a particular case for performing or directing the examination of the items submitted, interpreting the findings, writing the report, and providing evidence of fact and opinion for the court.

Representation Expense The costs for hosting guests, i.e., lunches, dinners, coffees offered by the lab, and presents given to guests or during visits abroad, etc.

Running Operational Expense Other costs than investment costs, personnel costs and facilities costs, e.g., consumables, traveling, QA, literature, contracting, representation, service and maintenance, information and advertisement.

Sample An item of evidence or a portion of an item of evidence that generates a reported result.

Scientist in Training An individual with no reporting rights being trained to become a reporting scientist.

Student Hours The sum of teaching hours in a course multiplied by the number of students attending the particular course.

Support Personnel Forensic laboratory staff providing various internal support services. Management and administration personnel not belonging to the operational units are included.

Teaching Hours Time spent teaching in the lecture room in hours (60 min).

Test An analytical process including, but not limited to, visual examination, instrumental analysis, presumptive evaluations, enhancement techniques, extractions, quantifications, microscopic techniques, and comparative examinations. This does not include technical or administrative reviews.

Total Expense The sum of the direct expenses (personnel, operating, and investment) and any administrative or other overhead expenses.

Total Funding The sum of all funding sources including jurisdictional budgeting, grant awards, bequests, and revenue sources.

Total Items Includes all items to which the laboratory assigns an item or tracking number. This is different than the number of items the laboratory receives. (The lab may split items up for analysis.)

Workload Total time spent on all work related to job, including overtime.

Appendix B

Definitions for FORESIGHT Investigation Area Definitions, Derived From QUADRUPOL and ILAC

Definitions in italics were created or redefined for FORESIGHT.

Accident Investigation All non-traffic accident investigations, such as work-related accidents.

Biology (Non-DNA) *The detection, collection, and non-DNA analysis of biological fluids.*

Computer Analysis The analysis of computers, computerized consumer goods, and associated hardware for data retrieval and sourcing.

Crime Scene Investigation The collection, analysis, and processing of locations for evidence relating to a criminal incident.

Digital Evidence—Audio & Video The analysis of multimedia audio, video, and still-image materials, such as surveillance recordings and video enhancement.

DNA Casework *Analysis of biological evidence for DNA in criminal cases.*

DNA Database Analysis and entry of DNA samples from individuals for database purposes.

Document Examination The analysis of legal, counterfeit, and questioned documents, excluding handwriting analysis.

Drugs-Controlled Substances The analysis of solid dosage licit and illicit drugs, including precursor materials.

Entomology Forensic entomology is the application of the study of arthropods, including insects, to criminal or legal cases.

Evidence Screening & Processing *The detection, collection, and processing of physical evidence in the laboratory for potential additional analysis.*

Environmental Analysis [See QUADRUPOL]

Explosives The analysis of energetic materials in pre- and post-blast incidents.

Fingerprints The development and analysis of friction ridge patterns.

Fire Analysis *The analysis of materials from suspicious* fires *to include ignitable liquid residue analysis.*

Firearms & Ballistics The analysis of firearms and ammunition, to include distance determinations, shooting reconstructions, NIBIN, and toolmarks.

Forensic Engineering & Material Science Failure and performance analysis of materials and constructions.

Forensic Pathology Forensic pathology is a branch of medicine that deals with the determination of the cause and manner of death in cases in which death occurred under suspicious or unknown circumstances.

Gun Shot Residue *The analysis of primer residues from discharged* firearms *(not distance determinations).*

Hairs & Fibers The analysis of human and animal hairs (non-DNA) and textile fibers as trace evidence.

Handwriting The evaluation of handwritten materials to categorize or identify a writer.

Marks & Impressions The analysis of physical patterns received and retained through the interaction of objects of various hardness, including shoeprints and tire tracks.

Odontology The identification of human remains through dental materials, for example by postmortem X-rays of the teeth compared to antemortem X-rays. Some forensic odontologists also analyze and compare bitemarks.

Other Specialties Other forensic science applications not covered by the other categories.

Paint & Glass The analysis of paints—generically, coatings—and glass as trace evidence.

Road Accident Reconstruction Analysis of criminal incidents involving vehicles and accidents (hit and run, for example).

Speech & Audio The analysis of live and recorded vocalizations in criminal investigations.

Trace Evidence The analysis of materials that, because of their size or texture, transfer from one location to another and persist there for some period of time. Microscopy, either directly or as an adjunct to another instrument, is involved.

Toxicology, Antemortem Toxicology involves the chemical analysis of body fluids and tissues to determine if a drug or poison is present in a **living** individual, to include blood alcohol analysis (BAC). Toxicologists are then able to determine how much and what effect, if any, the substance might have had on the person.

Toxicology, Postmortem Toxicology involves the chemical analysis of body fluids and tissues to determine if a drug or poison is present in a **deceased** individual. Toxicologists are then able to determine how much and what effect, if any, the substance might have had on the person.

Appendix C

The Laboratory Reporting and Analysis Tool (LabRAT) forms

LabRAT Template

Welcome to the Laboratory Reporting and Analysis Tool (LabRAT), part of the FORESIGHT project. The following worksheets address a variety of questions regarding the allocation of resources within your laboratory. With each worksheet, the requested items are highlighted in the yellow shaded cells. Other cells, with white background, may have data in the form of automated calculations. Do not worry about completion of items in those white background cells.

Organization (Data Source)	
Fiscal Year (e.g., 2007)	
Fiscal Year Period (e.g., Jul 1-Jun 30)	

Laboratory Detail		Laboratory LabRAT project manager	
		Contact Information:	
Jurisdiction (federal, state, local, private)		Phone:	
Size Population Served		Fax:	
Geographic Size Served		email:	
Number of separate facilities		Mailing:	
		Address:	
LIMS Provided / Version			
Standard Work Week (hrs)		City:	
Facilities - Square Feet		State/Prov:	
Administrative Office Area		Country:	
Laboratory Area		Zip:	
Other Area			

Please provide a brief descriptive profile of your organization and laboratories:

Please provide a brief descriptive of any anomolies in the reported year versus a "normal" year of operation.

In the box below, add any special notes regarding your completion of this worksheet.

Investigation area	case areas	total items	items outsourced	items examined internally	total samples from items examined internally	tests on samples examined internally	area reports	Median turn around time, days from last submission (internal)	Median turn around time, days from first submission (internal)	Median turn around time for analyst, days	open area cases at end of year	open area cases older than 30 days at end of year (backlog)	number of institutional cases in one area only	Floor area (sq ft)	Hrs in casework	Operational Staff FTE	Support Staff FTE
Accident Investigation																	
Blood Alcohol																	
Computer Analysis																	
Crime Scene Investigation																	
Digital evidence - Audio & Video																	
DNA Casework																	
DNA Database																	
Document Examination																	
Drugs - Controlled Substances																	
Entomology																	
Evidence Screening & Processing																	
Environmental analysis																	
Explosives																	
Fingerprints																	
Fire analysis																	
Firearms and Ballistics *																	
Forensic engineering and material science																	
Forensic Pathology																	
Gun Shot Residue (GSR)																	
Hairs & Fibers																	
Handwriting																	
Marks and Impressions **																	
Odontology																	
Paint & Glass																	
Road accident reconstruction																	
Serology/Biology																	
Speech & Audio																	
Toxicology ante mortem (excluding BAC)																	
Toxicology post mortem (excluding BAC)																	
Trace Evidence																	
Other Specialties (describe below)																	
TOTAL	-	-	-	-		-	-	-	-	-	-	-	-	-	-	0.00	0.00

*Includes toolmarks
**Includes tires and shoes

In each cell, enter the requested number for that row's investigation area

		In each cell, indicate the actual dollar expenditure for the Fiscal Year.															0							
Amount Budgeted (from all sources of funding) for the Investigation area for the year	Investigation area	Capital Expenditures current year	Capital Expenditures previous four years	Operational Staff Salary	Support Staff Salary	Benefits	Overtime and Temporary	Chemicals, reagents, consumables, gases	Travelig (non-training related)	QA and accreditation	Literature purchased	Subcontracting lab. investigations (cases)	Other subcontracts (e.g. consulting)	Representation costs	Service of instruments	Information / publications produced	Advertise-ments (for vacancies)	Non-instrument Repairs & maintenance	Equipment leasing	Facilities Rent - Leasing	Utilities	Telecommunic ation	Other	Total Expenditures
	Accident Investigation																							0
	Blood Alcohol																							0
	Computer Analysis																							0
	Crime Scene Investigation																							0
	Digital evidence - Audio & Video																							0
	DNA Casework																							0
	DNA Database																							0
	Document Examination																							0
	Drugs - Controlled Substances																							0
	Entomology																							0
	Evidence Screening & Processing																							0
	Environmental analysis																							0
	Explosives																							0
	Fingerprints																							0
	Fire analysis																							0
	Firearms and Ballistics *																							0
	Forensic engineering and material science																							0
	Forensic Pathology																							0
	Gun Shot Residue (GSR)																							0
	Hairs & Fibers																							0
	Handwriting																							0
	Marks and Impressions **																							0
	Odontology																							0
	Paint & Glass																							0
	Road accident reconstruction																							0
	Serology/Biology																							0
	Speech & Audio																							0
	Toxicology ante mortem (excluding BAC)																							0
	Toxicology post mortem (excluding BAC)																							0
	Trace Evidence																							0
	Administration and Support																							
	Other Specialties (describe below)																							
	0																							0
	0																							0
	0																							0
	0																							0
	0																							0
	0																							0
	0																							0
	0																							0
$0	TOTAL																							0
	TOTAL	$0							$0	$0	$0	$0	$0	$0	$0	$0	$0	$0	$0	$0	$0	$0	$0	$0
*Includes toolmarks	**Includes tires and shoes	#DIV/0!	#DIV/0!	#DIV/0!	#DIV/0!	#DIV/0!	#DIV/0!	#DIV/0!	#DIV/0!	#DIV/0!	#DIV/0!	#DIV/0!	#DIV/0!	#DIV/0!	#DIV/0!	#DIV/0!	#DIV/0!	#DIV/0!	#DIV/0!	#DIV/0!	#DIV/0!	#DIV/0!	#DIV/0!	$0

FORESIGHT: A Business Approach to Forensic Science Services.

Appendix D

Two example sets of tallies for items, cases, and tests in FORESIGHT

Biology/Serology case: *In a case of sexual assault, the kit, the victim's underwear, and a bedsheet are submitted; due to the context of the crime, the bedsheet is not examined. The laboratory does not perform microscopic hair examinations.*

	Total Items	Total Samples From Items Examined Internally	Tests Conducted Total Examined	Tests Conducted			
				Visual	AP	P30	Microscopic
Biology/Serology case	4	4	17				
1. Sexual assault kit							
1.1 Vaginal swab							
1.2 Oral swab					1	1	1
1.3 Anal swab					1	1	1
1.4 Hair sample (not tested)					1	1	1
1.5 Victim known sample							
2. Underwear (Two stains: one +, one –)				2	2	2	2
3. Bedsheet (not tested)							
4. Suspect known sample							

Fingerprints case: Two soda cans and three latent lifts are submitted from the scene as well as two sets of tenprint cards from the suspects.

Total Samples From Items	Total Items	Total Samples from items Examined Internally	Tests on Samples Examined Internally	Visual	Tests Conducted Develop (2 @)	Comparison Photography
Fingerprints case	7	7	18			
1. Two (2) soda cans (1 print @)		*1 print @*		2	4	2
2. Three (3) latent lifts		3			6	
3. Two (2) tenprint cards*		2				

*Each tenprint set counts as 1 item, to include major case prints, palms, etc.

Key Performance Indicators

3

Now that you know that benchmarking, comparing your performance to that of others in your industry, is a critical component of a successful operation, the next question becomes how to evaluate performance. Measurements were mentioned in the last chapter and in this chapter you'll learn about what measures to use and why. FORESIGHT has followed a classic economics approach to these measures and adapted them to the environment of the FSP. You could jump right in and start measuring but it is important to know where these measurements come from and how they work. Stick with it because FORESIGHT will make more sense once you finish this chapter. Let's go.

What Is a "Key Performance Indicator"?

How do you know your FSP is going in the right direction? To solve a problem, are people or equipment the answer? Do you need a bigger facility? To answer these questions, you need to know the critical measurable indicators of working toward an intended goal. Those critical measurable indicators are **key performance indicators** or **KPIs**. KPIs give you a tangible analytic foundation to make decisions and pay attention to what will get you where you want to go. Where you want to go is called a **target**, a desired level of performance. Remember the fingerprint unit that was having trouble getting reports out in the last chapter? Let's say you want to improve that performance: What should the target be? The rest of the units seem to be at parity or better, that is, one case submitted matches, another case being reported out. The unit gets just over 2,000 cases submitted each year, so production needs to more than double from less than 1,000 reports per year. That's a big ask! But now you have a target of 2,000 reports per year (given the current submission rate).

To be useful, KPIs should:

- Provide objective measurements of progress toward a target.
- Be relevant measures to help improve decision making.
- Compare measurements that show the degree of change over time.
- Track important factors like having the intended outcome, use of resources, quality, turnaround time, compliance or oversight, personnel, and money.

KPIs are connected to each other and to the rest of the FSP's efforts. Those measures that help run the FSP's operation track money, people, processes, and outputs. While important, they drive the FSP's larger goals (intermediate and end outcomes), like growth, trust, recognition, or other strategic goals.

DOI: 10.4324/9780429318528-6

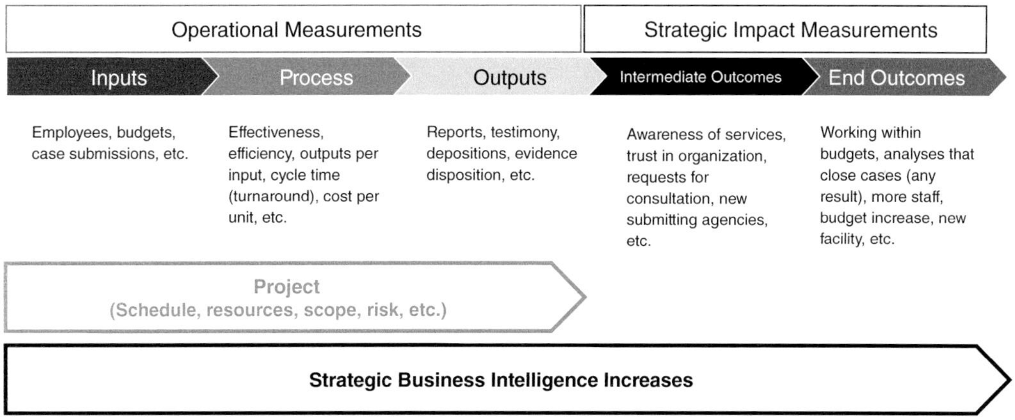

Figure 3.1 Operational measurements, like inputs, process, and outputs, should feed Strategic Impact Measurements, like intermediate outcomes and end outcomes, to provide gains to your overall Strategic Business Intelligence.

Your Coffee Shop

Step away from the mindset of your FSP for a minute and consider KPIs in another organization: A coffee shop. Imagine you own a coffee shop and write down answers to the following questions:

- What are your **inputs**, the amount, type, and quality of resources consumed in your processes that produce outputs?
- What are your **processes**, the things you do to control the costs, quality, or consistency of your outputs, like production or training?
- What are your **outputs**, the measures that show how much work has been done?
- What are your **intermediate outcomes**, those accomplishments that demonstrate the value or effect of what you have done?
- What are your **end outcomes**, the final tallies of your production or your ability to set new targets?

Now, coffee shops are generally for-profit, unlike FSPs, but think through this example. What about inputs? Coffee beans, certainly. What about cups? Is your place take-out only or do you offer a place to sit down? Only hot coffee or what about iced drinks? Sizes? Lids? What about zarfs (the cardboard sleeves that keep you from burning your fingers on the paper cup)? Water? Yes, water. Spoons, sugar, milks of various types, not to mention the capital investments, like chairs, tables, equipment, aprons, towels, and on and on. You could add to the list as you consider payment mechanisms, clean up, staff amenities, etc.

The processes include everything from welcoming customers, to brewing coffee and other drinks, completing the sale, cleaning up, and replenishing stock, among many other things. Other aspects include all the human resource stuff, like advertising for, interviewing, and hiring new employees or getting rid of poor performers, and training the new hires and criteria for raises and promotions, among others.

Figure 3.2 A coffee shop is a good non-forensic example to discuss systems thinking (Courtesy Jonathan Borba/Unsplash.com).

Outputs seem simple: Cups of coffee. Is that all, though? What strength or quality is your coffee? Dirty cups, used disposable cups, foamy milk pitchers, and more are also outputs of your processes. It's important to think through the entirety of your processes and identify *everything* that goes into, through, and out of them. As Drucker noted, "an integrated process that starts when supplies, materials, and parts arrive...and continues even after the finished product reaches the end user...there is only one cost: that of the total system."

On to outcomes. Intermediate outcomes would be sales, of course, especially repeat sales. Customer satisfaction means continued revenue. End outcomes would include turning a profit, adding products (maybe tea or chai drinks), or expanding from a take-out only to a place with enough room for seating. Using your KPIs for each of these aspects of your coffee shop will increase your intelligence about the coffee business, your coffee shop, and your caffeinated goals.

Back to Forensic Land

FORESIGHT has developed an entire family of KPIs tailored for the forensic industry. The hard work (it was but also fascinating). They tend toward the operational or project side but those can be combined to produce more strategic metrics. These metrics are based on standardized definitions, critical to benchmarking, agreed upon by the forensic industry. In the early days of FORESIGHT, this was the central issue because each FSP had its own language by practice or sometimes by statute. Paul Speaker was so right when he said, "Gather together a group of forensic laboratory directors and ask each to define what they mean by a case, a sample, an item, a test, or a host of other terms and you will hear as many different responses as the number of individuals who are quizzed."

The standard terms allow for the use of a method from accounting and another from economics. The first, called **common-size analysis** for budgeting, divides the funds allocated to each budget category by the total budget. Each line item is then expressed as a percentage of the total; this allows for comparing budget allocations between FSPs regardless of the organization's size or across time for a specific FSP.

The second method is called a **DuPont analysis** or a **DuPont expression**. Routinely taught in accounting and finance classes, the DuPont expression is a framework for analyzing an organization's performance. The formula was developed in 1914 by F. Donaldson Brown, an employee of the DuPont Corporation, hence the name. The formula is used to track financial performance by incorporating earnings, investment, and working capital together into a single number, called the **return on investment** or **ROI**. The DuPont expression is used to disassemble the various drivers of an organization's financial health. It delves into the why behind the ROI. In FORESIGHT, a DuPont expansion is used to break down overall performance metrics into meaningful KPIs that help to explain the ROI contribution from internal policy, productivity, and market forces, among others. Ratios permit an FSP to measure its performance over time, against specific other laboratory operations, and against industry averages or standards.

Why This Matters

The discussion by Speaker in the "Key Performance Indicators" and "Decomposition of return on investment" papers that follow this chapter is worth the time. It explains how FORESIGHT goes from simple metrics, like cases, tests, and costs, to DuPont-based ratios, like cost per case, testing intensity, and personnel productivity, making them KPIs. Those two papers offer a good in-depth discussion of how to use KPIs and why ROI is meaningful for FSPs.

Go Deeper

Drucker, P. 1996. *Managing in a time of great change*. Plume.

Latham, J. and Vinyard, J. 2010. *Organization diagnosis, design, and transformation*. John Wiley and Sons.

Speaker, P.J. 2008a. Key performance indicators and managerial analysis for forensic laboratories. *Forensic Science Policy & Management*, 1(1), pp.32–42.

Speaker, P.J. 2008b. The decomposition of return on investment for forensic laboratories. *Forensic Science Policy & Management*, 1(2), pp.96–102.

Speaker, P.J. 2010. A new twist on old tricks: DuPont expansions for public entities. *Journal of the Academy of Business Education, Proceedings*, pages 1–7.

The Father of ROI: Donaldson Brown: https://www.hagley.org/librarynews/father-roi-donaldson-brown

Article: Key Performance Indicators and Managerial Analysis for Forensic Laboratories

3.1

PAUL J. SPEAKER
West Virginia University, College of Business and Economics, P.O. Box 6025, Morgantown, WV 26506-6025, USA

Introduction

Forensic laboratories face the classic economic problem— how to allocate limited resources with increasing demand for services while maintaining high-quality standards. Questions abound with respect to performance of the laboratory, given budgeted resources (Dale and Becker, 2004; Koussiafes, 2004). How does the laboratory director know:

- If resources are appropriately allocated?
- If the laboratory performance is efficient?
- Whether alternative practices might result in improved, high-quality services?
- Whether sufficient safeguards are in place to assure the quality of analysis?
- Whether there is adequate investment for equipment, training, and development to enhance future analytical technique?
- Whether the laboratory is optimizing the return on investment for its constituency?

Similar questions face every organization, whether it is a manufacturing plant choosing the optimal level of production, a hospital deciding upon its level of emer-gency care, or a county government selecting which roads to repair. All organizations must address the economic problem and determine how to allocate scarce resources. And once a strategy has been implemented, how does the organization know whether its choices are appropriate? That is, how does the laboratory measure and monitor its attempts to make optimal decisions in meeting the mandates of its mission?

Fortunately, other industries have examined this economic problem (Kaskinen, 2007; Love et al., 2008; Smith, 2007). The answers in other industries to similar questions have left us with some powerful tools that can be applied to forensic laboratories (Bielski, 2007; Brooks, 2007; Petaschnick, 2008; Reilly, 1997). From accounting we learn that common-size financial statements can transform the financial statements of individual entities into size-adjusted analysis tools that can be easily compared against industry standards. From business finance we learn that transforming absolute dollar-performance measures into size-adjusted ratio measures permits any business to evaluate its allocation of resources into measures of liquidity, efficiency, risk, return, and market assessment.

DOI: 10.4324/9780429318528-7

In the sections that follow, these lessons are applied to forensic laboratories. In the first section, the technique of common-size statements is introduced for the budget allocations of laboratories. A common-size budget statement will permit a laboratory to consider the allocation of resources as a percentage of the total budget and compare to the industry average or to a select group of the most successful laboratories for signs of improvement. The third section addresses issues of language and measurement standards across the forensic laboratory industry. Unfortunately, there is no universally accepted terminology for forensic laboratories. The common language adopted in the European Quadrupol and U.S. Foresight studies is presented as the standard by which the remaining measures will be defined. In the fourth section, key performance indicators are developed that address issues in several areas of concern for forensic laboratories: return on investment, efficiency, risk/quality assurance, analytical process, and measures of market condition. Concluding comments follow with some warnings signs regarding data-driven management.

Common-Size Statements

Within any industry we find organizations of varying size, and forensic laboratories are no different. How do we compare a statewide forensic laboratory with a smaller county forensic laboratory, or a large metropolitan forensic laboratory with a forensic laboratory with a nationwide mandate? In the financial-services industry, a similar problem exists: the method in which we compare the community bank to the regional bank to the international bank; in healthcare, the need to compare the community health center to the regional hospital to the major teaching hospital; in food service, comparing the neighborhood restaurant to the national chain restaurant. How do you compare the performance of two organizations with similar missions, but unequal size?

To compare the decisions and performances of different-sized organizations, accountants introduced the concept of common-size statements. As applied to forensic laboratories, consider the allocation of funding within a forensic laboratory. With a common-size budget, the funds allocated to each budget category are divided by the total budget, and each line-item budgeted area is expressed as a percentage of the total, thus enabling comparison across laboratories or across time for a specific laboratory. When the common-size budget reveals significant differences in the allocation of funding across expense areas, it raises a red flag for the laboratory manager to look more closely at that area to see if practices and policies are moving the laboratory into the desired strategic direction. Consider Tables 1 and 2, examples of two forensic laboratories of different size. Laboratory Small has total funding and expenditures of less than $10 million annually, while Laboratory Large has total funding in excess of $50 million annually. Small spends $5.7 million for salaries while Large spends $26.5 million for the same expense category. Is Small spending too much on personnel (all else being equal), or perhaps, not enough? What about the personnel expense allocation of Large? Is it too much, too little, or just right? And what about the other expense categories or the source of funds—are there concerns that jump out?

Table 1 illustrates the dollar budgets for Laboratory Small and Laboratory Large. As presented, it would take a very experienced laboratory director to immediately draw some perspective about these two scenarios. Laboratory Large is nearly six times the size of Laboratory Small; its personnel expenses are roughly 4.5 times that of Small, while operational costs are more than 7.5 times the size. Large invests over $5.5 million in capital

Table 1 Dollar Budgets for Laboratory Small and Laboratory Large

Funding	Laboratory Small	Laboratory Large
Budget - Jurisdiction	$7,700,000	$52,700,000
Grants	$2,050,000	$0
Other funding sources (including revenue)	$50,000	$3,800,000
Total Funding Personnel costs Direct salaries	$9,385,500	$56,500,000
Direct salaries	$5,725,000	$26,500,000
Overtime & Temps	$605,000	$285,000
Benefits	$1,850,000	$9,200,000
Personnel development & training	$65,000	$805,000
Total Personnel	$8,245,000	$36,790,000
Running operational costs Chemicals, reagents, consumables, gases	$530,000	$3,450,000
Travel	$4,000	$310,000
Subcontracting	$25,000	$1,525,000
Leasing	$290,000	$940,000
Service of Instruments	$455,000	$810,000
Repairs & maintenance	$35,000	$750,000
Utilities	$47,500	$610,000
Telecommunications	$24,000	$305,000
Other operational costs	$12,500	$2,250,000
Total Operational Cost	$1,423,000	$10,950,000
Capital Expenditures (Investment costs) DNA Laboratory instruments/equipment	$0	$2,525,000
Non-DNA Laboratory instruments/equipment	$0	$1,100,000
Furniture	$15,000	$275,000
Information technology	$17,500	$1,400,000
Facilities > $10,000	$0	$99,900
Vehicles	$0	$127,500
Total Investment	$30,000	$5,527,400
Surplus (Deficit)	$102,000	$3,232,600

expenditures while Small only invests $30,000; and Large boasts a yearend surplus that is nearly one third the size of Small's entire budget. The direct comparison of the dollar funding and expenditures is very difficult to critically assess.

A common-size presentation delivers that immediacy as it converts each dollar category to a percentage of the entire budget. To make the conversion for Laboratory Small, which has total funding of $9.8 million, every line item for funding source or expense category is divided by the total funding level for the entire laboratory. Similarly for Laboratory Large, every line item is divided by its total funding level of $56.5 million. The results are illustrated in Table 2 for Laboratory Small and Laboratory Large. For additional analytical value, the corresponding percentages of funding and expense allocation for the average comparable forensic laboratory are presented. (Note that Table 2 is fabricated for purposes of illustration and is not the result of an industry study.)

In this cross-sectional presentation, the common-size statement serves to wave red flags at areas that require further investigation. If a laboratory's budget categorization differs significantly from the norm, laboratory management should dig further to develop

Table 2 Common-Size Budgets for Laboratory Small and Laboratory Large

Funding	Laboratory Small common Size Budget	Laboratory Large common Size Budget	Average Forensic Laboratory
Budget – Jurisdiction	76.71%	93.25%	87.32%
Grants	21.84%	0.00%	9.92%
Other funding sources (including revenue)	1.44%	6.75%	2.76%
Total Funding	**100%**	**100%**	**100%**
Personnel expenses			
Direct salaries	56.74%	47.07%	58.08%
Overtime & Temps	6.45%	0.51%	2.68%
Benefits	19.71%	16.34%	17.37%
Personnel development & training	0.69%	1.43%	0.92%
Total Personnel	**83.59%**	**65.35%**	**79.05%**
Running operational expenses			
Chemicals, reagents, consumables, gases	5.65%	6.13%	6.12%
Travel	0.04%	0.55%	0.38%
Subcontracting	0.27%	2.71%	4.39%
Leasing	3.09%	1.67%	4.83%
Service of Instruments	4.85%	1.44%	1.59%
Repairs & maintenance	0.37%	1.33%	0.71%
Utilities	0.51%	1.08%	1.34%
Telecommunications	0.26%	0.54%	0.46%
Other operational costs	0.13%	4.00%	0.97%
Total Operational Cost	**15.16%**	**19.45%**	**20.79%**
Capital Expenditures (Investment expenses)			
DNA Laboratory instruments/equipment	0.00%	4.48%	1.47%
Non-DNA Laboratory instruments/equipment	0.94%	1.95%	1.43%
Furniture	0.16%	0.49%	0.37%
Information technology	0.19%	2.49%	0.89%
Facilities > $10,000	0.00%	0.18%	0.24%
Vehicles	0.00%	0.23%	1.04%
Total Investment	**1.28%**	**9.82%**	**5.45%**
Surplus (Deficit)	**1.04%**	**5.72%**	**3.37%**

an understanding and explanation. Recognize, however, that merely being different from the average does not indicate that a problem exists; it may be the case that the difference is expected due to the nature of the missions of the two laboratories. Also, note that the industry average includes all organizations in the industry, both the efficient and the inefficient, and does not necessarily represent the ideal budget splits—just an average.

Nevertheless, consider the red flags raised by Table 2. In an examination of the source of funds, laboratories Small and Large both differ from the average laboratory. Small has a very high percentage of its funding (relative to the average laboratory funding) that comes from grants, while Large lacks the grants but has a relatively high percentage of funding that arises from its jurisdictional allotment. What might these red flags indicate? In the case of laboratory Large, further investigation reveals that it is a national laboratory that is not eligible for most of the available grant funding; from this investigation, the difference is easily explained. For laboratory Small, there is some concern as the jurisdictional budget (78.57%) is smaller than the total personnel expense of 84.13%, and may threaten long-term viability of some employment and laboratory performance.

Looking deeper into personnel expenses, Small devotes a much higher percentage of its budget towards direct salaries, overload, and temporary workers. This difference suggests that Small take a look at the salary levels and distribution of its workforce to look for any possible inefficient allocation of labor resources. Large, on the other hand, has a relatively lower total expense for benefits and may want to investigate what impact the reduced benefits may have on its workforce.

The long-term expense for capital sends some signals for additional investigation. In most environments, the miniscule investment in plant and equipment by Small would not bode well for the future. This percentage should be compared with recent years to see if this is a trend or merely an anomaly found in the current year. Further, when coupled with the high percentage spent in the service of equipment category, a red flag raises some initial concern for the long-term health of the laboratory—the high service expense possibly indicates older equipment. Laboratory Large, however, is making relatively large investments in DNA laboratory equipment and information technology, and is investing nearly twice that of the average laboratory. This may bode well for the future, but may also signal a need to monitor the use of equipment for efficiency versus overfunding.

Finally, note the final line category of laboratory "surpluses." Laboratory Small has little excess; Laboratory Large has a much higher percentage of unused funds compared to both Small and the industry average. Further investigation is warranted. Is Large overfunded, too big to be run efficiently, or even too slow to operate? Could the seemingly lower funding for personnel be masked by having too many unfilled positions? The benefit of commonsize statements is in raising these types of red flags to help direct the laboratory manager's inquiries.

A Common Nomenclature

Gather together a group of forensic laboratory directors and ask each to define what they mean by a case, a sample, an item, a test, or a host of other terms and you will hear as many different responses as the number of individuals who are quizzed. Each laboratory jurisdiction has developed its own language by practice or perhaps by statute. Unfortunately, disparate language in the same science makes comparison and the development of industry standards difficult; as yet universally accepted industry measures for forensic laboratories are lacking. To overcome this difficulty, the 2003 European Quadrupol study began to put a common language in place when it studied the national forensic laboratories of four countries. That common set of definitions and measures from Quadrupol was adopted and enhanced in the 2006–2009 U.S. Foresight study. In the present analysis, this language is adopted for the key performance indicators that follow.

The complete Quadrupol/Foresight glossary may be found in Appendix A. That glossary includes the following definitions presented in Table 3, which will be used in the next section.

Additionally, the European Quadrupol and U.S. Foresight studies separated functions of forensic laboratories into a series of investigative areas. Individual forensic laboratories will likely have some differences in their organizational chart with the standard definitions from these two studies. However, some separation was required to permit the detailed information that can be learned and shared with the industry, and these investigative areas are adopted in the present paper. Appendix B details the definitions of 'investigation areas' from the Quadrupol/Foresight Projects.

Ratio Metrics

While laboratories may have similar missions, there are differences such as populations served, geographic coverage, jurisdiction, legislative mandates, breadth of investigative areas, staffing, and existing facilities, that make direct comparisons of productivity across laboratories difficult. To find meaningful metrics across the range of forensic laboratories, it is helpful to make adjustments for resources and responsibilities through the creation of ratios. Similar to the common-size analysis tool discussed above, ratios provide relative measures that make for easier comparisons across disparate operational entities. Ratios permit a laboratory to measure its performance over time, against specific other laboratory operations, and against industry (averages) standards. The ability of a laboratory to assess performance over time permits management to relate performance to changes in either scientific advancement or managerial prowess, for example.

There is a seemingly endless array of potential metrics that might be devised from the data in any forensic laboratory. Consider the various sources of data from a laboratory, which include information on casework, budgets, and personnel. In the analysis that follows, some of the possible ratios are discussed, but the list is not exhaustive. Individual laboratories will likely have other items that are of particular interest to the management team, the laboratory director, or to those with oversight responsibilities.

For discussion purposes, the ratios are broken down into five categories: efficiency, quality or risk management, analytical process, market conditions, and return on investment.

Efficiency Measures

Not all forensic laboratories are built, provisioned, or staffed in the same manner. However, given the resources available, we can ask the question: How was a laboratory or an investigative area within a forensic laboratory able to perform relative to the resources at its disposal? Consider some ratios that provide an indication of efficiency. One measure of the resources for a particular laboratory is the physical size of the facility. While it is natural to expect that a larger facility will be able to process a higher volume, we can track how well it performs relative to its available facility and make comparisons across the industry or across time.

Consider one such measure, CASE/SQ.m (area cases processed/laboratory area). Although each laboratory will differ according to personnel, plant and equipment, this measure provides a simple comparative measure across facility space allocations. We proxy those resources via the size of the laboratory (or the size dedicated to a given investigative area) and consider the output: production/analyses performed. An individual laboratory can then compare similar investigative areas within the laboratory as well as their performance compared to laboratories of any size. As a rough generalization, a higher ratio suggests some greater efficiency. When this ratio is compared to the same measure for

Table 3 Selective Quadrupol/Foresight Definitions

Term	Definition	Notation
Backlog	Open cases older than 30 days as measured at the end of the year	BACK30
Case - Area Case	A request for examination in one forensic investigation area. An area case is a subset of an institute case.	CASE
Case - Institute Case	A request from a forensic laboratory "customer" that includes forensic investigations in one or more investigative areas.	ICASE
Full-Time Equivalent (FTE)	The work input of a full-time employee working for one full year.	FTE
Investment Expense	Sum of purchases of equipment, etc., with a lifetime longer than three years and a cost above $1,000 (alternatively capital expenses).	KEXP
Item	A single object for examination submitted to the laboratory. Note: one item may be investigated and counted in several investigation areas.	ITEM
Laboratory Area	Floor area used for forensic investigation, including sample and consumable storage rooms.	SQFT or SQ.m
Overhead Time	Total FTEs in hours in the investigation area subtracted by the total hours of casework, research & development, education & training, and support & service given to external partners.	OVRHD
Personnel Expense	Sum of direct salaries; social expenses (employer contribution to FICA, Medicare, workers comp, and unemployment comp); retirement (employer contribution only towards pensions, 401k plans, etc.); personnel development and training (internal or external delivery, including travel); and occupational health service expenses (employer contribution only).	LEXP
Report	A formal statement of the results of an investigation, or of any matter on which definite information is required, made by some person or body instructed or required to do so.	REPORT
Sample	An item of evidence or a portion of an item of evidence that generates a reported result.	SAMPLE
Test	An analytical process, including but not limited to visual examination, instrumental analysis, presumptive evaluations, enhancement techniques, extractions, quantization, microscopic techniques, and comparative examinations. This does not include technical or administrative reviews.	TEST
Total Expense	The sum of the direct expenses (personnel, operating, and investment) and any administrative or other overhead expenses.	TOTEXP
Total Funding	The sum of all funding sources including jurisdictional budgeting, grant awards, bequests, and revenue sources.	TOTFUND

peer laboratories, the measure offers the laboratory and its stakeholders an opportunity to assess the general productivity against an appropriately selected benchmark. For example, benchmarks might be specifically identified laboratories with similar missions or services, or with industry averages.

Other productivity measures of great value compare the productivity of capital or labor. Consider the largest single expenditure in most laboratories—investment in personnel. The measure TEST/FTE (tests performed/fulltime equivalent employees) offers a quick measure of the output per person in the laboratory. This is an easily calculated measure that provides both comparison to peer laboratory benchmarks and offers a regular (monthly, quarterly, or annual) measure for trend analysis within a laboratory.

The efficiency measures and benchmarks a forensic laboratory manager should choose are a local decision to meet the informational needs of the various stakeholders of the laboratory. In addition to those above, consider the following possible ratio measures:

ITEM/FTE = Items processed/FTE employees
SAMPLE/FTE = Samples processed/FTE employees
TEST/FTE = Tests completed/FTE employees
REPORT/FTE = Reports completed/FTE employees
CASE/SQFT = Area cases processed/laboratory square feet area
ITEM/SQFT = Items processed/laboratory square feet area
SAMPLE/SQFT = Samples processed/laboratory square feet area
TEST/SQFT = Tests completed/laboratory square feet area
REPORT/SQFT = Reports completed/laboratory square feet area
ITEM/SQ.m = Items processed/laboratory square meter area
SAMPLE/SQ.m = Samples processed/laboratory square meter area
TEST/SQ.m = Tests completed/laboratory square meter area
REPORT/SQ.m = Reports completed/laboratory square meter area
FTE/FTEBUD = Actual FTE employees/budgeted FTE employees

Each of these measures is created from easily obtained laboratory information. The periodic calculation of these measures, whether annually, quarterly, or monthly, provides an immediate data-analysis tool for management. Each investigative area within a forensic laboratory can track ratios of interest and either compare that period's performance to past performance or make comparisons to an industry average or to specific peer labs to gain a sense of productivity.

These measures of efficiency are the type of measures that are followed in most industries. For example, capacity utilization measures are usually important to track. Since resources are limited whereas the potential output is unlimited, a laboratory may want to keep track of how well it uses the resources at its disposal. How quickly does management respond to vacancies and fill those vacancies? The last of the listed measures compares the allocation of resources versus the budgeting of those resources. Further, given each laboratory's uniqueness, it may have additional efficiency measures that may be followed to track other concerns, issues, or points of emphasis.

Quality/Risk Management

The regular calculation and reporting of ratios can help monitor the laboratory in other areas besides efficiency. A second such category relates to measures of qualityor risk-management measures. A laboratory faces multiple problems with its processing of cases, given limited resources. Errors may occur from either commission or omission. The formermay receive the bulk of the attention as errors may occur from poor calibration of equipment, human error, insufficient samples, less sophisticated techniques, etc. Likewise, errors of omission may result from delays in processing evidence. These delays may result in contamination, recidivism, or other losses.

One quality measure is the separate testing of the accuracy of a laboratory's normal output in which a random selection of prior test results is selected and retested to verify the accuracy of the original testing. At the conclusion of the retesting of this random sample of procedures, an error rate can then be calculated and used as a sample measure of quality. While this independent verification of results is a normal part of most quality-assurance programs in any forensic laboratory, it does represent a statistical sample, is not a census of the entire laboratory, and is subject to individual laboratory choices on allowable error in the sampling process. Going back to the efficiency metrics discussed above, some labs are required to retest every sample. Such double-testing assures high quality but would most likely result in relatively lower return-oninvestment measures when compared across the industry

There are proxies that address quality and do not require the collection of any additional data. Consider the measure TEST/CASE (tests completed/cases processed). The quality-assurance processes of the previous paragraph would be partially revealed in this measure. That is, a quality-assurance program within a laboratory that increased the sample size for retesting would be revealed as a higher numerator (tests completed) in this measure. This would also be true for any case in which a greater number of separately revealing examinations were conducted. Both the test error rate and the ratio TEST/CASE provide some monitoring of the errors of commission.

For errors of omission, a reasonable proxy would attend to the number of cases not yet processed. A ratio of BACK30/CASE (backlog cases in excess of thirty days/total cases processed) provides a useful trend measure that can be monitored from regularly collected data and used on a regular basis to assess the work not being completed.

As with efficiency, there are a variety of other ratio measures that are easily constructed and may have pertinent information content to an individual forensic laboratory. These include the following:

ERROR/SAMPLE = Commission sample error rate
ERROR/ITEM = Commission item error rate
ERROR/CASE = Commission case error rate
TEST/ITEM = Tests completed/items processed
TEST/SAMPLE = Tests completed/samples processed
SAMPLE/CASE = Samples processed/case processed

Analytical Process (Production Function)

Over time, new scientific achievements, new instrumentation, and new techniques are introduced that affect the processes by which samples are analyzed. Not every organization will select the same equipment or the same techniques for analysis. This is true in other industries as well as with forensic laboratories. For other industries, attention is placed with these choices through a look at the allocation of resources for capital equipment, human resources (labor), and other inputs into the production process.

A simple comparative ratio to reflect a laboratory's choices is the dollar allocation of resources from the common-size statements. If a laboratory has substituted more equipment for personnel, then the manager's question becomes one of assessing whether this change has a positive impact on the laboratory's output. A measure LEXP/TOTEXP (personnel [labor] expense/total expense) will have a variety of implications that may be tracked on a regular basis.

Other analytical process measures include a look at the other expenditure categories as a percentage of total expenditures, the capital expenditure to labor expense ratio analyzed, time management measures, or even the allocation of labor resources across investigative areas. Dale and Becker (2004) suggest alternative hiring practices to optimize the use of limited budgets and change the mix of scientists versus other technical workers in the laboratory for which these ratios offer some use.

KEXP/TOTEXP = Capital expense/total expense
OPEXP/TOTEXP = Operational expenses/total expense
KEXP/LEXP = Capital expense/labor expense
CASEWK/HRS = Casework hours/FTE hours
COURTHRS/HRS = Testimony hours/FTE hours
FTE_{sup}/FTE = Support staff FTE/total staff FTE
FTE_{Op}/FTE = Operational staff FTE/total staff FTE
FTE_{sci}/FTE_{Op} = Scientists FTE/operational staff FTE

Market Conditions

Local economic conditions may strongly influence the performance of an individual forensic laboratory. A forensic laboratory may be highly efficient, adopt the most productive analytical procedures, and optimize quality, yet fall short comparatively to other laboratories simply because of the local economic conditions, particularly in the labor market. For example, low-unemployment/highincome geographical areas may skew analytical comparisons to geographical areas with high-unemployment or lower wage rates. Further, changes in socio-economic conditions may affect the relative rates of forensics over time; this may in turn affect operational efficiency. To illustrate, consider a situation with declining economic performance and increasing forensic rates; such a socioeconomic condition may cause the forensic laboratory to reach its capacity, causing some operational dysfunction that was not present when forensic rates were lower and the forensic laboratory operated at optimal volumes. Or across countries, there may be social/cultural choices that have an effect on individual laboratory performance that need to be part of operational comparisons.

The Bureau of Labor Statistics (www.bls.gov), the Census Bureau (www.census.gov), state economic authorities, and other public sources (www.usa.gov) provide individual market data for comparison. But the data from individual laboratories also reflect information about these market conditions and are readily available to the forensic laboratory director and managers. A combination of data from the budgetary process and from a laboratory's personnel database and case volume metrics can permit some regular monitoring of the local market effects on performance.

Consider the average worker compensation as one such indicator of market conditions. The ratio of LEXP/FTE (personnel expense/full-time equivalent employees) provides the average expense across all employee compensation. Because the cost of living will vary greatly across locations, it will be more expensive to hire in high cost-of-living markets as the forensic laboratory is competing with organizations in many industries for the skilled workers it must employ. Similarly, a laboratory could track the expense attributed to other variable costs in the analytical process, VC/CASE (non-labor operational expenses/cases processed).

Return on Investment (ROI)

Ultimately, the questions to be answered come down to the overall performance of the forensic laboratory.

- How does the laboratory compare to its peers?
- Is the forensic laboratory taking sufficient care of the funds provided by its stakeholders?
- Does management have the laboratory on a path of continuous improvement?

To evaluate the return, we first must address what the laboratory is attempting to accomplish in such a way that comparisons across laboratories may yield meaningful results.

"If you don't know where you are going, then any path will get you there." It becomes paramount that organizations take a careful look at their mission, vision, and values, and establish goals that are consistent with their charge. In the case of forensic laboratories, there are some commonalities, but the underlying goals and affiliated optimization decision will differ across organizations. There are three general categories into which organizations tend to fall, and all three categories apply to the broadly defined industry of forensic laboratories. The categories include for-profit, not-for-profit, and government entities.

When economists evaluate an organization in a forprofit industry, the optimization problem is relatively simple. The decisions of the business are guided by a wealth-maximization criterion and actions become relatively easy to judge based upon that standard. And, although we don't know what the highest attainable level of wealth might actually be, we can determine whether managerial actions have improved the performance of the business or resulted in decline. That is, either an action served to increase stock price (or some other appropriate measure of wealth such as profits) or it did not.

When it comes to the not-for-profit or government entity, the goals are a little more disparate. However, we can make some generalizations that extend to the industry of forensic laboratories. Each nonprofit organization has a goal (mission) that is targeted towards some sense of "greater good" that for-profit market solutions do not adequately address. The nonprofit organization tries to alleviate as much pain and suffering or to achieve as much of the "greater good" as it can with the resources at its disposal. And, for these organizations, the resources rarely, if ever, are adequate enough to declare, "Mission accomplished." There is always more to be done. For the nonprofit organization, the optimization problem still exists but is operationalized by maximizing resources, funding, or revenues, and then using these funds to achieve as much "good" as possible.

For the government entity, leadership faces a problem similar to that of the not-for-profit organization. Its sights are set on some "greater good" that pure for-profit market solutions have failed to adequately address. Managerial decisions in this environment differ from that of the nonprofit manager in that funding/resources/revenues are predetermined by some extra-organizational (e.g., legislative) process, and the manager is left with an exogenous budget constraint. Given a budget for funding, the optimization problem for management becomes one of getting the most output from the limited funds. However, the other side of the coin is that managers, by maximizing output for a given budget, are actually minimizing the cost per unit of output.

Summarizing, the generalization of objectives for each type of organization:

- Not-for-profits maximize revenue;
- Government entities minimize cost; and
- For-profit organizations maximize the difference between revenue and cost.

Because the government organization dominates the forensic laboratory industry, the ROI measures that are most appropriate address the mission of government forensic laboratories. Consider then the mandate to correctly process as many cases as possible for a given budget allocation. The measure CASE/TOTEXP (area cases processed/total expenditures) is a tractable measure that has direct comparison across laboratories or across time for trends within an individual laboratory, and may be appropriate for comparisons across investigative areas that are similar in nature. The metric is easily computed from existing databases within the laboratory and can be a regularly (monthly, quarterly, annually) produced measure at the laboratory level or constructed by investigative area within the laboratory

Note that the inverse of this ratio, cost per case (TOTEXP/CASE), is also an item of interest to the laboratory and all of its stakeholders. In fact, the prior discussion on maximization of the cases processed for a given budget allocation may be turned around as the problem of minimizing the average cost of processing a case. Additionally, objectives may be measured in terms of reducing backlog as a measure of return.

There are other related ROI measures that may be computed and tracked. Some of these alternatives:

ITEM/TOTEXP = Items processed/total expenditures
SAMPLE/TOTEXP = Samples processed/total expenditures
TEST/TOTEXP = Tests completed/total expenditures
REPORT/TOTEXP = Reports completed/total expenditures
BACK30/TOTEXP = Backlog cases/total expenditures

Cautions and Conclusions

Forensic laboratories routinely generate data from casework performance across investigative areas, personnel and budget allocations, and corresponding expenses. This article considers ways in which laboratories can make data-driven managerial decisions through the regular extraction of data to create key performance indicators. The techniques borrow from the experience in other industries and are modified for the specific needs of forensic laboratories.

From accounting we learn how common-size financial statements can be easily created to transform the "financial statements" of individual laboratories into sizeadjusted analytical tools that can be easily compared against peer labs and industry standards. From business finance we learn that transforming absolute dollarperformance measures into size-adjusted ratios permits an evaluation of the allocation of resources into measures of efficiency, quality/risk management, analytical process choices, market conditions, and return on investment.

These key performance indicators can then be compared to peer laboratory performance and/or be used to determine in-house trends for the proper management of the scarce resources at its disposal or to provide quantitative support for the acquisition of additional resources. As the leading organizations in the industry begin to assess and adopt a common set of measures through the determination of industry-wide standards for data collection, then the stories may be told by the leaders in the industry in an effort to identify best practices. Dissemination of those success stories and adoption of similar practices offers a winning opportunity for all.

Some cautionary tales must also be borrowed from the application of these measures in other industries. It is a natural tendency to lose sight of the organizational goals and begin to manage (i.e., influence) the measures. Remember that the ratios serve to proxy performance in the individual areas of concern. No single ratio should be evaluated in isolation. There is a natural temptation to play to the measures and lose sight of the goals. For example, with a goal of a lower cost per case, one way for forensic laboratory managers to improve the metric is to assume more risk, for example, reducing quality-assurance activities. But that risk may prove dangerous and even devastating in the long run.

Implementation of the ratios into the management process can be tremendously beneficial to laboratories, but stakeholders must be reminded to maintain the balance between return, risk, and efficiency to achieve the organizational goals and avoid the pitfalls that have confronted other industries.

References

Bielski, L. 2007. KPI: Your metrics should tell a story. *ABA Banking J.* 10: 66–68.

Brooks, M.G. 2008. Getting a handle on analytics for strategic success. *Healthcare Finan. Manag.* 7: 100–104.

Dale, M.W., & W.S. Becker. 2004. A case study of forensic scientist turnover. *Forensic Sci. Commun.* 6: 1–9.

Kaskinen, J. Creating a best-in-class KPI program. 2007. *Strategic Finance.* 10: 29–33.

Koussiafes, P.M. 2004. Public forensic laboratory budget issues *Forensic Sci. Commun.* 6: 1–7.

Love, D., L. Revere, & K. Black. 2008. A current look at the key performance measures considered critical by health care leaders. *J. Health Care Finance* 34: 19–33.

Petaschnick, J. 2008. Using key performance indicators wisely. *Health Care Collector*, 22: 7–9.

Reilly, F.K. 1997. The impact of inflation on ROE, growth and stock prices. *Finan. Serv. Rev.*, 6: 1–17.

Smith, F.O. 2007. Asset management and the search for actionable intelligence. *Control Eng.*, 12: 40–44.

Appendix A

Nomenclature Established by Quadrupol and U.S. Foresight Projects

Assistant/Analyst An individual carrying out general casework examinations or analytical tests under the instruction of a reporting scientist or reporting analyst and who is able to provide information to assist with the interpretation of the tests.

Backlog Open cases older than 30 days as measured at the end of the year.

Case - Institute Case A request from a forensic laboratory "customer" that includes forensic investigations in one or more investigative areas.

Case - Area Case A request for examination in one forensic investigation area. An area case is a subset of an institute case.

Casework All laboratory activities involved in examination of cases.

Casework time Total FTEs for the operational personnel in the investigation area (in hours) subtracted by the hours of R & D and E & T, and support and service given to external partners.

Crime Perceived violation of the law that initiates a case investigation.

Direct Salary Total salary paid to employees, including overtime compensations, vacation salary, bonuses, etc.

Examinations (Exams) The word QUADRUPOL used for "test"; see both "test" and "sample" in this glossary for the changes adopted by U.S. Foresight.

Facility Expense Sum of rents, cleaning, and garbage collection, security, energy, water, communication, ICT infrastructure, and facility maintenance.

Floor Area Total of all floor area including office, laboratory, and other.

Full-Time Equivalent (FTE) The work input of a full-time employee working for one full year.

Full-time Researcher A forensic scientist whose primary responsibility is research and who is not taking part in casework.

Investigation Area Area limited by item type and methods as they are listed in the benchmarking model.

Investment Expense Sum of purchases of equipment, etc. with a lifetime longer than three years and a cost above $1,000 (alternatively, capital expenses).

Item A single object for examination submitted to the laboratory. Note: One item may be investigated and counted in several investigation areas.

Laboratory Area Floor area used for forensic investigation, including sample and consumable storage rooms.

Non-reporting Manager An individual whose primary responsibilities are in managing and administering a laboratory or a unit thereof and who is not taking part in casework.

Office Area Floor area of offices (square feet).

Operational Personnel Personnel in operational units providing casework, research and development (R & D), education and training (E & T), and external support services. Non-reporting unit heads are included.

Other Area Floor area of space not belonging to laboratories or offices, i.e., corridors, lunch corners, meeting rooms etc. (square feet).

Overhead Time Total FTEs in hours in the investigation area subtracted by the total hours of casework, R & D, E & T, and support and service given to external partners.

Personnel Expense Sum of direct salaries; social expenses (employer contribution to FICA, Medicare, workers comp, and unemployment comp); retirement (employer contribution only towards pensions, 401k plans, etc.); personnel development and training (internal or external delivery, including travel); and occupational-health service expenses (employer contribution only).

Report A formal statement of the results of an investigation, or of any matter on which definite information is required, made by some person or body instructed or required to do so.

Reporting Analyst An analyst responsible in non-complicated cases (e.g., simple drugs analysis) for performing the examination of the items submitted, interpreting the analysis results, writing the analysis report and, if necessary, providing factual evidence for the court.

Reporting Scientist The forensic scientist responsible in a particular case for performing or directing the examination of the items submitted, interpreting the findings, writing the report, and providing evidence of fact and opinion for the court.

Representation Expense The costs for hosting guests: **lunches, dinners, coffees offered by the lab, and presents given to guests or during visits abroad, etc.**

Running Operational Expense Others cost than investment costs, personnel costs, and facilities costs, e.g. consumables, traveling, QA, literature, contracting, representation, service and maintenance, information, and advertisement.

Sample An item of evidence or a portion of an item of evidence that generates a reported result. Scientist in Training An individual with no reporting rights being trained to become a reporting scientist.

Student Hours The sum of teaching hours in a course multiplied by the number of students attending the particular course.

Support Personnel Forensic laboratory staff providing various internal support services. Management and administration personnel not belonging to the operational units are included.

Teaching Hours Time spent teaching in the lecture room in hours (60 minutes).

Test An analytical process, including but not limited to visual examination, instrumental analysis, presumptive evaluations, enhancement techniques, extractions, quantifications, microscopic techniques, and comparative examinations. This does not include technical or administrative reviews.

Total Expense The sum of the direct expenses (personnel, operating, and investment) and any administrative or other overhead expenses.

Total Funding The sum of all funding sources including jurisdictional budgeting, grant awards, bequests, and revenue sources.

Total Items Includes all items to which the laboratory assigns an item or tracking number. This is different than the number of items the laboratory receives (the laboratory may split items up for analysis).

Workload Total time spent on all work related to job, including overtime. Accident Investigation All non-traffic accident investigations, such as work-related accidents. Biology (Non-DNA) The detection, collection, and non-DNA analysis of biological fluids.

Blood Alcohol The analysis of blood or breath samples to detect the presence of and quantify the amount of alcohol.

Computer Analysis The analysis of computers, computerized consumer goods, and associated hardware for data retrieval and sourcing.

Crime Scene Investigation The collection, analysis, and processing of locations for evidence relating to a criminal incident.

Digital Evidence - Audio & Video The analysis of multimedia audio, video, and still-image materials, such as surveillance recordings and video enhancement.

DNA Casework Analysis of biological evidence for DNA in criminal cases.

DNA Database Analysis and entry of DNA samples from individuals for database purposes.

Appendix B

Definition of Investigation Areas from the Quadrupol and U.S. Foresight Projects

Document Examination The analysis of legal, counterfeit, and questioned documents, excluding handwriting analysis.

Drugs - Controlled Substances The analysis of solid dosage licit and illicit drugs, including precursor materials.

Entomology Forensic entomology is the application of the study of arthropods, including insects, to criminal or legal cases.

Evidence Screening & Processing

The detection, collection, and processing of physical evidence in the laboratory for potential additional analysis.

Environmental Analysis The analysis of naturally occurring materials, such as soil or water, for foreign substances with criminal implications.

Explosives The analysis of energetic materials in pre- and post-blast incidents.

Fingerprints The development and analysis of friction ridge patterns.

Fire Analysis The analysis of materials from suspicious fires to include ignitable liquid residue analysis.

Firearms and Ballistics The analysis of firearms and ammunition, to include distance determinations, shooting reconstructions, NIBIN, and toolmarks.

Forensic Engineering and Material Science

Failure and performance analysis of materials and constructions.

Forensic Pathology Forensic pathology is a branch of medicine that deals with the determination of the cause and manner of death in cases in which death occurred under suspicious or unknown circumstances.

Gun Shot Residue (GSR) The analysis of primer residues from discharged firearms (not distance determinations). Hairs & Fibers The analysis of human and animal hairs (non-DNA) and textile fibers as trace evidence. Handwriting The evaluation of handwritten materials to categorize or identify a writer.

Marks and Impressions The analysis of physical patterns received and retained through the interaction of objects of various hardness, including shoeprints and tire tracks.

Odontology The identification of human remains through dental materials, for example by postmortem X-rays of the teeth compared to antemortem X-rays. Some forensic odontologists also analyze and compare bite marks.

Paint & Glass The analysis of paints—generically, coatings—and glass as trace evidence.

Road Accident Reconstruction Analysis of criminal incidents involving vehicles and accidents (hit and run, for example). Speech & Audio The analysis of live and recorded vocalizations in criminal investigations.

Toxicology, Ante-Mortem Toxicology involves the chemical analysis of body fluids and tissues to determine if a drug or poison is present in a living individual, to include blood alcohol analysis (BAC).

Toxicology, Post-Mortem Toxicology involves the chemical analysis of body fluids and tissues to determine if a drug or poison is present in a deceased individual.

Trace Evidence The analysis of materials that, because of their size or texture, transfer from one location to another and persist there for some period of time. Microscopy, either directly or as an adjunct to another instrument, is involved.

Article: The Decomposition of Return on Investment for Forensic Laboratories

3.2

PAUL J. SPEAKER
West Virginia University, College of Business and Economics, P.O. Box 6025, Morgantown, WV 26506-6025, USA

Introduction

Leaders across all types of organizations seek means for continuous improvement in the performance of their organizations. When it comes to practicing methods that efficiently tie daily performance to success of the organization, managers of forensic laboratories face similar problems to those faced by hospital administrators or corporate chief executive officers. Constant feedback is necessary for timely action. The determination of the metrics most appropriate for each industry is the challenge.

To determine which metrics will be most valuable, it is critical to start with an examination of the mission, vision, and values of an organization. What are the goals of an organization and what does it hold to be most valuable? Good metrics will point toward success along the way, toward important goals, and toward change in policy or practice to better reach organizational goals.

Unfortunately, the service industry of forensic laboratories does not follow a standard practice with respect to the collection and publication of data that might assist in the management of resources. Recently, two studies of-fered some hope for the standardization of language and the development of tools to meet the goals of forensic laboratories. The first study, QUADRUPOL—Development of a Benchmarking Model for Forensic Laboratories (2003), offered a standardized definition set for measurement of the inputs and non-financial outcomes of forensic laboratories in Europe. The second study, FORESIGHT NIJ— Measure, Preserve, Improve, adopted the foundation created in the QUADRUPOL study and expanded it to include connections between casework, operational expenditures, and personnel detail for laboratories across North America.

In this paper, the language of these studies is preserved as various metrics are considered. The next section examines a simple analytical technique that breaks down return on investment goals into component parts of tractable metrics for comparison for a laboratory over time or for comparison across the industry.

Connecting Goals to the Optimization Problem

Consider the performance of the forensic laboratory attempting to successfully process as many cases as possible given its level of funding. The ratio of cases processed to total expenditures provides one possible description of the return on investment (ROI) for a forensic

 DOI: 10.4324/9780429318528-8

laboratory, and it is a measure that is comparable to other laboratories. Additionally, the inverse of the measure represents a much-requested measure: average cost per case. In the presentation of various metrics for forensic laboratories, Speaker (2009) cautions about the potential abuse that can occur when attention is limited to a single measure. Fortunately, authors in several disciplines (see Burns, Sale, & Stephan, 2008; Pares, 1980; and Soliman, 2008, for some examples) have presented the mechanisms by which that caution can be mitigated.

To see how a laboratory might be able to extract valuable information from some common metrics, begin with the ROI measure above. By measuring the cases processed to the total laboratory expenditures, we construct a metric for ROI that can be compared across laboratories and across time.

$$\frac{CASE}{TOTEXP} = \frac{AreaCasesProcessed}{TotalExpenditures} \tag{1}$$

where the variables above follow the definitions from the QUADRUPOL/FORESIGHT studies, namely:

- An area case refers to a request for examination in one forensic investigation area.
- Total expenditures include the sum of the direct expenses (personnel, operating, and investment) and any administrative or other overhead expenses.

(Note that a full listing of the common definitions adopted by the European Quadrupol and U.S. Foresight studies is compiled in Appendix A.)

Once a laboratory begins to track this ROI measure, there are several ways in which forensic laboratory managers might increase this ratio over time. Strategies to increase efficiency or to modify processes may have a very positive effect on performance; but it is also possible to improve this ROI measure by undertaking practices that might not be so desirable, particularly via increases in risk taking. A decomposition of the ROI measure improves the information content of the return ratio by breaking it down into a variety of components to express managerial decisions and other conditions that contribute to the return. That is, can we devise a usable relationship between ROI and the various components that influence the rate of return?

$$ROI = f\left(efficiency,\ quality\,/\,risk\ management,\ analytical\ process,\ market\ conditions\right) \tag{2}$$

Examples of alternative ratios in each of these categories are described by Speaker (2009). These include amongst others:

- Efficiency measures such as cases processed per fulltime equivalent employee;
- Quality/risk management measures such as tests performed per area case;
- Analytical process measures such as personnel expenditures/total expenditures; and
- Market condition measures such as the average compensation per employee.

The key for such a breakdown was developed in 1919 by DuPont executive F. Donaldson Brown as he attempted to analyze another company in which Dupont had made a substantial investment. Brown realized that higher ROI resulting from improvements in efficiency were highly desirable, while higher ROI that was merely accompanied by higher risk was to be avoided. The technique, a DuPont expansion, calls for the transformation of the ROI measure into component parts via an expansion from multiplication by the number 1 in various forms.

To demonstrate, first consider some additional notation and the definitions from the QUADRUPOL/FORESIGHT nomenclature. The present ROI expansion will occur when we multiply the ROI measure by the following forms of the number 1.

$$1 = \frac{LEXP}{LEXP} = \frac{FTE}{FTE} = \frac{TEST}{TEST}$$

where the notation and QUADRUPOL/FORESIGHT definitions are:

- LEXP refers to personnel (labor) expenditures and includes the sum of direct salaries, social expenses (employer contribution to FICA, Medicare, workers' comp, and unemployment comp), retirement (employer only contribution toward pensions, 401k plans, etc.), personnel development and training (internal or external delivery, including travel), and occupational health service expenses (employer contribution only).
- FTE or full-time equivalent employee is the work input of a full-time employee working for one full year.
- TEST represents tests completed. A test is an analytical process including, but not limited to, visual examination, instrumental analysis, presumptive evaluations, enhancement techniques, extractions, quantization, microscopic techniques, and comparative examinations. This does not include technical or administrative reviews.

Consider the insertion of these alternative expressions of the number 1 as multiplicative terms in the relationship defined in (1).

$$\frac{CASE}{TOTEXP} = \frac{CASE}{TOTEXP} \times \frac{LEXP}{LEXP} \times \frac{FTE}{FTE} \times \frac{TEST}{TEST} \tag{3}$$

By rearranging the order of the terms in the numerator and denominator, we unveil other metrics that have meaning with respect to the arguments in the functional relationship described in expression (3).

$$\frac{CASE}{TOTEXP} = \frac{CASE}{TEST} \times \frac{LEXP}{TOTEXP} \times \frac{FTE}{LEXP} \times \frac{TEST}{FTE} \tag{4}$$

$$\frac{CASE}{TOTEXP} = \frac{\dfrac{TEST}{FTE} \times \dfrac{LEXP}{TOTEXP}}{\dfrac{LEXP}{FTE} \times \dfrac{TEST}{CASE}} \tag{5}$$

Or, using the ratio definitions provided in Speaker (2009), this becomes:

$$ROI = \frac{LaborProductivity \times LaborExpenseRatio}{AverageCompensation \times TestingIntensity} \tag{6}$$

While this represents only one of many potential breakdowns of the return on investment measure, it has an advantage in that it addresses each of the categories from the functional form in (2) and has an appealing message to laboratory managers and stakeholders. Consider first the two items in the numerator. As either ratio increases, holding the other ratios constant, ROI increases. The first measure, labor productivity, is an efficiency measure. As workers become more productive and successfully complete more tests per person, the overall case output is increased. The second numerator ratio captures part of the analytical process for the laboratory. For most laboratories, personnel expense represents the highest category of expenses and provides more immediate returns in the present than expenditures toward capital. This suggests a mixed blessing. While the present return is higher from greater dedication of resources toward immediate application to casework (rather than investment in equipment for future application), there may be longer-term consequences that could reduce future ROI.

The two terms in the denominator capture market conditions and quality considerations. The higher the average compensation, the lower will be the return on investment. This proxy for market conditions is an important qualifier for inter-laboratory comparisons. The ROI for a laboratory is dependent upon local market conditions and, in markets with high cost of living, the forensic laboratory may have no choice in what it must pay as it competes with all other local industry for skilled employees. A forensic laboratory can't just decide to ignore the market or it will not be able to hire adequately. Alternative explanations for the size of this ratio then fall to other changes that may be related to the average compensation. See Dale and Becker (2004) for an examination of a change in the analytical process that has an impact on average compensation.

The other item in the denominator, the ratio of tests performed to cases analyzed, provides some indication of risk/quality management. Generally, when a laboratory conducts more tests for a given case, regardless of whether the number of tests is dictated by statute, policy, or practice, the increased testing will reduce the overall return on investment unless the additional testing can overly influence productivity. This last comment regarding the ROI expansion permits a more detailed analysis of proposed changes in risk management and begs the question as to whether a statutory requirement or policy shift in testing is worth the tradeoff on performance.

Notice that the inverse of expression (5) addresses a critical concern to forensic laboratories. The maximization of the return on investment (CASE/TOTEXP) is alternatively expressed as the minimization of the average cost per case for a given budget constraint. Inverting (5) we obtain:

$$\frac{TOTEXP}{CASE} = \frac{\frac{LEXP}{FTE} \times \frac{TEST}{CASE}}{\frac{TEST}{FTE} \times \frac{LEXP}{TOTEXP}} \tag{7}$$

Removing the notation and replacing with the definitions, expression (7) becomes:

$$\frac{AverageCost}{PerCase} = \frac{AverageCompensation \times TestingIntensity}{LaborProductivity \times LaborExpenseRatio} \tag{8}$$

This expression should appeal to the manager's common sense. When market conditions dictate that personnel expenses will be higher for a given service, average costs rise. As more tests for a case are conducted, costs will also rise. Turning to the denominator we also expect costs to fall as employees are more productive or when immediate resources are dedicated toward current period output rather than future investment.

ROI Expansion Applications

Because the ROI expansion is expressed entirely in ratios, forensic laboratories of any size may be compared via metrics easily collected on a regular (weekly, monthly, quarterly, or annually) basis. To illustrate, consider three fictional forensic laboratories and their information from records of casework, personnel, and budgets. Note that the analysis presented in the previous section can be applied to the entire laboratory or to individual investigative areas within a forensic laboratory. Because the mission, location, or other choices of each forensic laboratory may call for differences in services provided and in the intensity of those activities within the overall laboratory, the analysis for a single investigative area is highlighted in the example to follow.

The three laboratories, S, M, and L, each have some common investigative areas, but differ in the total investigative services rendered. Because of the different scope of services, a laboratory-wide comparison of the ratios would be difficult to interpret. However for a common investigative area for each laboratory, a meaningful comparison can be obtained. Suppose for this example that each of these three forensic laboratories has a fingerprint identification section. (Appendix B defines each of the investigative areas from the QUADRUPOL and FORESIGHT studies and includes the select area, fingerprint identification.) For the fingerprint identification investigative area, the three laboratories' data are reported in Table 1.

Direct comparison is very difficult, given the different sizes for the three forensic laboratories, but conversion to ratios can enable a worthwhile comparison. Although laboratory L processes more than eight times as many cases laboratory S, it does so with a substantially larger budget. Using the ROI measure defined in (1), we find:

ROIS = 0.0066 cases per dollar or an average cost per case of $152.17;
ROIM = 0.0043 cases per dollar or an average cost per case of $234.43; and
ROIL = 0.0025 cases per dollar or an average cost per case of $404.15.

Using either the return on investment or the average cost measure suggests a much higher level of performance for laboratory S versus its larger counterparts. It is somehow able to do much more with its limited resources.

Table 1 Key Laboratory Metrics

Fingerprint Identification	Laboratory S	Laboratory M	Laboratory L
Cases Processed (CASE)	1,150	3,050	9,650
Total Expenditures (TOTEXP)	$175,000	$715,000	$3,900,000
Personnel Expenditures (LEXP)	$155,000	$445,000	$3,075,000
Full-time Equivalent Employees (FTE)	2.00	5.35	33.00
Tests Performed (TEST)	4,100	14,300	58,300

The ROI measures indicate for these three laboratories that the larger the laboratory, the lower the return. Is this an indication of economies of scale or is it something else? Breaking down the data into the components described by (8) for average cost per case reveals more useful information for the comparison of the laboratories (Table 2).

The decomposition of the average cost per case ratio into the four component costs begins to reveal the reasons for the differing performances for the three laboratories. First, when average compensation (LEXP/FTE) is higher, so is average cost per case. This is partly a product of market conditions. That is, higher cost of living environments tend to also have a higher salary scale, and the laboratory cannot ignore local economic conditions, hence higher costs for operations.

Of course market conditions alone do not fully explain the disparity in the average compensation. Other supporting explanations for higher compensation include greater productivity or a different mix of scientists, analysts, and support staff. The greater productivity argument appears to be at play in this case for some of the laboratories. Consider the ratio for personnel productivity (TEST/FTE) in the fifth column. The average compensation for laboratory M is higher than the average compensation for laboratory S, but this seems to be justified by the relatively higher productivity of the staff at laboratory M. For laboratory L, however, these two columns raise a red flag as Laboratory L pays the highest compensation but has the least productive workers of the three laboratories. Here the decomposition points to an area for greater inspection.

The testing intensity ratio (TEST/CASE) also appears to be directly related to the average cost per case. Higher testing is used as a proxy for risk—the more tests for a case, the lower the risk of an error of commission and the higher the quality assurance. That additional testing comes at a cost and increases the average cost per case. The increased testing also has indirect costs, potentially increasing errors of omission, as backlog is increased ceteris paribus with additional testing. As with the market measure, the full explanation likely involves more than just risk. Greater testing may also be reflective of greater complexity to an average case or to other variables. There is a large disparity in the testing behavior of these three laboratories. Again, a red flag is raised with the substantially lower testing level (and resultant lower average cost) for laboratory S and substantially higher testing level of laboratory L. Is the former laboratory cutting corners or is it simply more efficient? And is the latter laboratory testing too much? Are there statutory reasons, policies, or practices that call for the testing levels? The differences beg explanation and should be investigated.

The labor expense ratio (LEXP/TOTEXP), coupled with the other ratios, helps to complete the picture of the performance differentials. This measure suggests that laboratory S dedicates most of its expenditures for the here and now, and reserved very little in this year for investment in the future via capital expenditures. Laboratory M appears to be doing just the opposite with much smaller allocations to current period expenditures in the form of personnel expenses. While it could indicate some differential in analytical process via

Table 2 Ratio Expansion

	Average Cost per Case	*Average Compensation*	*Testing Intensity*	*Personnel Productivity*	*Labor Expense Ratio*
Laboratory S	$152.17	$77,500	3.57	2,050	88.57%
Laboratory M	$234.43	$83,178	4.69	2,673	62.24%
Laboratory L	$404.15	$93,182	6.04	1,767	78.85%

a more capital-intensive practice by laboratory M, it is more likely there are differences in capital expenditures. Since investment in equipment occurs at discrete times, while personnel expenditures are continuous, this ratio should always be checked alongside the average cost measure. Laboratory M might just be better at writing grants for equipment, but since it doesn't spread out the capital expense over time, it gives an appearance of more costly behavior.

In summary, if a review of a laboratory was limited to the ROI or average cost per case measure, then it might be concluded that laboratory S was the top performer among the three laboratories analyzed. However, the decomposition of these measures into the four component ratios points to specific areas for greater investigation before anointing laboratory S the best performer. While laboratory S seems reasonably efficient in its operations, it has been able to reduce costs by investing less in its future than the other laboratories and testing at a lower rate, perhaps sacrificing quality. Laboratory L is the most expensive laboratory and the decomposition offers several reasons why that is so. Laboratory L offers exceptionally high salary and benefits, has a less productive staff than the other laboratories, and may be testing more than necessary for the average case. All of these contribute to a poor relative performance and calls for a close review of policies, practices, and procedures. Laboratory M, on the other hand, appears to be the laboratory whose practices are exemplary. Personnel are paid a little more than the smaller laboratory, but those payments seem to be justified with a high level of productivity. Laboratory M also appears to be investing in its future with greater capital expenditures, which should enable a lowering of average costs in future periods.

Concluding Remarks

Ratio metrics provide a quick and convenient mechanism by which to track performance for a laboratory and a means by which to compare that performance with best practices within the industry. The ease of use comes with a caution to not rely solely on a single performance measure. Rather, a series of measures may be used to paint the full performance picture. The decomposition of the return on investment ratio or the average cost per case ratio shows that these performance metrics can be presented as a combination of metrics relating outcomes to efficiency, risk, analytical process, and market conditions.

The calculation of these ratio metrics can be performed on a regular basis across each investigative area within the laboratory and compared over time and across the industry. As such, they offer a simple, yet highly informative tool for a laboratory manager to use on a regular basis. The measures also point to areas of future research as information from high-performing laboratories is cataloged and the performance stories are shared with the rest of the industry. The decomposition presented above offers a foundation to share those stories.

References

Bielski, L. 2007. KPI: Your metrics should tell a story. *ABA Banking J.*, 99(10): 66–68.

Brooks, M.G. 2008. Getting a handle on analytics for strategic success. *Healthcare Financial. Manage.*, 62(7): 100–104.

Burns, D.C., J.T. Sale, & J.A. Stephan. 2008. A better way to gauge profitability: Systematic ratio analysis using the advanced DuPont model. *J. Accountancy*, 206(2): 38–42.

Dale, W.M., & W.S. Becker. 2004. A case study of forensic scientist turnover. *Forensic Sci. Commun.* 6(3): 1–9.
Kaskinen, J. 2007. Creating a best-in-class KPI program. *Strategic Finance*, 89(4): 29–33.
Koussiafes, P.M. 2004. Public forensic laboratory budget issues. *Forensic Sci. Commun.* 6(3): 1–7.
Love, D., L. Revere, & K. Black. 2008. A current look at the key performance measures considered critical by health care leaders. *J. Health Care Finance*, 34(3): 19–33.
Pares, A. 1980. The return on equity decomposition (ROED) and its importance to financial statement analysis. *J. Business Finance & Accounting*, 7(3): 365–375.
Petaschnick, J. 2008. Using key performance indicators wisely. *Health Care Collector*, 22(3): 7–9.
QUARUPOL. 2003. European Network of Forensic Science Institutes.
Reilly, F.K. 1997. The impact of inflation on ROE, growth and stock prices. *Financial. Serv. Rev.* 6 (1): 1–17.
Smith, F.O. 2007. Asset management and the search for actionable intelligence. *Control Eng.* 54(12): 40–44.
Soliman, M.T. 2008. The use of DuPont analysis by market participants. *The Accounting Rev.* 83(3): 823–853.
Speaker, P.J. 2009. Key performance indicators and managerial analysis for forensic laboratories. *Forensic Sci. Policy Manage.* 1(1): 32–42.

Appendix A

Nomenclature Established by QUADRUPOL and U.S. Foresight Projects

Assistant/Analyst An individual carrying out general casework examinations or analytical tests under the instruction of a Reporting Scientist or Reporting Analyst and who is able to provide information to assist with the interpretation of the tests.

Backlog Open cases that are older than 30 days as measured at the end of the year.

Case - Institute Case A request froma forensic laboratory "customer" that includes forensic investigations in one or more investigative areas.

Case - Area Case A request for examination in one forensic investigation area. An area case is a subset of an institute case.

Casework All laboratory activities involved in examination of cases.

Casework Time Total FTEs for the operational personnel in the investigation area (in hours) subtracted by the hours of R&D, and E&T and support and service given to external partners.

Crime A perceived violation of the law that initiates a case investigation.

Direct Salary Total salary paid to employees, including overtime compensations, vacation salary, bonuses, etc. **Examinations (Exams)** The word QUADRUPOL used for "test"; see both "test" and "sample" in this glossary for the changes adopted by U.S. Foresight.

Facility Expense Sum of rents, cleaning and garbage collection, security, energy, water, communication, ICT infrastructure, and facility maintenance.

Floor Area Total of all floor area including office, laboratory and other.

Full-Time Equivalent (FTE) The work input of a full-time employee working for one full year.

Full-Time Researcher A forensic scientist whose primary responsibility is research and who is not taking part in casework.

Investigation Area Area limited by item type and methods as they are listed in the benchmarking model.

Investment Expense Sum of purchases of equipment, etc., with a lifetime longer than three years and a cost above $1,000 (alternatively capital expenses).

Item A single object for examination submitted to the laboratory. Note: One item may be investigated and counted in several investigation areas.

Laboratory Area Floor area used for forensic investigation, including sample and consumable storage rooms. **Non-Reporting Manager** An individual whose primary responsibilities are in managing and administering a laboratory or a unit thereof and who is not taking part in casework.

Office Area Floor area of offices (square feet).

Operational Personnel Personnel in operational units providing casework, research and development (R & D), education and training (E & T), and external support services. Non-reporting unit heads are included.

Other Area Floor area of space not belonging to laboratories or offices, i.e., corridors, lunch corners, meeting rooms, etc. (square feet).

Overhead Time Total FTEs in hours in the investigation area subtracted by the total hours of casework, R&D, E&T, and support and service given to external partners. **Personnel Expense** Sum of direct salaries, social expenses (employer contribution to FICA, Medicare, workers' comp, and unemployment comp), retirement (employer contribution only towards pensions, 401(k) plans, etc.), personnel development and training (internal or external delivery, including travel), and occupational health service expenses (employer contribution only).

Report A formal statement of the results of an investigation, or of any matter on which definite information is required, made by some person or body instructed or required to do so.

Reporting Analyst An analyst is responsible in non-complicated cases (e.g., simple drugs analysis) for performing the examination of the items submitted, interpreting the analysis results, writing the analysis report and, if necessary, providing factual evidence for the court.

Reporting Scientist The forensic scientist responsible in a particular case for performing or directing the examination of the items submitted, interpreting the findings, writing the report, and providing evidence of fact and opinion for the court.

Representation Expense The costs for hosting guests, i.e., lunches, dinners, coffees offered by the lab, and presents given to guests or during visits abroad, etc.

Running Operational Expense Other costs than investment costs, personnel costs, and facilities costs, e.g., consumables, traveling, QA, literature, contracting, representation, service and maintenance, information and advertisement.

Sample An item of evidence or a portion of an item of evidence that generates a reported result.

Scientist in Training An individual with no reporting rights being trained to become a reporting scientist.

Student Hours The sum of teaching hours in a course multiplied by the number of students attending the particular course.

Support Personnel Forensic laboratory staff providing various internal support services. Management and administration personnel not belonging to the operational units are included.

Teaching Hours Time spent teaching in the lecture room in hours (60 minutes).

Test An analytical process including, but not limited to, visual examination, instrumental analysis, presumptive evaluations, enhancement techniques, extractions, quantifications, microscopic techniques, and comparative examinations. This does not include technical or administrative reviews.

Total Expense The sum of the direct expenses (personnel, operating, and investment) and any administrative or other overhead expenses.

Total Funding The sum of all funding sources including jurisdictional budgeting, grant awards, bequests, and revenue sources.

Total Items Includes all items to which the laboratory assigns an item or tracking number. This is different than the number of items the laboratory receives (the laboratory may split items up for analysis).

Workload Total time spent on all work related to job, including overtime.

Appendix B

Definition of Investigation Areas From the QUADRUPOL and U.S. FORESIGHT Projects

Accident Investigation All non-traffic accident investigations, such as work-related accidents.

Biology (Non-DNA) The detection, collection, and non-DNA analysis of biological fluids.

Blood Alcohol The analysis of blood or breath samples to detect the presence of and quantify the amount of alcohol.

Computer Analysis The analysis of computers, computerized consumer goods, and associated hardware for data retrieval and sourcing.

Crime Scene Investigation The collection, analysis, and processing of locations for evidence relating to a criminal incident.

Digital Evidence - Audio & Video The analysis of multi-media audio, video, and still-image materials, such as surveillance recordings and video enhancement.

DNA Casework Analysis of biological evidence for DNA in criminal cases.

DNA Database Analysis and entry of DNA samples from individuals for database purposes.

Document Examination The analysis of legal, counterfeit, and questioned documents, excluding handwriting analysis.

Drugs - Controlled Substances The analysis of solid dosage licit and illicit drugs, including pre-cursor materials.

Entomology Forensic entomology is the application of the study of arthropods, including insects, to criminal or legal cases.

Evidence Screening & Processing The detection, collection, and processing of physical evidence in the laboratory for potential additional analysis.

Environmental analysis The analysis of naturally occurring materials, such as soil or water, for foreign substances with criminal implications.

Explosives The analysis of energetic materials in pre- and post-blast incidents.

Fingerprints The development and analysis of friction ridge patterns.

Fire Analysis The analysis of materials from suspicious fires to include ignitable liquid residue analysis.

Firearms & Ballistics The analysis of firearms and ammunition, to include distance determinations, shooting re-constructions, NIBIN, and toolmarks.

Forensic Engineering & Material Science Failure and performance analysis of materials and constructions.

Forensic Pathology Forensic pathology is a branch of medicine that deals with the determination of the cause and manner of death in cases in which death occurred under suspicious or unknown circumstances.

Gun Shot Residue (GSR) The analysis of primer residues from discharged firearms (not distance determinations).

Hairs & Fibers The analysis of human and animal hairs (non-DNA) and textile fibers as trace evidence.

Handwriting The evaluation of handwritten materials to categorize or identify a writer.

Marks & Impressions The analysis of physical patterns received and retained through the interaction of objects of various hardness, including shoeprints and tire tracks.

Odontology The identification of human remains through dental materials, for example by postmortem X-rays of the teeth compared to antemortem X-rays. Some forensic odontologists also analyze and compare bite marks.

Paint & Glass The analysis of paints—generically, coatings—and glass as trace evidence.

Road Accident Reconstruction Analysis of criminal incidents involving vehicles and accidents (hit and run, for example).

Speech & Audio The analysis of live and recorded vocalizations in criminal investigations.

Toxicology, Antemortem Toxicology involves the chemical analysis of body fluids and tissues to determine if a drug or poison is present in a living individual, to include blood alcohol analysis (BAC).

Toxicology, Postmortem Toxicology involves the chemical analysis of body fluids and tissues to determine if a drug or poison is present in a deceased individual.

Trace Evidence The analysis of materials that, because of their size or texture, transfer from one location to another and persist there for some period of time. Microscopy, either directly or as an adjunct to another instrument, is involved.

4 Budgets

A budget is essentially a financial statement which lays out how your FSP plans to spend its money. Think of it as a monetary plan for how you'll meet your organizational goals. Budgets have two basic components: Sources of revenue (taxes, income, grants) and operating expenses (costs, like people and consumables). Revenue and expenses are plotted over time (your budgetary cycle or calendar), including drafting the budget, getting it approved, and receiving the funds. FORESIGHT helps in the budgeting process by providing realistic and measurable metrics that are aligned with the defined activities in your FSP's budget. Each activity you perform should have a specific dollar amount, both expense and revenue sources, associated with it.

Your budget is a tool. It should be stained, worn, reviewed, and used; it is part of a process, not an event. Mike Tyson, the professional boxer said, "Everybody has a plan until they get punched in the mouth." This echoes the old military saying of no plan survives first contact with the enemy. A corollary to this notion is "All budgets are wrong," that is, you never expend funds exactly according to the approved plan. Budgets are not holy documents to be shelved and never touched again. They are a living, changing part of your FSP's management plan and in reality are one of the most dynamic parts of your operations.

Maximize Quality Throughput

As discussed earlier, FSPs have a number of competing demands:

- Process as much evidence as possible
- In a timely fashion
- Given limited resources
- At or under budget
- While maintaining quality

Not an easy task. The KPIs that are most relevant and useful, therefore, are those that relate to increasing cases processed/reports issued while maintaining quality for the given budget. These metrics provide proof for process or policy changes. The measure of average cost per case (report issued) provides the most robust measure for how an FSP is doing.

Remember the struggling fingerprint unit in the previous chapters? Take a look at some metrics from that group.

Full-Time Equivalent Employees	7.5
Total Tests Performed	15,300
Total Cases Evaluated	875
Total Expenditures (overhead, labor, etc.)	$750,000

DOI: 10.4324/9780429318528-9

For 875 cases, doesn't 15,300 tests seem to be a lot? Looking at the FORESIGHT typical testing capability (median), it's 8.75 tests per case; based on that, this unit should be around 7,656 tests. At a cost per case of $882.35, this unit is below the average of $996 but they could be much, much lower. They are definitely **not** doing as much good as they could be with their budget. What's going on?

It's Important to Remember

The FORESIGHT numbers are flags on issues, not the issues themselves. The real answer requires inquiries, interviews, and possibly more data before a cause can be found. In talking with the fingerprint staff, it seems there is an unofficial "policy" that a suspect's ten-print card is compared to every questioned latent print from the crime scene. Even if the first latent print is associated with a known print, the staff feel they must continue to compare every other print from the scene. The staff feel this is providing as much service as possible in every case. Management, however, feels that cost minimization and throughput maximization are more important. That $996 industry average cost per case includes both the best and worst performers; the next question is, where do the best performers sit in regard to cost per case? How about cases per employee? You check the latest FORESIGHT data and find that the best performers have a cost of only $711 per case and produce 170 reports per employee. Your fingerprint unit is costing about $111 *more* but producing 54 *fewer* cases per employee. Something has to change, yes?

In discussions with the fingerprint staff, perhaps a compromise could be reached: Find a suitable stopping rule that provides enough service to the clients that also reduces the cost per case and allows the unit to do more good in more cases. Using the unit's metrics as the current state and the FORESIGHT data as targets, you can now propose several options and test them objectively.

That leads to getting your data into FORESIGHT and evaluating performance, the topics of the next section.

Laboratory Analysis and Reporting Tool (LabRAT)

The Laboratory Analysis and Reporting Tool, or LabRAT (see what we did there?) is the data entry form for FORESIGHT. It is basically a spreadsheet in Microsoft Excel but there is a good deal of formulas and programming embedded in it to make its use easier for you. The form captures a holistic view of personnel, operations, and costs. It provides a preliminary view of laboratory metrics, so it is useful as soon as you input data. Ultimately, the completed LabRAT is submitted to FORESIGHT for a report specifically for the submitting FSP and a cumulative annual report for all submitting FSPs. The former is the FSP's tool for using FORESIGHT to evaluate their operations and the latter is a valuable resource for where the FSP industry is for that year.

Figure 4.1 The Laboratory Analysis and Reporting Tool, or LabRAT, is the data entry form for FORESIGHT; some LIMS may have features that allow for direct data input.

Don't Panic

It may seem like a lot at first but the following step-by-step discussion will show you exactly what to do.

Some terms you should be familiar with:

- **Capital**: Durable produced goods that are, in turn, used as productive inputs for further production; that is, instrumentation, equipment, and furniture.
- **Full-time equivalent (FTE)**: The full-time equivalent or FTE definition refers to the number of hours considered full-time; for example, one FSP may have a 40-hour work week, while another might have a 37-hour work week.
- **Consumables**: A commodity that is intended to be used up relatively quickly, like chemical reagents, gasses, PPE, swabs, pipette tips, and so on.
- **Salary**: A fixed regular payment, typically paid on a monthly or biweekly basis but often expressed as an annual sum, made by an employer to an employee, especially a professional or white-collar worker.
- **Wages**: A fixed regular payment, typically paid on a daily or weekly basis, made by an employer to an employee, especially to a manual or unskilled worker.
- **Investigative Area**: In FORESIGHT, the operational services offered by an FSP, like DNA, Fingerprints, and Firearms. The complete list and definitions are at the end of this section.
- **Cases**: A request for a forensic service.

First of All, Levels

It's important to have as much data as possible so that you get the most accurate information for your FORESIGHT profile. FSPs can have a hard time finding data, especially financial data, because of the structure of various jurisdictions. Again, don't panic. The LabRAT can accept data at a variety of levels and still give you good information. But, as with any analysis, the more and better data you can provide, the more accurate and useful the results will be.

Let's start with the easy stuff that helps FORESIGHT and you to get the most accurate answers. First of all, information about your FSP:

- **Calendar year versus Financial year**: What is your fiscal year? Some FSPs are on a calendar fiscal year, some mirror the federal government, etc. The main point is be consistent and provide the data for the year—whatever year that is—you choose. Match the time period for the financial data with the time period of the casework.
- **Demographic, descriptive information**:
  - Jurisdiction (national, state, metro, regional, private)
  - Number of separate facilities
  - LIMS Provider/Version
  - Standard Work Week (hrs)
  - Population served
- **Contact information**: Who do we call with questions?
- **Facility descriptive information:**
  - What currency do you use? (FORESIGHT is international; other currencies are converted to yours)
  - Do your total expenditures include costs for **utilities** (electricity, gas, etc.) or **telecommunications** (phone, internet, etc.)?
- **Other data or exceptions**: Anything else that you think would help FORESIGHT understand your FSP.

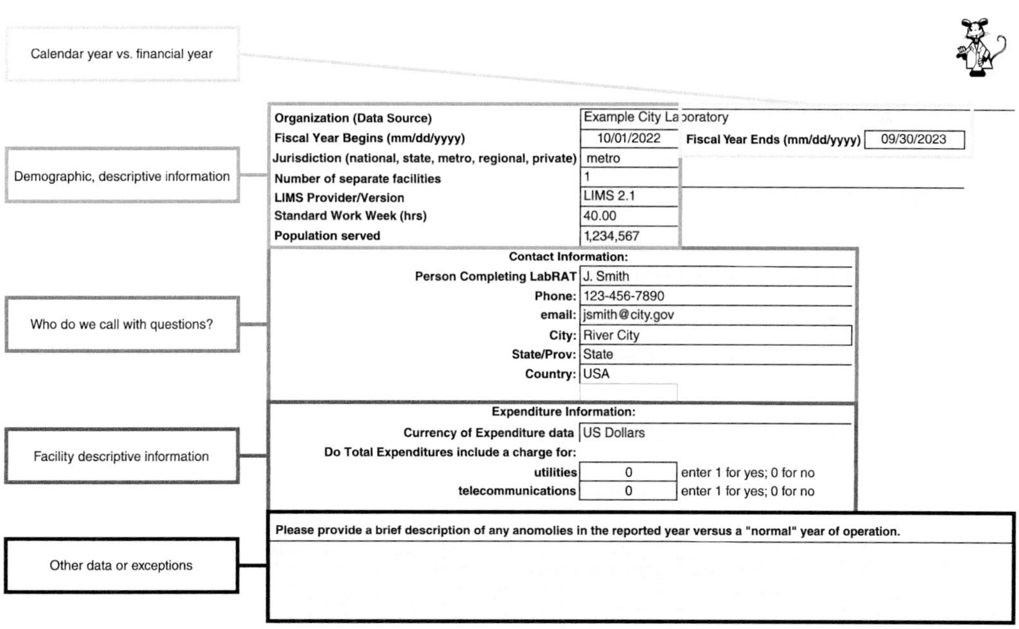

Figure 4.2 The demographic and description portion of LabRAT.

The LabRAT levels data collection sheet separates the minimum data necessary from the maximum data possible. **At a minimum, do the following:**

- "Open Me First" on worksheet 2
- "Casework Level 1" worksheet, which ask for cases submitted for each investigative area and FTE employees in each area,
- "Expenditure Level 1" worksheet, which asks for personnel expenditures (salaries, wages, benefits, temporary employees, and overtime) for each area of investigation and laboratory total expenditures for capital, consumables, and the lump sum of other expenditures.

Investigation area	Cases	FTE
Blood Alcohol	828	0:45
Crime Scene Investigation	5,116	18.00
Digital evidence	576	2.00
DNA Casework	1,052	15.00
DNA Database (including CODIS)		
Document Examination including handwriting)		
Drugs-Controlled Substances	5,251	10.45
Evidence Screening & Processing		
Explosives		
Fingerprints	2,068	3.50
Fingerprints Database (including IAFIS)	1,752	0.50
Fire analysis	15	0.10
Firearms and Ballistics	252	2.50
Firearms Database (including NIBIN)	225	0.50
Forensic Pathology		
Gun Shot Residue (GSR)	154	1.00
Marks and Impressions		
Serology/Biology	1,179	2.00
Toxicology ante mortem (excluding BAC)		
Toxicology post mortem (excluding BAC)		
Trace Evidence (includes Hairs & Fibers, Paint & Glass)	63	1.00
Administration and Support		5.00
Other Specialties (describe below)		
TOTAL		62.00

Figure 4.3 Level I Cases and FTEs (Full-Time Equivalents, better known as employees) on LabRAT.

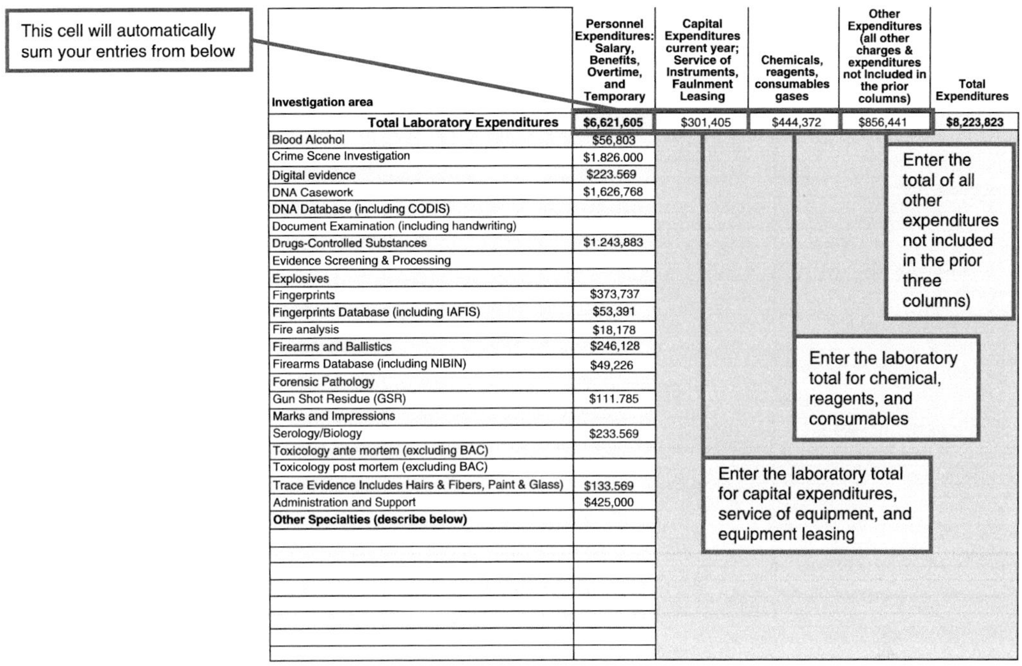

Investigation area	Personnel Expenditures: Salary, Benefits, Overtime, and Temporary	Capital Expenditures current year; Service of Instruments, Faulnment Leasing	Chemicals, reagents, consumables gases	Other Expenditures (all other charges & expenditures not Included in the prior columns)	Total Expenditures
Total Laboratory Expenditures	**$6,621,605**	$301,405	$444,372	$856,441	**$8,223,823**
Blood Alcohol	$56,803				
Crime Scene Investigation	$1.826.000				
Digital evidence	$223.569				
DNA Casework	$1,626,768				
DNA Database (including CODIS)					
Document Examination (including handwriting)					
Drugs-Controlled Substances	$1.243,883				
Evidence Screening & Processing					
Explosives					
Fingerprints	$373,737				
Fingerprints Database (including IAFIS)	$53,391				
Fire analysis	$18,178				
Firearms and Ballistics	$246,128				
Firearms Database (including NIBIN)	$49,226				
Forensic Pathology					
Gun Shot Residue (GSR)	$111.785				
Marks and Impressions					
Serology/Biology	$233.569				
Toxicology ante mortem (excluding BAC)					
Toxicology post mortem (excluding BAC)					
Trace Evidence Includes Hairs & Fibers, Paint & Glass)	$133.569				
Administration and Support	$425,000				
Other Specialties (describe below)					

Figure 4.4 The Level I laboratory expenditures portion of LabRAT.

With this data, FORESIGHT can provide you with a report, the **Summary Measures Worksheet**, and the immediate feedback you get is based on the data you enter. Partial data entry under the Level II worksheets is fine. Don't worry about having all the data. *How can FORESIGHT provide information with so little data entered?* Easy: FORESIGHT has over a decade of data about hundreds of FSPs from around the world. That data allows FORESIGHT to proxy (or estimate) the information based on a minimum of data; however, the more specific the data you enter, the more accurate the information you get out.

Personnel expenditures are broken out by Investigative Area to better determine the costs for each, as they will vary considerably. Capital expenditures, consumables, and other expenditures are bundled together for ease of entry.

The Data Entered at Level I Gives Immediate Feedback to an FSP

The information provided here is awesome! You now have a good idea of how much each kind of case costs, how much it costs for your FSP to handle each item and sample, and how much each test costs. And that's just a part of what LabRAT returns with the minimum of data.

Investigation area	Expenditures Total	Personnel Expenditures	Capital Expenditures	Consumables Expenditures	Other Expenditures	FTE	Area Cases	Items	Samples	Tests	Reports
Blood Alcohol	$74,564	$60,158	$2,380	$3,508	$8,518	0.49	828	802	NA	NA	830
Crime Scene Investigation	$2,536,443	$1,960.,211	$95,181	$140,328	$340,724	19.58	5,116	NA	NA	NA	NA
Digital evidence-Audio & Video	$302,507	$238,481	$10,576	$15,592	$37,858	2.18	576	1,285	NA	NA	558
DNA Casework	$2.218.803	$1.738,610	$79,317	$116.940	$283.937	16.32	1.052	1.563	4,958	44.816	995
DNA Database	$0	$0	$0	$0	$0	0.00	0	NA	NA	NA	NA
Document Examination	$0	$0	$0	$0	$0	0.00	0	NA	NA	NA	NA
Drugs Controlled Substances	$1,656,334	$1.321,799	$55,258	$81,468	$197,809	11.37	5,251	5,522	5,764	NA	5.138
Evidence Screening & Processing	$0	$0	$0	$0	$0	0.00	0	NA	NA	NA	NA
Explosives	$0	$0	$0	$0	$0	0.00	0	NA	NA	NA	NA
Fingerprints	$511,879	$399,833	$18,507	$27.286	$66.252	3.81	2.068	2,701	4,931	NA	1.844
Fingerprints Database (including IAFIS)	$73.126	$57,119	$2.644	$3,898	$9.465	0.54	1,752	NA	NA	NA	1,752
Fire analysis	$22,125	$18.924	$529	$780	$1.893	0.11	15	55	55	NA	11
Firearms and Ballistics	$344,801	$264,768	$13.220	$19,490	$47.323	2.72	252	798	1,405	NA	249
Firearms Database (including NIBIN)	$68.960	$52.954	$2.644	$3,898	$9,465	0.54	225	225	NA	NA	185
Forensic Pathology	$0	$0	$0	$0	$0	0.00	0	NA	NA	NA	NA
Gun Shot Residue (GSR)	$151,254	$119,241	$5.288	$7,796	$18.929	1.09	154	294	294	NA	136
Marks and Impressions	$0	$0	$0	$0	$0	0.00	0	NA	NA	NA	NA
Serology/Biology	$312,507	$248,481	$10.576	$15,592	$37,858	2.18	1,179	2.235	2,296	NA	893
Toxicology ante mortem (excluding BAC)	$0	$0	$0	$0	$0	0.00	0	NA	NA	NA	NA
Toxicology post mortem (excluding BAC)	$0	$0	$0	$0	$0	0.00	0	NA	NA	NA	NA
Trace Evidence	$173,038	$141,025	$5.288	$7,796	$18,929	1.09	63	104	104	NA	73
Other Specialties (describe below)											
	$0	$0	$0	$0	$0	0.00	0	NA	NA	NA	NA
	$0	$0	$0	$0	$0	0.00	0	NA	NA	NA	NA
	$0	$0	$0	$0	$0	0.00	0	NA	NA	NA	NA
	$0	$0	$0	$0	$0	0.00	0	NA	NA	NA	NA
	$0	$0	$0	$0	$0	0.00	0	NA	NA	NA	NA
	$0	$0	$0	$0	$0	0.00	0	NA	NA	NA	NA
	$0	$0	$0	$0	$0	0.00	0	NA	NA	NA	NA
	8,446,340.57	6,621,604.57	301,405.00	444,372.00	1,078,959.00	62.00					

Figure 4.5 The Level I Summary Measures tab gives you immediate feedback on the data you entered.

On to Level II

The rows on the main LabRAT sheet are the Investigative Areas. The columns are the **Metrics** FORESIGHT uses, like cases, items, items outsourced, test, and samples. The full definitions for these are at the end of this section. This level of detail defines Level II and helps with being more accurate about your FSP's specific operational information. Like the data entered at Level II, you'll get the same information out of LabRAT but it's more accurate because it's based directly on your input.

The case information for Level II would look like the grid depicted in Figure 4.6.

How to enter all this data for Level II? Some Laboratory Information Management Systems (LIMS) participated in a project funded by the Laura and John Arnold Foundation and managed by the American Society of Laboratory Directors (ASCLD) to create a way to export data from the LIMS directly into LabRAT. Check with your LIMS provider to see if they can do this. Otherwise, that data will need to be pulled from your LIMS (or whatever system you use) and entered into LabRAT.

How to *count all this information?* The Item-Sample-Test Examples tab in LabRAT gives you examples, vetted by forensic FSP managers, on how to count items, samples, and tests for the various Investigative Areas (Figure 4.7).

As best you can, follow the counting "rules" listed in LabRAT. This keeps the data consistent and more precise when compared between other FSPs. In an early effort, one FORESIGHT participant seemed to be doing ten times the fingerprinting tests of other participants…until it was discovered that they were counting ten-print cards as "ten tests" while the others were counting it as one. Thus, the counting "rules."

The financial information at Level 2 looks something like Figure 4.8.

Investigation area	cases	items	items outsourced	items examined internally	total samples from items examined internally	tests on samples examined internally	reports	Median turn around time, days from last submission (internal)	Median turn around time, days from first submission (internal)	open area cases at end of year	open area cases older than 30 days at end of year (backlog)	Operational Staff FTE	Digital Evidence gigabytes examined	DNA Casework Sexual Assault Kits examined	DNA Database offender entries	DNA Database attestee entries
Blood Alcohol	**828**	828		802			830		7	64	2	0.45				
Crime Scene Investigation	**5,116**								32			18.00				
Digital evidence	**576**	1,352		1,285			558		127	122	82	2.00	1,244,712			
DNA Casework	**1,052**	1,584		1,563	4,958	44.816	995		136	1,078	944	12.75		421		
DNA Database (including CODIS)	**0**															
Document Examination (including handwriting)	**0**															
Drugs Controlled Substances	**5,251**	5,522		5,522	5,764		5,138		49	866	287	10.45				
Evidence Screening & Processing	**0**															
Explosives	**0**															
Fingerprints	**2.068**	2,701		2.701	4.931		1,844		72	423	349	4.00				
Fingerprints Database (including IAFIS)	**1,752**						1,752									
Fire analysis	**15**	55		55	55		11		146	8	7	0.10				
Firearms and Ballistics	**252**	798		798	1405		249		63			2.50				
Firearms Database (including NIBIN)	**225**	225		225			185		63			0.50				
Forensic Pathology	**0**															
Gun Shot Residue (GSR)	**154**	294		294	294		136		90	19	13	1.00				
Marks and Impressions	**0**															
Serology/Biology	**1,179**	2,235		2,235	2,296		893		90	343	280	1.50				
Toxicology ante mortem (excluding BAC)	**0**															
Toxicology post mortem (excluding BAC)	**0**															
Trace Evidence (includes Hairs & Fibers, Paint & Glass)	**63**	104		104	104		73		192	22	14	1.00				
Administration and Support	**0**															
Other Specialties (describe below)																
	0															
	0															
	0															
	0															
	0															
	0															
	0															
	0															
TOTAL												54.25				

Figure 4.6 The Level II data entry portion of LabRAT. Because this is your granular data, the information you get will be more accurate.

Investigative Area	Total items	total samples from items examined internally	tests on samples examined internally	Tests conducted					
Scenario 2	1	1	4	1	1		1		
Scenario 3	1	5	11	1	1	1	2	1	2
Fingerprints	*Two soda cons and three latent lifts are submitted from the scene as well as two sets of tenprint cards from the suspects.*								
	Items	**Samples**	**Tests**	**Visual**	**Develop (2 @)**	**Comparison**	**Photography**		
Two (2) soda cans (1 print @)	2	2		2	4	4	2		
Three (3) latent lifts	3	3				6			
Two (2) tenprint cards	2	2							
TOTAL	**7**	**7**	**18**						
Fire analysis	3 cans of fire debris recovered from the scene are submitted to the lab								
	Items	**Samples**	**Tests**	**Extraction + GC/MS**					
	3	3	3	3					
Firearms and Ballistics	*Scenario 1: A weapon is seized from a suspect in a homicide case; two bullets are removed from the victim's body.* *Scenario 2: A weapon and 10 bullets from the scene are submitted.*								
	Items	**Samples**	**Tests**	**Function Tests**	**Test firing**	**Comparison**			
Scenario 1	3	3	4	1	1	2			
Scenario 2	11	11	12	1	1	10			
Forensic Pathology	Decedent submitted to ME with five histology slides prepared and examined.								
	Items	**Samples**	**Tests**	**Autopsy**	**Histology slides**				
	1	5	6	1	5				
Gun Shot Residue (GSR)	*One GSR kit submitted for examination.*								
	Items	**Samples**	**Tests**	**Microscopic**	**SEM**				
	1	3	6	3	3				
	One GSR kit and one garment submitted for examination; one stub from kit or created from stubbing garment								
	Items	**Samples**	**Tests**	**SEM/EDS**					
	2	1	1	1					

Figure 4.7 The counting examples for FORESIGHT in LabRAT.

Investigation area	Personnel Expenditures: Salary, Benefits, Overtime, and Temporary	Capital Expenditures current year; Service of instruments, Equipment Leasing	Capital Expenditures, Service of Instruments, Equipment Leasing for the prior four years	Chemicals, reagents, consumables, gases	Other Expenditures (all other charges & expenditures not included in the prior columns)	Total Expenditures (Personnel, Capital, Consumables, & Other Expenditures)
Total Laboratory Expenditures	$6,621,605	$301,405		$444,372	$856,441	$8,223,823
Blood Alcohol	$56,803					
Crime Scene Investigation	$1,826,000					
Digital evidence	$223,569					
DNA Casework	$1,626,768					
DNA Database (including CODIS)						
Document Examination (including handwriting)						
Drugs-Controlled Substances	$1,243,883					
Evidence Screening & Processing						
Explosives						
Fingerprints	$373,737					
Fingerprints Database (including IAFIS)	$53,391					
Fire analysis	$18,178					
Firearms and Ballistics	$246,128					
Firearms Database (including NIBIN)	$49,226					
Forensic Pathology						
Gun Shot Residue (GSR)	$111,785					
Marks and Impressions						
Serology/Biology	$233,569					
Toxicology ante mortem (excluding BAC)						
Toxicology post mortem (excluding BAC)						
Trace Evidence (includes Hairs & Fibers, Paint & Glass)	$133,569					
Administration and Support	$425,000					
Other Specialties (describe below)						
TOTAL	$6,621,605	$0	$0	$0	$0	$0

Figure 4.8 The Level II Expenditures portion of LabRAT.

The capital, consumables, and other costs are broken out and, depending on how your FSP handles the budget structure (which may be out of your control), you can either bundle the costs together, as shown above, or break them out by each Investigative Area. Some Areas will be more "instrument heavy" than others, while others will be more "consumable heavy," so it may be useful to have that level of data to consider the Areas separately.

Now We Get to Benchmarking!

Each year, all of the LabRAT submissions get gathered together to produce an annual report. Annual reports go back to 2008 but the reports starting in 2011 are more complete. The annual reports go to each FORESIGHT FSP and are then posted publicly (see where under Go Deeper). These reports provide a useful summary of where the forensic industry is for any one year; after enough data had been collected, trend analysis was added. The reports are very comprehensive and cover the main metrics that FSP managers are interested in, like cost per case, cases per FTE, turnaround times, and so on.

As an example, showing cases per FTE (that is, submissions), such data would look like that shown in Figure 4.9.

Cases per FTE by Investigative Area

Area of Investigation	Example City Lab	25th percentile	Median	75th percentile
Blood Alcohol	1,691.6	318.8	605.4	983.4
Crime Scene Investigation	261.3	15.7	45.6	75.7
Digital evidence	264.8	26.2	42.7	87.5
DNA Casework	64.5	78.3	101.0	133.6
DNA Database	NA	1,204.9	2,515.7	3,702.9
Document Examination	NA	16.5	21.0	26.6
Drugs - Controlled Substances	462.0	300.3	360.5	481.5
Evidence Screening & Processing	NA	105.5	144.9	174.1
Explosives	NA	5.6	7.6	11.2
Fingerprints	543.2	94.4	133.0	162.2
Fingerprints Database (including IAFIS)	3,221.4	216.7	327.2	549.6
Fire analysis	137.9	26.4	43.4	70.4
Firearms and Ballistics	92.7	45.0	63.8	112.4
Firearms Database (including NIBIN)	413.7	336.7	684.4	1,092.2
Forensic Pathology	NA	104.4	155.5	207.1
Gun Shot Residue (GSR)	141.6	26.5	34.1	57.8
Marks and Impressions	NA	14.2	19.1	27.2
Serology/Biology	542.0	58.5	112.7	144.9
Toxicology ante mortem (excluding BAC)	NA	139.9	185.5	266.8
Toxicology post mortem (excluding BAC)	NA	138.4	169.1	202.6
Trace Evidence	57.9	29.8	33.7	38.5

Figure 4.9 Cases per FTE by Investigative Area for a fictitious FSP.

Tests per Sample by Investigative Area				
Area of Investigation	**Example City Lab**	**25th percentile**	**Median**	**75th percentile**
Blood Alcohol	NA	1.65	1.75	1.83
Crime Scene Investigation	NA			
Digital evidence	NA	4.13	4.31	4.48
DNA Casework	9.04	4.01	4.17	4.37
DNA Database	NA	0.97	1.00	1.04
Document Examination	NA	1.88	2.66	2.76
Drugs - Controlled Substances	NA	2.11	2.21	2.35
Evidence Screening & Processing	NA	4.75	5.03	5.31
Explosives	NA	4.10	4.28	4.44
Fingerprints	NA	2.13	2.30	2.49
Fingerprints Database (including IAFIS)	NA	0.41	0.41	0.41
Fire analysis	NA	1.48	1.54	1.64
Firearms and Ballistics	NA	1.16	1.21	1.29
Firearms Database (including NIBIN)	NA	3.00	3.00	3.00
Forensic Pathology	NA			
Gun Shot Residue (GSR)	NA	1.42	1.54	1.62
Marks and Impressions	NA	1.42	1.47	1.50
Serology/Biology	NA	1.11	1.16	1.20
Toxicology ante mortem (excluding BAC)	NA	6.54	7.05	7.50
Toxicology post mortem (excluding BAC)	NA	2.50	2.65	2.86
Trace Evidence	NA	1.91	2.04	2.18

Figure 4.10 Tests per sample by Investigative Area for a fictitious FSP.

What does this mean? This data show the values for the number of submissions for each Investigative Area. For example, the median (middle value) number of cases submitted to a fingerprint unit is 152. The upper 75th percentile (the value at which 25% of the answers lie above that value and 75% of the answers lie below that value) is 216 and the lower 25th percentile is 104. If your fingerprint unit receives 174 per year, you now know where your unit stands in terms of all FORESIGHT FSPs.

Figure 4.10 shows another example looking at tests per sample.

Looking at fingerprints again, the median is 2.3 tests per sample. Remember, a test is an analytical process (that contributes to a reportable result) and a sample is an item of evidence or a portion of an item of evidence that generates a reportable result. This metric is a way of evaluating testing intensity. Note that the range of tests per sample for fingerprinting doesn't vary much (2.16 to 2.38, meaning that only a few FSPs do more than two tests per sample). Document examination (0.9 to 3.98) and explosives (3.05 to 5.19), however, vary more, meaning that the type of evidence they are dealing with requires more latitude in choosing methods to get results. This can be useful for FSPs with those services to help decide if they are over or under testing. Under testing may mean a number of things, like cutting corners, a lack of instrumentation, or not enough training to exploit the evidence. Over testing might mean a too-cautious approach to testing (repeating exams or rerunning samples) or retaining out-of-date policies and procedures simply because you have "always done it that way."

Reports per FTE by Investigative Area				
Area of Investigation	**Example City Lab**	**25th percentile**	**Median**	**75th percentile**
Blood Alcohol	1,695.7	339.4	548.8	982.9
Crime Scene Investigation	NA	21.2	50.4	82.4
Digital evidence	256.5	23.1	39.0	86.0
DNA Casework	61.0	76.9	99.9	132.6
DNA Database	NA	1,838.8	2,954.5	3,884.6
Document Examination	NA	18.5	20.8	25.2
Drugs - Controlled Substances	452.0	290.9	355.1	476.7
Evidence Screening & Processing	NA		87.8	
Explosives	NA	5.4	7.0	11.4
Fingerprints	484.4	96.2	123.2	159.0
Fingerprints Database (including IAFIS)	3,221.4	263.5	453.9	556.5
Fire analysis	101.1	24.4	38.6	61.1
Firearms and Ballistics	91.6	45.5	65.0	101.2
Firearms Database (including NIBIN)	340.2	318.8	476.4	838.6
Forensic Pathology	NA	207.9	209.3	210.8
Gun Shot Residue (GSR)	125.0	25.2	32.7	47.0
Marks and Impressions	NA	15.3	18.4	31.9
Serology/Biology	410.5	46.5	95.5	132.7
Toxicology ante mortem (excluding BAC)	NA	135.4	168.4	247.2
Toxicology post mortem (excluding BAC)	NA	133.8	156.8	191.7
Trace Evidence	67.1	26.8	31.2	34.6

Figure 4.11 Reports per FTE by Investigative Area for a fictitious FSP.

Checking on reports per FTE, that is, productivity, we find the results, as shown in Figure 4.11.

In fingerprinting, the average employee produces 140 reports per year. How do yours stack up? Let's say each of your fingerprint examiners produce 183 reports per year. Sounds great, right? With the upper 75th percentile at 191, it seems like your staff are killing it in productivity. **But remember, the numbers are *not* the answer, they lead you to the answer.** That high productivity could be because none of them are making associations and reporting out a lot of inconclusive results; those are obviously easier and quicker than proper examinations. Now, to be fair, they could be a tight, highly productive team and should be commended for their hard work. As an FSP manager, however, it is critical to know which it is.

This is the power of benchmarking.

Benchmarking is an objective evaluation that provides FSPs with a set of metrics that can be used as a point of reference for evaluating performance and suggest a level of quality. Data in isolation is meaningless or has limited value. We need to compare our data to our peers. Your FSP's past performance sets the baseline to monitor progress. The real value comes from comparisons with the forensic industry. Is your FSP up to par? Excelling? Lagging? The main goals of a benchmarking exercise like FORESIGHT are:

- Find out where your FSP is in the industry
- Inform the development of your FSP's strategy
- Evaluate better practices by setting targets and testing against them
- Understand your trends over time
- Communicate and support your FSP's performance to the decision makers

The first four are internal where FORESIGHT can drive improvements, create a sense of common goals, and inspire innovation. **The last one is external** and FORESIGHT can help demonstrate to your stakeholders how your FSP compares with others, supporting your initiatives, and validating requests for resources. Instead of asking for an additional 10% budget each year, only to be met with reductions or rejections, a budget that is strategically laid out with supporting data turns the request into an "if, then" statement: If you provide a X% increase, we will do Y, and show you by delivering Z.

Benchmarking has its limits, however. Good benchmarking evaluates performance and defines better practices. Bad benchmarking uses data to justify mediocre performance. This is one of the reasons FORESIGHT added the metric of "X per 100,000 population." By comparing against the population served, the metric becomes a per capita type of measure that reduces the influence of size. Also, just because other FSPs perform at a particular level, doesn't mean that that's where you should be. Every FSP has a backlog, but that doesn't absolve your FSP from having one. FSPs that gain the most value from FORESIGHT get creative and borrow ideas from others, looking at who is the best at a particular metric and asking why that is the case and what are they doing differently. Because other FORESIGHT FSPs use the same definitions, they can have this conversation with minimum confusion.

Go Deeper

Houck, M., Riley, R., Speaker, P. and Witt, T. 2009. FORESIGHT: A business approach to improving forensic science services. *Forensic Science Policy & Management*, 1(2), pp.85–95. (See full article in Chapter 2)

Latham, J. and Vinyard, J. 2010. *Organization diagnosis, design, and transformation*. New York: John Wiley and Sons.

FORESIGHT Website:
https://business.wvu.edu/research-outreach/forensic-business-studies/foresight

FORESIGHT Investigative Area Definitions

Blood Alcohol	The analysis of blood or breath samples to detect the presence of and quantify the amount of alcohol.
Computer Analysis	The analysis of computers, computerized consumer goods, and associated hardware for data retrieval and sourcing.

Crime Scene Investigation	The collection, analysis, and processing of locations for evidence relating to a criminal incident.
Digital evidence	The analysis of multimedia audio, video, and still image materials, such as surveillance recordings and video enhancement. Includes computer analysis as defined above.
DNA Casework	Analysis of biological evidence for DNA in criminal cases.
DNA Database	Analysis and entry of DNA samples from individuals for database purposes.
Document Examination	The analysis of legal, counterfeit, and questioned documents, including handwriting analysis.
Drugs - Controlled Substances	The analysis of solid dosage licit and illicit drugs, including pre-cursor materials.
Evidence Screening & Processing	The detection, collection, and processing of physical evidence in the laboratory for potential additional analysis.
Explosives	The analysis of energetic materials in pre- and post-blast incidents.
Fingerprints	The development and analysis of friction ridge patterns.
Fire analysis	The analysis of materials from suspicious fires to include ignitable liquid residue analysis.
Firearms and Ballistics	The analysis of firearms and ammunition, to include distance determinations, shooting reconstructions, NIBIN, and toolmarks.
Forensic Pathology	Forensic pathology is a branch of medicine that deals with the determination of the cause and manner of death in cases in which death occurred under suspicious or unknown circumstances.
Gun Shot Residue (GSR)	The analysis of primer residues from discharged firearms (not distance determinations).
Hairs & Fibers	The analysis of human and animal hairs (non-DNA) and textile fibers as trace evidence.
Marks and Impressions	The analysis of physical patterns received and retained through the interaction of objects of various hardness, including shoeprints and tire tracks.
Paint & Glass	The analysis of paints—generically, coatings—and glass as trace evidence.
Serology/Biology	The detection, collection, and non-DNA analysis of biological fluids
Toxicology, Ante-mortem	The chemical analysis of body fluids and tissues to determine if a drug or poison is present in a living individual, excluding blood alcohol analysis (BAC).

Toxicology, Post-mortem	The chemical analysis of body fluids and tissues to determine if a drug or poison is present in a deceased individual, excluding blood alcohol analysis (BAC).
Trace Evidence	The analysis of materials that, because of their size or texture, transfer from one location to another and persist there for some period of time. Microscopy, either directly or as an adjunct to another instrument, is involved. Includes Hairs & Fibers and Paint & Glass as defined above.
Other Specialties	Other forensic science applications not covered by the other categories.

FORESIGHT Definitions

backlog	Open cases that are older than 30 days after submission to the laboratory.
capital expense	Purchases of equipment, instruments, etc., with a lifetime longer than three years and a cost above $1,000.
Case	A request for examination in one forensic investigation area.
Casework time	Total for operational personnel in an investigation area (in hours) subtracted by the hours of R&D and T&E and support and service given to external partners.
full-time equivalent (FTE)	The work input of a full-time employee working for one full year.
investigation area	Area limited by item type and methods as they are listed in the "definitions of investigative areas" tab.
Item	A single object for examination submitted to the laboratory. Note: One item may be investigated and counted in several investigation areas.
non-reporting Manager	An individual whose primary responsibilities are in managing and administering a laboratory or a unit thereof and who is not taking part in casework.
operational Personnel	Personnel in operational units providing casework, research and development (R & D), education and training (E & T) and external support services. Non-reporting unit heads are included.
personnel expense	Sum of direct salaries, social expenses (employer contribution to FICA, Medicare, Workers Comp, and Unemployment Comp), retirement (employer contribution only toward pensions, 401K plans, etc.), personnel development and training (internal or external delivery, including travel), and occupational health service expenses (employer contribution only).

Report	A formal statement of the results of an investigation, or of any matter on which definite information is required, made by some person or body instructed or required to do so.
Request	A request for examination in one forensic investigation area.
Sample	An item of evidence or a portion of an item of evidence that generates a reportable result.
support personnel	Forensic laboratory staff providing various internal support services. Management and administration personnel not belonging to the operational units are included.
Test	An analytical process, including but not limited to visual examination, instrumental analysis, presumptive evaluations, enhancement techniques, extractions, quantifications, microscopic techniques, and comparative examinations. This does not include technical or administrative reviews.
Turn-around time	The number of days from a request for examination in an investigative area until issuance of a report. (Note that a case may have multiple requests and each new request has a separate turn-around time.)
Workload	Total time spent on all work related to job, including overtime.

Go Deeper

Kobus, H., Houck, M., Speaker, P., Riley, R. and Witt, T., 2011. Managing performance in the forensic sciences: Expectations in light of limited budgets. *Forensic Science Policy & Management: An International Journal*, 2(1), pp.36–43.

Speaker, P. and Fleming, A.S., 2009. Monitoring financial performance: An approach for forensic crime labs. *The CPA Journal*, 79(8), p.60.

Speaker, P.J., 2010. The KPI of CSI: A business application in the public sector. *Journal of the Academy of Business Education*, 11, pp. 1–12.

Speaker, P.J. and Fleming, A.S., 2010. Benchmarking and budgeting techniques for improved forensic laboratory management. *Forensic Science Policy and Management*, 1(4), pp.199–208.

Speaker, P.J., 2015. Financial management of forensic science laboratories: lessons from project FORESIGHT 2011–2012. *Forensic Science Policy & Management: An International Journal*, 6(1–2), pp.7–29.

Article: Managing Performance in the Forensic Sciences

Expectations in Light of Limited Budgets

4.1

HILTON KOBUS,[1] MAX HOUCK,[2] PAUL SPEAKER,[3] RICHARD RILEY,[4] AND TOM WITT[5]

[1]*Flinders University, School of Chemistry, Physics and Earth Sciences, Adelaide, Australia*

[2]*West Virginia University, Forensic Science Initiative, Morgantown, West Virginia, USA*

[3]*West Virginia University College of Business and Economics, Finance, Morgantown, West Virginia, USA*

[4]*West Virginia University, Division of Accounting, Morgantown, West Virginia, USA*

[5]*West Virginia University, Bureau of Business and Economic Research, Morgantown, West Virginia, USA*

Introduction

The common experience for most forensic science laboratories worldwide is that service demand outstrips resources, resulting in a steady climb in case backlogs with a subsequent impact on the timeliness of investigative and court processes (Peterson & Hickman 2005; Durose 2008). The continuing challenge for forensic science managers is to balance these competing forces and provide outcomes that meet justice system expectations, as well as the expectations of parent agencies and the public. Forensic laboratories are not immune to the classic economic puzzle—that of finding a solution to the problem of unlimited demands in the face of limited resources—that is, demand management (Crum & Palmatier 2003). This puzzle is of course characteristic to the public sector economy. Some form of rationing must be adopted and preferably that rationing is derived from a decision-making process consistent with the mission of the laboratory. In a for-profit capitalist market, price serves as the rationing mechanism, a goal consistent with the mission of the for-profit business. The public laboratory, however, faces a different objective, output maximization for a given budget (Speaker 2009a), which correlates into a cost minimization problem for the processing of laboratory submissions.

The effectiveness of forensic science is a function of both the quality of the scientific analysis and the timeliness with which that analysis is provided. The gap between the demands for forensic science services and the budget-limiting supply of those services has implications for the efficiency and effectiveness of forensic science. Consider a service

 DOI: 10.4324/9780429318528-10

mandate forced on a forensic science laboratory and the laboratory reaction to that mandate. The laboratory is charged with providing quality analysis of evidence in a timely manner. In an environment of unlimited resources, the question of effectiveness of forensic science services would be reduced to an issue of timeliness. However, because laboratory resources are insufficient to satisfy the demand for laboratory services, effectiveness faces a dual-edged sword. In order to improve the quality of analysis of cases under review, more resources (both time and money) are devoted to a given case. This additional investment in quality comes at the cost of cases not yet reviewed, an opportunity cost.1

Thus, effectiveness requires an evaluation of the tradeoff between cases reviewed and cases delayed (backlog). Greater investment in cases under review increases the quality (one dimension of effectiveness) of reviewed cases (reducing sins of commission), but simultaneously reduces effectiveness via increased time to process those cases as well as delays in the processing (backlog) of other cases (an increase in sins of omission).[2]

Laboratories continually make decisions that affect the tradeoff between errors of commission and errors of omission. However, laboratories can address questions of efficiency that may improve their performance with respect to both types of errors. Speaker (Speaker 2009b) offers a series of alternative key performance indicators that might be addressed by a laboratory as it attempts to improve processes to more effectively and efficiently allocate its resources. In a later presentation (Speaker 2009a), these key performance indicators are connected to highlight the relationship between efficiency, quality, and economies of scale.[3]

Efficiency, Cost, and Timeliness: Perceived Value

When it comes to efficiency and effectiveness, two methods of evaluation are commonly referenced, cost effectiveness and cost-benefit analysis. In actuality, cost effectiveness is merely a subset of cost-benefit analysis. Cost effectiveness is generally described as a comparison of alternative methods to achieve a similar, but not identical result. Mathematically, cost-effectiveness is the ratio of the perceived benefit divided by the cost of that activity. A comparison of alternatives is reduced to a comparison of the ratio of the benefits to the costs.

More broadly, cost-benefit analysis is a comparison of the incremental benefit[4] to the incremental cost[5] from taking a particular course of action, but does not require that there be a host of alternatives under review as found with cost-effectiveness studies. In general, a particular action should be taken as long as the additional gain (marginal benefit) exceeds the additional cost (marginal cost). This is the guiding principal behind the allocation of scarce resources in addressing the economic problem; that is, to allocate scarce resources in light of more demand than resources can meet. An implication of actions for which the marginal benefit exceeds the marginal cost is improvements in both efficiency and effectiveness.

Active Process Management

One of the problems that forensic science laboratories face is that case backlogs can grow to the point where court proceedings are delayed and public criticism in the media can result. This has occurred in Victoria and Queensland recently in both DNA and drug matters and in South Australia, where Forensic Science South Australia (FSSA) was publicly criticized for delays in pathology reports.[6] This sort of criticism is not helpful to the judicial process if it fails to consider the realities of the economic problem that the unlimited wants are being met with limited resources. However, it did have a positive outcome in Queensland and Victoria when the laboratories were given significantly increased funding. The disadvantage of providing resources when the situation reaches a crisis is that it takes time to recruit, train, and certify new staff, and therefore there is a significant lead time before any impact on backlogs occurs.

To avoid this crisis management approach, improved service delivery to meet user expectations needs to be a combination of three activities: continual improvement in laboratory systems, planned staff growth, and a proactive demand management strategy.

Active management requires close interaction between users and laboratory so that each party understands the others' challenges and that joint concerns are focused on the overall mission. The provision of forensic science services needs to work as a partnership between laboratory and user and cannot function successfully as a traditional provider/customer model where unlimited demand ignores finite resources. An example of such a partnership is the relationship between the FSSA Biology and DNA Database Groups and the South Australian Police (SAPOL) DNA Management Section to help manage both sample and information flow between the two organizations.

An important outcome of an effective forensic science service is that it has value at the end-user stage. Information from the UK has indicated that the detection rate in crime increased from 26% to 40% when DNA profiling intelligence was available (Wallace 2006). There is opportunity for laboratories and users of forensic science services to work together to gather data that will assist managing demand and improving the effectiveness of forensic science outcomes. The forensic science provider should not be left alone to cope with demand pressure and must have the support of the legal system to achieve the level of service expected.

Automation

An important internal strategy to address rising backlogs is to continually improve laboratory processes and systems to achieve higher efficiency. In a scientific organization, automation has long been a logical path to follow in the quest for efficiency (Nelson 1969). A perceived relationship exists between cost (in time or money) and the evidentiary value of the examination or analysis (Figure 1). Economists refer to this cost as the opportunity cost, which is measured as the value of the next best alternative. The choice to spend time and money on a particular endeavor implies that an alternative use of the time and money has been forsaken. Improvements in efficiency may result by minimizing the number of examinations that are labor intensive (and thus costly) and low value in the lower right quadrant and maximize those in the upper left quadrant. Shifting processes toward the upper left quadrant presumably would improve efficiency and cost effectiveness.

Thirty years ago, ABO typing was the routine method for sourcing body fluids, using techniques such as absorption/elution and absorption/inhibition (Gaensslen 1983). They were labor-intensive and gave (relatively) low value results since groups such as A and O are common to a large proportion of the population (Butler 2005). The introduction of electrophoretic iso-enzyme and protein systems such as phospho-gluco mutase and haptoglobin analysis, though still labor intensive, began to increase the resolution—and therefore the

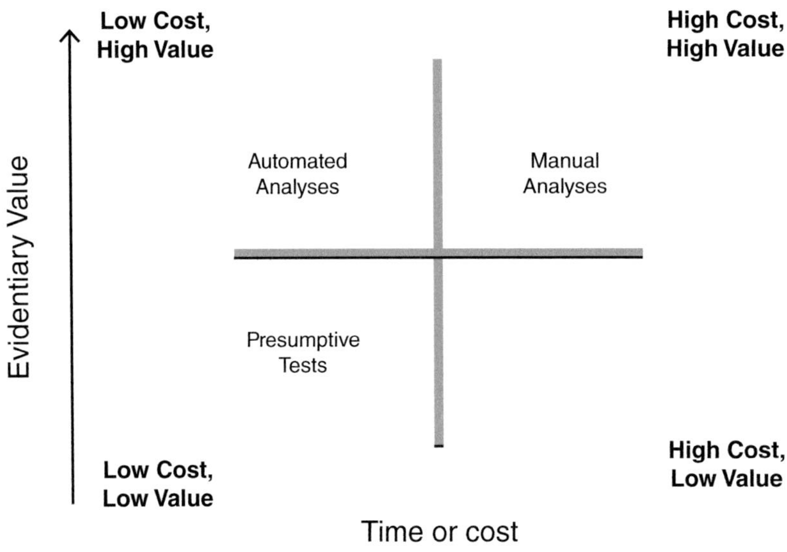

Figure 1 The perceived relationships between evidentiary value and time or cost of analysis.

value—of body fluid analysis; this moved the process toward the upper right quadrant. This is an example of the classic marginal cost versus marginal benefit analysis in economics, that is, only adopt the process if the additional expected benefit exceeds the additional expected cost. The first generation of DNA profiling systems in the late 1980s significantly increased the value of the analysis by greatly increasing the resolution of sourcing, but it was also labor-intensive, slow, and expensive. The key factor was the specificity of the analysis and how the resultant evidentiary value (upper quadrants) outweighed the costs in time and money. The current generation of multiplexed polymerase chain reaction-based (PCR) systems, such as Profiler Plus, combined with parallel column capillary electrophoresis and robotic liquid handling platforms have transformed body fluid analysis into a highly automated rapid throughput operation (Gaensslen 1983; Butler 2005).

There are many other processes in forensic science laboratories that, through improvements in technology, have made the transition to an upper quadrant; gunshot residue (GSR) analysis is a notable example. Early GSR anal-ysis involved making wax (paraffin) casts of a suspect's or deceased's hands, followed by chemical spot tests on the cast with a color change being presumptive for the presence of nitrates (Meng & Caddy 1997). The wax casting process was highly labor intensive and the spot tests had a low resolution (*any* nitrates, such as common garden fertilizer, would provide a positive result), placing the paraffin method clearly in the lower right quadrant. The modern technology that performs fully automated searching of samples using a scanning electron microscope/energy dispersive x-ray analysis (SEM/EDS) system provides definitive identification of GSR particles and totally unattended operation (Romolo & Margot 2001).

Many forensic laboratories are continually seeking opportunities to widen and improve automated systems. For example, the Forensic Science South Australia (FSSA) has recently introduced robotic platforms to handle DNA sample preparation and manipulation. Various automated reporting systems have been established, including online communication to Southern Australian Police (SAPOL) to provide DNA database intelligence, an automated illicit drug reporting system, and an automated analytical system to process data under the new using drugs while driving (drug driving) legislation. However, automation is a capital-intensive endeavor, often involving significant initial examination of internal processes and information technology (IT) development to integrate processes, train employees, and transfer sample information. Again, a full understanding of the costs is expressed through the opportunity cost of these actions.

Some processes, however, do not lend themselves to automation and will remain high input activities. The canonical example of this is the post mortem examination. Autopsies require the hands-on involvement and expertise of a forensic pathologist for every examination undertaken and automated unattended operation is not yet possible. These examinations are high input and of high value and therefore fall into the upper right quadrant. This is further emphasized by the worldwide shortage of forensic pathologists; even if a budget allowed unlimited resources, the supply of new pathologists is lacking. The review of medical case files following an apparent natural cause death plays an important part in providing the coroner with information on whether a post mortem should be conducted. This allows informed decisions to be made that ensure the number of post mortems that do not provide information beyond that available from the medical history are minimized. It is interesting to note that of 1,365 post mortems conducted at FSSA in 2005/2006 almost 900 were natural cause deaths. The file reviews help ensure an information-based approach to this demand. It is possible that such an approach could go further but adds to the "hands-on" nature of this method.

Another example of an activity that is complex beyond automation is crime scene investigations. Evidence recovery operations at crime scenes to search, locate, secure, and package evidence is labor intensive and a critical high-value operation. A final example of labor-intensive, high-value work is the production of reports by the scientists and pathologists. Reporting the findings from post mortem examinations is most likely the major factor contributing to service delivery delays in pathology.[7] While significant progress in automation has been made and likely will continue in analytical technologies, there will always be rate-determining steps caused by essential upper right quadrant activities that effect forensic service delivery. The key to efficient use of resources committed to the upper right quadrant category cases, therefore, is to minimize those that might fall into the lower right quadrant.

Triage

The low-input, low-value examinations located in the lower left serve a useful purpose. Rapid presumptive tests, such as color tests for blood and drugs, are a good example. These tests do not provide positive identification of a substance but indicate whether it is worth proceeding to further examinations; in this sense, they are a type of triage. These types of tests are low value because of their low resolution for sourcing but make a useful contribution to laboratory decision making in eliminating the need for further testing.

The critical factor for a successful forensic triaging system is for all decisions to be based on sound information and on case-based context; the exclusion of samples for reasons of cost or value has to be justified. It should be a marginal cost versus marginal benefit decision. The simplest strategy to address the economic problem in this regard is where valid sampling regimens reduce sample numbers for analysis. For example, in a packet containing a large number of ecstasy (MDMA) tablets, a statistically relevant and properly derived representative sample can be selected: the entirety of the tablets do not need to be analyzed (Aitken & Taroni 2004). Another way to triage is through presumptive testing (Ponce & Pascual 1999). A range of preliminary analytical processes can be used to reduce or eliminate the need for further, more definitive, and costly analysis. Using screening tests for blood and drugs, as mentioned above, can eliminate unnecessary analyses and provide direction for further testing. FSSA has recently introduced an automated real-time PCR procedure for quantifying DNA in extracts from crime scene samples. If no DNA is detected, the sample does not proceed to full DNA profiling; this saves both time and money. The roadside drug testing under the new drug driving legislation in South Australia is another good example of analytical screening—only samples from positive tests are sent to FSSA for confirmatory analysis.

Many forensic science cases involve multiple items collected from crime scenes and the prioritization of examinations and triaging of samples needs to be based on a combination of laboratory experience and user requirements that reflect the economic realities of the benefits versus the opportunity cost of any examination. Good communication with the person requesting the examination is the starting point. The relevance of the examinations requested and the significance of the range of items submitted for examination should be established. Then testing is performed on relevant items with mutual understanding of the economic problem and the realities of the scientific investigation under consideration. In major crime cases, FSSA holds case management meetings that involve investigators, crime scene examiners, and prosecution staff to address these issues.

A range of examinations exist that allow laboratory personnel to advise users about the potential utility and success of specific samples. For example, the FSSA has gathered extensive data on the success rates for obtaining DNA profiles from contact (so-called "touch") samples on different surfaces. The analysis figures for a twelve-month period show that from 2,700 contact DNA samples analyzed, usable profiles were only obtained from approximately 300 (FSSA, personal communication). Clearly a more efficient service can be provided if the number of negative samples can be reduced. The research was able to identify those sample types that consistently gave negative results, and this has allowed a better approach to crime scene sampling and exhibit prioritizations based on firm intelligence. Another example is gunshot residue sampling, for which there is extensive information that shows that samples not collected from hands of live subjects within four hours of the shooting do not produce a result (Meng & Caddy 1997).

Benchmarking

The perceived value of forensic tests is a goal: Moving any one test or method from a "high cost" quadrant of Figure 1 to the "low-cost, high-value" quadrant takes time, research, testing, and evaluation. This is essentially an iterative process, the Shewhart cycle for continuous improvement (Shewhart 1939; Deming 1986) used throughout the industry for quality improvements. Notwithstanding the benefits of effective case and sample triage, automation is typically the way most analytical methods would be moved to the upper left quadrant, reasoning that people are an organization's greatest expense (Senge 1994; Speaker 2009a) and capital expenditures can be amortized over time. Effectiveness and efficiency are laudable goals, as are cost savings, but where to start?

Collaborative benchmarking is one approach. The FORESIGHT Project of the West Virginia University College of Business and Economics in conjunction with the Forensic Science Initiative provides benchmarking data to participating laboratories worldwide (Houck et al. 2009)[8]. An analysis of data from FORESIGHT laboratories paints an interesting picture (see Figure 2) in contrast to the idealized image of Figure 1. Using the mean number of cases submitted per full-time employee and the mean cost per case in US$, the reality of cost effectiveness can be seen. Automated methods, such as blood alcohol, drugs, and toxicology, appear in the low-cost quadrant, while handson methods, such as trace evidence and questioned documents, appear in the high-cost section. This may align with previous qualitative expectations and historical assumptions, yet unexpected anomalies appear: DNA being a high cost method is not unexpected, but given the automation utilized in the practice, a more productive score would be anticipated.[9] The same is true for GSR. Conversely, with little to no automation, it would not have been expected that firearms would be in the middle of both axes. Interestingly, high-demand cases seem to cost

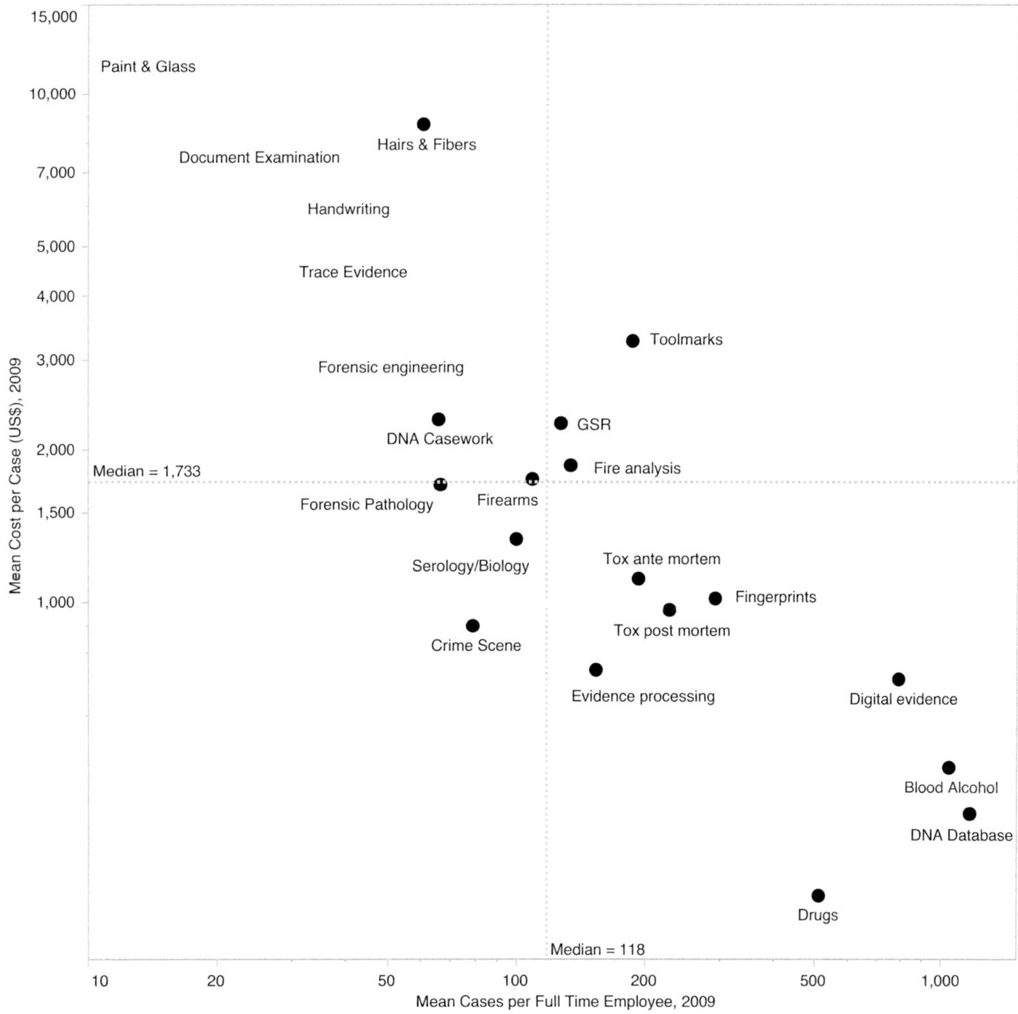

Figure 2 A plot of mean cases per full time employees at participating FORESIGHT laboratories against the mean cost per case in US dollars by analytical area of the laboratory for 2009.

less, perhaps indicating an optimization of some kind (Figure 3). Data such as this provides a starting point for the plan-do-check-act cycle described by Shewhart:

- **Plan**: Establish objectives necessary to achieve goals and intended deliverables and outcomes.
- **Do**: Implement the processes.
- **Check**: Measure the results of the new process and compare against previous and expected data.
- **Act**: Analyze any differences and look for causal connections.
- **Repeat.**

Essentially, the Shewhart cycle is a scientific approach to quality where a hypothesis is developed, tested, and refined (Shewhart 1939). Iteration is the key to success, and quality is a process, not a singular event (Deming 1986), hence the "repeat" phase of the cycle. Absent of baseline measurements and a methodology like the Shewhart cycle, however, quality improvements are limited to luck and coincidence.

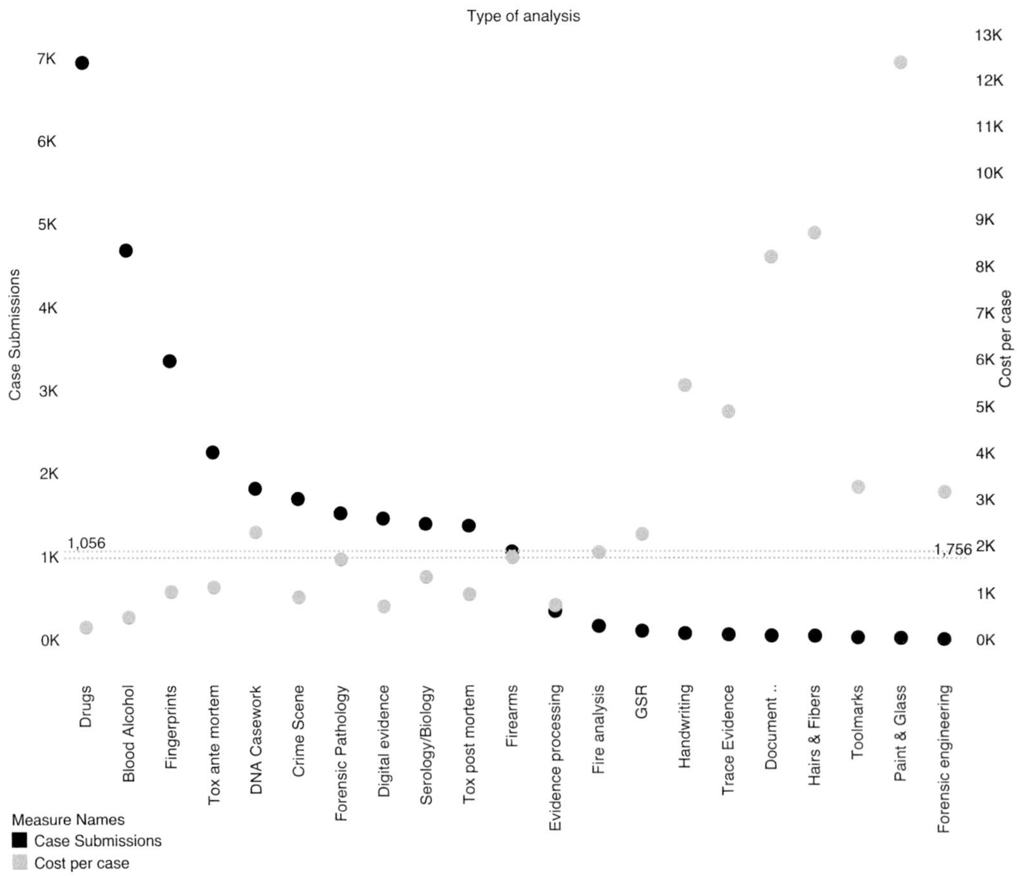

Figure 3 A plot of case submissions versus cost per case by type of analysis of participating FORESIGHT laboratories for 2009. The median values are shown next to the relevant axis and by the gray lines.

Managing the Economic Problem—Further Thoughts

The previous sections describe the economic justification for the adoption of the optimal production process within the laboratory. This optimization, in turn, will suggest comparable needs for resources—labor, capital, and materials—subject to a budget constraint, that provides a schedule of the supply of laboratory services to compare to the demands of the judicial system. It is not a matter *if* resources will be limited but rather *by how much*. Budget pressures are common in public service and, even if an agency is well resourced, matching demand continues to be a challenge. The management of demand is a strategy that laboratory managers must pursue. An efficient, information-based triaging system can at least minimize or at best eliminate unnecessary examinations. All sectors of the criminal justice system may not agree with a process that rationalizes or prioritizes samples submitted for examination (the moral implication being that one crime in a category is as much a priority as another). The perception, perhaps fueled by popular media, that all items collected at a crime scene should be examined is a quaint notion in light of the pressures facing forensic laboratories around the globe. As a recent author noted,

> The TV cliché of 'doing a full forensic examination' on a scene is simply not an option. Without a clear understanding of what is already known about an offence and the objective that an investigator is seeking to achieve, a forensic practitioner would end up applying random techniques in the hope that one of them turned up something useful. This is not only time consuming and expensive, it risks swamping the investigation with irrelevant material. What is required by all forensic disciplines is a clear objective and an understanding of how it is to be achieved. This requires a dialogue between the investigators and the forensic practitioner, thus enabling the latter to apply their specialist knowledge to the problem and identify the most appropriate techniques to achieve the objective. It is not necessary that the investigator understands the technique, so long as they set a clear objective and know what type of result can be expected (Stelfox 2009, p. 134).

Forensic knowledge can also be applied to focus forensic examinations on adding the most value to the investigation through "optimizing" examinations on certain types of crimes, for example, trace evidence in crimes of a close, personal, or intimate nature (Houck 2001; Houck 2003). Not conducting forensic examinations to investigate intimate associations, like microscopic or mitochondrial hair examinations (Houck & Budowle 2001), when the individuals are known to have had recent prior intimate contact makes sense to reduce laboratory time, effort, and costs. This is not to advocate that all of the examinations in the lower right quadrant should be eliminated from a laboratories' analytical portfolio of services, but only that they be applied *when appropriate* to answer investigative questions or provide lead value.

Conclusion

Forensic laboratories are not immune to the classic economic puzzle—unlimited requests for services that outstrip the resources allocated to provide those services. Proactive process management, automation, triage, and benchmarking are approaches that can assist in this process of planning, assessing, testing, and reviewing improvement strategies. Managing demand and balancing resources is central to the effective allocation of forensic science services.

In this article we examine these challenges and provide a discussion that can be used by forensic laboratory management for the purposes of developing a strategy for managing performance in light of limited budgets.

Endnotes

1 The term *opportunity cost* is a reference to the economic cost of a decision. Rather than limiting our view to accounting costs, the cost of any action may be much more than just the pecuniary expense. Opportunity cost refers instead to the value of the next best alternative. That is, choosing to take a particular action implies foregoing some alternative action. That forgone action is the true measure of cost.
2 Notice that this trade-off in effectiveness may have little impact on efficiency where efficiency is a measure of cost per scientific test.
3 The phrase *economies of scale* is a reference to another type of efficiency in which greater economies of scale refers to reductions in the average cost from providing a good or service. Perfect economies of scale occur when average cost is minimized.
4 Note that the additional benefit may be either a revenue-based (expense reduction) measure or a utility-based measure, a perception of the gain.
5 As noted above, the measure of cost is more appropriately an opportunity cost (economic cost) rather than an accounting cost.
6 Personal communication, Hilton Kobus. Several examples throughout are offered from one author's (Hilton Kobus) professional experience at operational forensic laboratories.
7 Reporting is not always and not necessarily labor intensive. The reporting requirements are at least partially dependent on legislation and/or jurisdiction. Law and court rules permitting, it may be possible to automate reporting of certain types of routine casework, for example, drugs and toxicological analysis and DNA database hits.
8 The FORESIGHT Project is funded through awards from the National Institute of Justice, 2003-RC-CX-K001 and 2008-DNBX-K223.
9 Externalities are also involved. In one author's (Max Houck) discussion with EU and U.S. DNA analysts, it was determined that the U.S. analyst testified more in *one month* than the EU analyst did in *one year*. This kind of burden on an analyst's time preparing for testimony, testifying, and travel certainly effects their productivity.

References

Aitken, C., and F. Taroni. 2004. *Statistics and the evaluation of evidence for forensic scientists*. New York: John Wiley and Sons.

Butler, J.M. 2005. *Forensic DNA typing: Biology and technology behind STR markers*. San Diego, CA: Academic Press.

Crum, C., and G. Palmatier. 2003. *Demand management best practices: Process, principles, and collaboration*. Fort Lauderdale, FL: J. Ross Publishing.

Deming, W.E. 1986. *Out of the crisis*. Boston, MA: MIT Center for Advanced Engineering Study.

Durose, M. 2008. *Census of publically funded crime laboratories*. Washington, DC: Bureau of Justice Statistics.

Gaensslen, R.E. 1983. *Sourcebook in forensic serology, immunology, and biochemistry*. Washington, DC: U.S. Government Printing Office. Houck, M.M. 2001. *Mute witnesses*. San Diego, CA Academic Press.

Houck, M.M. 2003. *Trace evidence analysis*. San Diego, CA: Academic Press.

Houck, M.M., and B. Budowle. 2001. Correlation of microscopic and mitochondrial DNA analysis of hairs. *Journal of Forensic Sciences* 45(5):1–4.

Houck, M.M., R. Riley, P.J. Speaker, and T.S. Witt. 2009. FORESIGHT: A business approach to improving forensic science services. *Forensic Science Policy and Management* 1:85– 95.
Meng, H., and B. Caddy. 1997. Gunshot residue analysis–a review. *Journal of Forensic Sciences* 42(4): 553–570.
Nelson, M. 1969. Automation in the laboratory. *Journal of Clinical Pathology* 22:1–10.
Peterson, J.L., and M.J. Hickman. 2005. *Census of publicly funded forensic crime laboratories, 2002*. Washington, DC: U.S. Department of Justice, Office of Justice Programs.
Ponce, A., and F. Pascual. 1999. Critical review of pre-sumptive tests for bloodstains. *Forensic Science Communications* 1(2). http://www.fbi.gov/about-us/lab/forensic-science-communications/fsc/archive/july1999/ponce.htm (accessed March 10, 2011).
Romolo, F., and P. Margot. 2001. Identification of gunshot residue: a critical review. *Forensic Science International* 119(2): 195–211.
Senge, P. 1994. *The fifth discipline: The art & practice of the learning organization*. New York: Doubleday Business.
Shewhart, W. 1939. *Statistical method from the viewpoint of quality control*. New York: Dover.
Speaker, P. 2009a. The decomposition of return on investment for forensic laboratories. *Forensic Science Policy and Management* 1:96–102.
Speaker, P. 2009b. Key performance indicators and managerial analysis for forensic laboratories. *Forensic Science Policy and Management* 1(1): 32–42.
Stelfox, P. 2009. *Criminal investigation*. Portland, OR: Willan Publishing.
Wallace, H. 2006. The UK national DNA database. *Nature: EMBO Reports* 7:S26–S30.

Article: Benchmarking and Budgeting Techniques for Improved Forensic Laboratory Management

4.2

PAUL J. SPEAKER AND A. SCOTT FLEMING
West Virginia University, College of Business and Economics, Morgantown, West Virginia, USA

Introduction

Economic realities have forced laboratories to re-examine their procedures and the associated costs at a time when the demand for forensic laboratory services have increased and simultaneously resources have been cut or at best remained stagnant. The identification and monitoring of reliable metrics in the examination of performance have never been more critical to a forensic laboratory's success. In this article we discuss two distinct yet related types of metrics, external benchmarking and internal budgeting, as avenues to successful laboratory management. The metrics discussed in this analysis will assist the management teams of forensic laboratories through improved planning, control, evaluation, and communication to interested parties.

Budgeting provides a tool in the planning process that connects an organization's mission to its strategic plan and offers specific metrics to highlight the laboratory's progress toward meeting that mission. Properly implemented, the budgetary process becomes an integral part of a forensic laboratory's business plan. The process helps to organize a measurable plan, offers control and monitoring mechanisms, provides the necessary metrics for internal and external communication of performance, and has a built-in feedback mechanism for continuous performance improvement.

In this paper we outline a process for a full budgetary feedback loop. We begin in the following section with a look at the link between a forensic laboratory's mission, strategy, and its financial statements. From those financial statements we demonstrate the incremental budget process and build the process around an example using the fingerprint identification investigative area for a hypothetical laboratory. To develop our metrics and examples we use data from two recent studies. The first study (QUADRUPOL 2003) established the groundwork and offered a standardized definition set for measurement of the inputs and outcomes of forensic laboratories in Europe. The second study, FORESIGHT (Houck et al. 2009), adopted the foundation of the European study and extended the standardization to include connections between casework, operational budgets, and personnel detail for forensic laboratories across North America. Next, we introduce the tools to compare a planned budget to actual performance via variance analysis. We demonstrate how that comparison enters into the feedback loop for the next budget cycle and what the information gleaned from the analysis indicates about the attention to the mission of the unit. Finally, the budgetary loop is connected to external benchmarks via an example using fingerprint identification to complete the feedback loop.

DOI: 10.4324/9780429318528-11

Linking Mission and Strategic Plan to Metrics

The mission of a laboratory may range in complexity from "process all of the forensic requests that investigators send" (Speaker & Fleming 2009, p. 61) to "provide timely and innovative evidence processing and analysis for any investigation that requires support to detect, enhance, or recover evidence, including on-site assistance such as search and seizure and expert testimony."[1] The strategy (and the challenge) is to match the capabilities of the laboratory in a manner to accomplish the mission, and the fiscal plan should be consistent with this objective. How does the laboratory compare to its peers in efficiency and effectiveness? What are the fiscal and operating plans for the laboratory to meet the mission? How do you connect the fiscal measures to a laboratory's professed strategy? The fiscal and strategic plans are symbiotically related; a change in one leads to a change in the other. The role of the budgetary process in corporate planning, goal formation, and the impact on strategic decisions has been outlined elsewhere (e.g., Lin 1979; Mintzberg 1979; Kahalas 1977; King & Cleland 1977). For an organization to meet its mission the budget must be integrated into the plan, and the two cannot be contradictory.

The use of external metrics in the form of benchmarking allows a laboratory to compare facilities with similar missions (Speaker 2009a), and the use internal metrics in the form of budgets allow a laboratory to monitor, control, and evaluate actual performance to expectations (Speaker & Fleming 2009). Describing and relaying these metrics and their link to the mission and strategic plan improve communications with stakeholders both internal and external to the laboratory.

The link between potential metrics and strategic planning follows the dictum that when it comes to selecting a strategy to meet goals, if you can't measure it, then you can't meet it. Thus, it becomes critical for the forensic laboratory to carefully voice its mission, its statutory mandates, and connect the associated goals to the internal management of the laboratory. One case study of a strategic planning change (Dale & Becker 2004) relates the performance goals of one laboratory to attempts to change the traditional personnel model. To evaluate the change, they carefully determine the measures and test these measures against the changes in a particular strategy. Another study (Speaker 2009b) demonstrates how a more generic mission of maximizing cases processed for a given budget, or minimizing the cost per case processed, leads to a series of metrics to explain how individual laboratories meet that goal, and how to compare their performance over time or across all laboratories.

In meeting their mission, laboratories will obtain data from a variety of sources—casework, personnel records, and financial records. As demands for laboratory services are increasing while the funding for laboratory services are somewhat stagnant, the ability to further stretch limited resources becomes the focus for many managers. A key tool for managers to optimize the output from those limited resources is from the planning and monitoring of resources through budget management.

The Budget Process

Budgeting is a management tool used to plan for both financial and operational purposes over a specified period of time. Hagen and Harden (1995, p. 772) describe the budget as "a list of revenues and expenses during a certain period of time . . . It is the answer to the question, who does what, when, and how in the preparation and the implementation of the budget." Additionally they note that the budget is a process to reduce uncertainty, can be used as a device for commitment to fiscal discipline, and that the rules used in the budgetary process affect fiscal performance. Often considered to be solely a planning device, the budget also assists in the control, evaluation, and communication of performance. While many forensic laboratories receive their financial support through legislative dictum, enhanced use of the budget and the budgetary process can improve efficiency and control over the resources provided. Organizations construct budgets to achieve a goal within a boundary of scarce resources that often include time, money, or even talent, and then use those budgets to maintain efficiency and to reduce fiscal surprises.

Different methods exist for developing a budget. *Incremental budgeting* uses the prior period, usually the prior fiscal year, as the foundation and incrementally adds resources to become the new budget. *Zero-based budgeting*, on the other hand, is a budgeting method where each budget is developed and justified from scratch each year. This method is generally considered to be more comprehensive but also more time consuming. Research into the effectiveness of each method is mixed. Wildavsky and Hammond (1965) examine zero-based and incremental budgeting methods and find benefits for each. Zero-based budgeting has educational benefit to those within the organization in the process and appears to have psychological benefits in improving self-esteem of those individuals involved. The researchers also note, though, that the failure to consider the budget as a whole may be characterized as irrational, that agencies generally receive legislative mandates so that no program would be expected to proceed without history or learning based on that history. Lin (1979) suggests that zero-based budgeting may be best utilized in a dynamic business model or industry, and similarly Wetherbe (1976) notes that incremental or traditional budgeting may best be utilized in an organization that is more traditional and manufacturing based. Wetherbe and Montanari (1981) suggest that zero-based budgeting would be effective when incorporated into the planning framework for service-oriented organizational units.

Generally speaking, public forensic laboratories receive their budgetary totals through a governmental allocation process while private laboratories possess greater self-direction and planning autonomy. This being the case, it is suggested that the optimal budgetary process for either type of laboratory would include aspects of both zero-based and traditional incremental budgeting methods. The learned history of laboratory processes and costs related to those processes need not be ignored. The cost of "reinventing the wheel" during each budgeting cycle in all likelihood will exceed the benefit to be gained. It may be beneficial, though, to periodically apply zero-based budgeting to a particular process or program as part of a comprehensive financial and operational review. This approach may allow for the efficiency and expediency of the traditional budgeting method in the budget as a whole while allowing for the benefits of zero-based budgeting to be utilized at a program level.

The budgetary period usually covers a year and is typically aligned with the funding agency's fiscal year or calendar year. The year is often sub-divided into months and/or quarters to allow for better planning and monitoring. The initial budget is often called the *static budget*. The static budget is established and remains unchanged during the year. A *rolling* or *continuous budget* is a budget that is continually updated or recast as new information become available. The budgeting period in a rolling budget is usually the next twelve months regardless of the fiscal or calendar year.

The advantages of budgets and the budgeting process within an organization include improving operational and fiscal coordination, providing a structure for gauging performance, motivating managers and employees, and improving operational and fiscal control. The budget links the purpose of the organization to plans necessary to achieve that purpose; it identifies resource needs, shortfalls, and excesses; it provides a basis for performance evaluation; and it becomes a mechanism from which effective communication can evolve. In essence the budget has four primary uses: plan, control, evaluate performance, and communicate results. Along with an appropriate feedback loop, Figure 1 illustrates the steps in the budgeting process.

The development of the budget flows from the top of the profit and loss statement to the bottom. The primary starting point is the revenue or services budget, as illustrated in Figure 2, and the most difficult item to budget is the number of tests or procedures, broken down in sub-periods of time such as months or quarters. Once the number of tests or procedures has been budgeted, then variable and fixed laboratory costs can be budgeted, such as materials, labor, and overhead to ascertain a total laboratory cost per test. Finally, nondirect general and administrative costs are budgeted, which leads to a budgeted profit/loss statement. These individual component budgets are compiled together in what is commonly known as the master budget.

There are certain steps to follow in developing, maintaining, and utilizing a budgetary process, and these are used for the overall budget, within each component budget, and within each procedure budget. Elaborating upon the steps depicted in Figure 1, consider what is included in each stage of the budgeting process. *Planning* references the identification of the problem and recognition of uncertainties, obtaining relevant information, and making predictions predicated upon chosen alternatives. The *control* stage includes the monitoring of costs and the motivation of employees. *Evaluation* of performance requires an in-depth analysis including a consideration of static budget variances and a breakdown of flexible variances. The fourth stage, *communicate,* references both the interpretation of those variances and the dissemination of the lessons learned. Finally, the feedback loop implies a constant flow of information through each stage of the process in a given budgetary cycle and across budgetary cycles to continually refine the process.

Within any organization there exist units or centers where similar products, tasks, or services may be grouped for ease of management, marketing, and/or accounting of results. A forensic laboratory is similar to firms in other industries in that there are separate investigative areas for DNA analyses, fingerprint analyses, hair and fiber analyses, etc. To demonstrate the steps in budgeting, we use the investigative area fingerprint identification as the subject center.

The first stage of the budgetary process involves the development of the budget plan. Consider this first step for our sample investigative area, fingerprint identification. Management begins by identifying the problem and uncertainties for fingerprint analysis. For problem identification this could be something as simple as: develop a budget

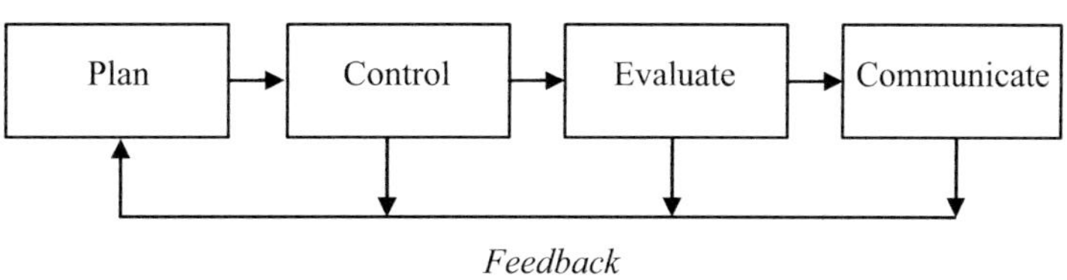

Figure 1 Budgetary Loop.

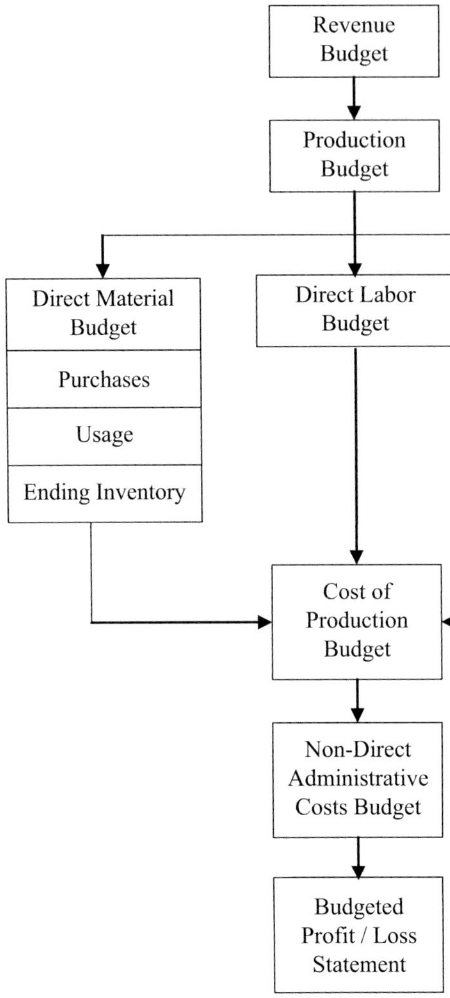

Figure 2 Master Budget and Components.

for fingerprint identification to cover the expected number of analytical requests (tests). The uncertainties might include items such as: the number of expected tests; the cost of reagents and materials per test; the time to process a test; the labor required to process the test; the cost of labor for processing; and non-personnel expenditures.

Once the issues have been identified, management begins to assemble supporting information related to the issues. Key information from historical records provides a basis for developing the answers to the uncertainties above. For this example, refer to Table 1 for some historical information regarding a hypothetical budget for the fingerprint identification section of a laboratory.

Table 1 provides a basis by which the incremental budgeting process will be explained. The column "Actual Y1" provides the starting point for our example. Suppose that the figures in this column represent the actual experience in the first year. From that actual experience, management builds its planned budget for year two with a consideration of changes anticipated for the coming year.

Table 1 Incremental Budgeting Example

Personnel Expenditures	Actual Y1	Planned Y2
Salary & Wages—base	$388,000	$388,000
Salary & Wages—overtime	$63,500	$63,500
Social costs (employer FICA, Medicare, etc.)	$32,500	$32,500
Retirement (Pensions, 401K, etc.)	$38,500	$38,500
Personnel development & training	$1,500	$1,500
Health care/insurance	$55,850	$55,850
Life Insurance	$770	$770
Total Personnel Expenditures	**$580,620**	**$580,620**
Non-Personnel Expenditures		
Chemicals/reagents	$15,000	$16,500
Traveling (non-training related)	$750	$750
QA and accreditation	$9,000	$9,000
Literature purchased	$350	$350
Service of instruments	$7,500	$7,875
Utilities	$37,500	$39,375
Telecommunications	$7,500	$7,875
Capital Expenditures	$25,000	$25,000
Other	$70,000	$73,500
Total Non-Personnel Expenditures	**$172,600**	**$180,225**
TOTAL EXPENDITURES	**$753,220**	**$760,845**

Make Predictions and Choose Alternatives

For forensic laboratories, the basis for predictions may be crime statistics and trends, economic analysis, state directives, or other indicators. For example, we may use federal crime statistics and trends for the estimated annual number of fingerprint tests, employment data and state budgetary figures for pay increases, and supplier figures or the producer price index (PPI) for chemicals and reagents. For this example, we find evidence at the federal level that the crime rate is expected to stay steady but locally the crime rate has been variable; the state has indicated that there will be no raises even though industry average pay increases have been around 5%; and suppliers indicate that cost of chemicals and reagents are expected to increase by 10%, although the inflation measures, such as the Producer Price Index (PPI), are relatively flat.

For this example (illustrated in Table 1), we assume that the laboratory management has projected that the crime rate, and thus the number of expected fingerprint tests, will hold steady; there will be no raises; and chemical and reagent cost will increase by 10%. As the column "Planned Y2" illustrates, the planned budgets include these changes to reflect the expected economic conditions from stagnant salaries to inflated costs for materials.

Figure 1 illustrates the second stage of the budgetary process as one of control. Many of the costs within a budget are considered to be *variable costs*, while others are *fixed costs*. Variable costs are driven by some factor that may or may not be controllable. In the working example, chemical and reagent costs are variable, that is, the total chemical and reagent cost is dependent upon the number of fingerprint tests. As the number of fingerprint tests increase, so does the total reagent cost. But fixed costs are not dependent upon these factors that affect variable costs (within a relevant range), and while some are not controllable, many are. Each cost type should be separated and analyzed. In examining select costs from the QUADRUPOL and FORESIGHT studies, we see that the expected year 2 cost of travel is $750, the cost of literature is $350, and the cost of utilities is $39,375. Each of these fixed costs can be monitored and controlled, and depending upon the laboratory, some are more controllable than other. Further, by communicating the fiscal goal of each item, employees may be motivated to help contain the costs.

The third step in the budget process calls for an evaluation of performance. A tool to help in that evaluation is budget variance analysis in which the *ex ante* planned budget items are compared to the *ex post* actual revenues and expenditures. The comparison of actual results to the budget is critical for evaluation of performance, particularly when the question is addressed as to whether or not the resulting experience could have been accurately anticipated. Variance analysis helps to determine where to focus managerial efforts.

A demonstration of variance analysis is presented in Table 2. The table begins with the column (1) information, planned Y2, which was initially presented in Table 1 as the budgeted amounts for each category of expenditure. In the next column we have added the actual year 2 experience in a side-by-side comparison to the planned amounts for each expenditure category. The variance is the difference between the actual experience and the budgeted amount for that category. A negative dollar entry in the third column indicates a line item where actual expenditures were less than the amount budgeted (a surplus), while a positive variance indicates expenditures in excess of those planned. The fourth column expresses the variance as a percentage of the planned expenditure.

Table 2 Budget Variance Report

Personnel Expenditures	(1) Planned Y2	(2) Actual Y2	(3) Variance	(4) Var %
Salary & Wages—base	$388,000	$370,000	($18,000)	−4.64%
Salary & Wages—overtime	$63,500	$109,183	$45,683	71.94%
Social costs (employer FICA, Medicare, etc.)	$32,500	$34,493	$1,993	6.13%
Retirement (Pensions, 401K, etc.)	$38,500	$42,361	$3,861	10.03%
Personnel development & training	$1,500	$2,750	$1,250	83.33%
Health care/insurance	$55,850	$58,269	$2,419	4.33%
Life Insurance	$770	$770	$0	0.00%
Total Personnel Expenditures	**$580,620**	**$617,826**	**$37,206**	**6.41%**
Non-Personnel Expenditures				
Chemicals/reagents	$16,500	$18,048	$1,548	10.32%
Traveling (non-training related)	$750	$1,875	$1,125	150.00%
QA and accreditation	$9,000	$8,852	($148)	−1.64%
Literature purchased	$350	$350	$0	0.00%
Service of instruments	$7,875	$9,250	$1,375	18.33%
Utilities	$39,375	$53,156	$13,781	36.75%
Telecommunications	$7,875	$7,925	$50	0.67%
Capital expenditures	$25,000	$32,500	$7,500	30.00%
Other	$73,500	$70,500	($3,000)	−4.29%
Total Non-Personnel Expenditures	**$180,225**	**$202,456**	**$22,231**	**12.34%**
TOTAL EXPENDITURES	**$760,845**	**$820,282**	**$59,437**	**7.89%**

While a manager might desire perfect foresight for budget planning, the reality is that not all circumstances will be perfectly predicted and therefore some variance is inevitable. Every laboratory director must decide what degree of accuracy is optimal, which in itself is a marginal cost versus marginal benefit problem. To evaluate a marginal cost vs. marginal benefit problem, the gain from increased accuracy must be weighed against the cost of making a more accurate prediction. In that vein, a laboratory should set some standards to trigger a more in-depth analysis for the next budget cycle.

In setting those standards for further investigation, a consideration must be given to both realistic expectations for the degree of accuracy of predictions and the potential gains from greater accuracy. While it is unrealistic to expect 100% accuracy in budget predictions, is it reasonable to expect actual expenditures to be within 99% or 95% or even 90% of the budget plan? That standard for accuracy must be considered in advance as it will lead to corresponding actions after the end of the budget cycle.

Column 4 of Table 2, variance percentage (Var%), is listed to provide a quick review of such a standard. Regardless of the dollar size of the budget category, if performance falls within some predetermined x% of the plan, then planning is deemed reasonable and no elaborate redesign of the planning is required. For demonstration purposes in this example, suppose that the hypothesized laboratory in Table 2 does not conduct a more in-depth budget planning analysis when the variance percentage is within (+/–) 5% of the plan. Using that criterion, the highlighted entries in column 4 indicate all category candidates for further review.

With respect to the second trigger, potential gains, we shift attention to column 3 of Table 2. It is quite possible that a large variance percentage merely reflects a line item that is so small relative to the entire budget that it does not bear much scrutiny, as in-depth analysis would cost more to conduct than the potential gains from the analysis. Again, laboratory management should establish a dollar threshold for the actual variance below which no further review is merited. For this demonstration, sup-pose the leadership of the laboratory in Table 2 have predetermined that any variance within (+/–) $1,500 of plan does not require more detailed review, but any variance outside those parameters is a candidate for further review.

With these two criteria in mind, the highlighted items in Table 2 can be reviewed. Each of the highlighted items for the variance and variance percentage columns have fallen into the potential investigation area because they are either outside the 5% forecasting accuracy rule on variance percentage or outside the $1,500 level for potential gain from greater accuracy. Notice, however, that personnel development & training, traveling, and service of instruments all have a variance percentage above 5%, but none of these have a dollar variance outside the $1,500 limit. As such, they will not be investigated further since the cost of investigation may well cost more than the potential gains from greater review. Similarly, salary & wages—base, health care/insurance, and other non-personnel expenditures—have dollar variances beyond the $1,500 threshold, yet none fall outside the 5% variance percentage criterion and thus are considered accurate enough to not bear detailed scrutiny. Thus, for the next budget cycle, salary & wages—overtime, social costs, retirement, chemicals/reagents, utilities, and capital expenditures—will all be reviewed in greater detail for potential improvements in forecasting the next budget. An example of such a review, using chemicals/reagents, is conducted below following the introduction of benchmarking standards to assist in that review.

External Metrics and Benchmarking

Public forensic labs generally have limited budgets, vary in size and complexity, and sometimes offer little in the way of direct comparison. One measure, though, lends itself to comparability across labs, and that is return on investment (ROI). One measure of ROI for the forensic industry is the total number of cases processed for a given level of funding, with the inverse being the level of funding divided by the total number of cases, or the average cost per case. The following formula can be used to extract this valuable information:

$$\frac{CASE}{TOTEXP} = \frac{\textit{Area cases processed}}{\textit{Total Expenditures}}$$

These measures are defined in the QUADRUPOL and FORESIGHT studies, where an area case refers to a request for examination in one forensic investigation area, and total expenditures is the sum of the direct expenses (personnel, operating, and investment) and any administrative or other overhead expenses.

We are able to gain greater insight into the formula using the DuPont expansion technique (Speaker 2009b). First developed in 1919 by DuPont executive F. Donaldson Brown, we transform the ROI measure into component parts. Brown noted that higher ROI was desirable, in general, but could result from too much risk rather than increased efficiency. Such undesirable results should be avoided. To detect such a potential problem, components to consider for addition include: efficiency measures such as cases processed per full-time-equivalent employee; quality/risk management measures such as tests performed per area case; analytical process measures such as personnel expenditures/total expenditures; and market condition measures such as the average compensation per employee

By adding the additional terms we expand the initial formula:

$$\frac{CASE}{TOTEXP} = \frac{CASE}{TOTEXP} \times \frac{LEXP}{LEXP} \times \frac{FTE}{FTE} \times \frac{TEST}{TEST}$$

Where, following the definitions from the QUADROPOL/FORESIGHT studies:

- LEXP refers to fully loaded labor expenditures and includes the sum of direct salaries, employer contributions (FICA, Medicare, workers comp, and unemployment comp), retirement (employer-only contributions toward pensions, 401K, 403b, etc.), personnel development and training (internal or external delivery, including travel), and occupational health service expenses (employer contribution only).
- FTE is the work input of a full-time employee working for one full year.
- TEST represents tests completed. A test is an analytical process, including but not limited to visual examination, instrumental analysis, presumptive evaluations, enhancement techniques, extractions, quantization, microscopic techniques, and comparative examinations. This does not include technical or administrative reviews.

By manipulating the above formula we can obtain the following:

$$\frac{CASE}{TOTEXP} = \frac{\frac{TEST}{FTE} \times \frac{LEXP}{TOTEXP}}{\frac{LEXP}{FTE} \times \frac{TEST}{CASE}}$$

Or, stated another way:

$$ROI = \frac{Labor\ productivity}{Average\ compensation} \times \frac{Labor\ Expense\ ratio}{Testing\ Intensity}$$

From labor productivity and labor expense ratio we gain a measure of efficiency and utilization, while average compensation is a market condition measure and testing intensity is a quality measure. These expressions provide a framework for the manager to gauge laboratory performance and to compare metrics to other laboratories.

To illustrate, consider our example laboratory and its fingerprint identification investigative area. Using the aggregate data from the QUADRUPOL and FORESIGHT studies to create a hypothetical laboratory, we use the following metrics for fingerprint identification for their planned activity associated with the planned budget cycle (Planned Y2) illustrated in Tables 1 and 2 above: Total expenditures of $753,220 and personnel expenditures of $580,620. Additionally, suppose that plans called for 2,400 cases to be processed and 28,000 tests to be performed by 7.70 fulltime equivalent employees. By using the inverse of ROI we can find the average cost per case, and by inserting the data provided we find:

$$\frac{Average\ Cost}{Case} = \frac{Average\ Compensation}{Labor\ productivity} \times \frac{Testing\ Intensity}{Labor\ Expense\ ratio}$$

Using the data above, we can substitute the following values into the components of this expression.

$$Average\ Compensation = personnelexpenditures\,/\,full - \ time\ equivalent\ employees$$

$$= \$580{,}620\,/\,7.70 = \$75{,}405$$

$$Testing\ Intensity = testsperformed\,/\,casesprocessed$$

$$= 28{,}000\,/\,2{,}400 = 11.67$$

$$Labor\ productivity = testsperformed\,/\,full\text{-}\ time\ equivalent\ employees$$

$$= 28{,}000\,/\,7.70 = 3.636$$

$$Labor\ expense\ ratio = personnelexpenditures\,/\ totalexpenditures$$

$$= \$580{,}620\,/\,\$753{,}220 = 77.09\%$$

Substituting these values into the expression for the average cost per case yields:

$$\frac{Average\ Cost}{Case} = \frac{\$75{,}405}{3{,}636} \times \frac{11.67}{77.09\%} = \$\ 313.84$$

The decomposition of the cost per case ratio into the four components allows managers to better evaluate the laboratory performance, particularly across the industry. Not only can we examine and compare ROI, but we can examine and compare average compensation, labor productivity, testing intensity, and the labor expense ratio. In turn, this can allow for improved goal setting, communication of performance, and potentially improved performance by allowing the manager to focus on specific areas that affect the ROI measure. Additionally, if a measure appears to be an outlier as compared to the industry, either above or below, then it may merit further investigation as to data completeness and/or accuracy.

The European forensic industry standards (QUADRUPOL 2003) and corresponding North American forensic industry standards in FORESIGHT (Houck et al. 2009) provide a starting point for the acquisition of industry metrics for external comparison. While laboratory managers may be victims of market conditions via the average compensation component of the average cost per case, comparison of the other three ratios to industry standards may suggest some internal goals for process improvement. Establishing such goals for improvement then enters the feedback loop in the budgetary cycle for the next fiscal year.

Return to Budget Review Performance Evaluation—Chemicals/Reagents Example

Recall that the budget variance suggested a few areas for which greater scrutiny was recommended based on the joint criteria of a variance percentage in excess of 5% and a dollar variance in excess of $1,500. Among these expenditure areas for greater investigation was the expenditure for chemicals/reagents (chemicals, reagents, consumables, and gases). Consider such an investigation into those expenditures as an example of how to conduct such a review. Table 3 highlights some additional detail on personnel and casework to be used along with the budget detail of Table 2 for this demonstration.

Assume the actual number of fingerprint tests was 32,000 (compared to the planned budget of 28,000 tests) and the actual reagent cost was $18,048 (compared to the planned budget of $16,500). Note that the volume exceeded expectations resulting in an increase in cost. The difference between the planned (static) budget and actual results is known as the static budget variance, which in this case is $1,548 unfavorable (Speaker and Fleming, 2009, p. 64).

Next, we determine how much the laboratory should have spent given the actual volumes. This is the flexible budget, and it shows the amount of spending given complete accuracy of budgetary volumes. To determine the flexible budget we take the actual number of tests and multiply it by the standard cost per test. This standard is based upon the number of tests and chemicals/reagents cost upon which we budgeted. In this case, the standard is the $16,500 reagent cost divided by the 28,000 tests for a standard cost of $0.589/test. To determine how much chemicals/reagents cost should have been expected, we simply multiply the actual number of tests (32,000) by the standard cost per test of $0.589, which equals $18,857.14.

From this analysis, the manager can now decompose the static budget variance into two components, the volume variance and the flexible budget variance. The volume variance illustrates the impact of the increase or decrease of the cost object volume, in this case the number of fingerprint tests; the flexible budget variance highlights the spending variance generally controllable by managing opera tions. In this case, the unfavorable static

Table 3 Non-Financial Data: Casework and Personnel Casework

Casework	2,400		
Total area cases			
Total items	11,318		
Items examined internally	11,318		
Tests on samples examined internally	28,000		
Area reports	2,304		
Median turnaround time, days from first submission (internal)	193		
Median turnaround time for analyst, days	63		
Open area cases at end of year	623		
Open area cases older than 30 days at end of year (backlog)	208		
Floor area (sq ft)	2,500		
Hrs in casework	8,525		
Personnel	Planned FTE	Actual FTE	Variance
Reporting Scientists/Analysts	4.95	4.25	(0.70)
Scientists in training	2.50	2.50	0.00
Support staff/Administration	0.25	0.25	0.00
Total Full-Time Equivalents	7.70	7.00	(0.70)

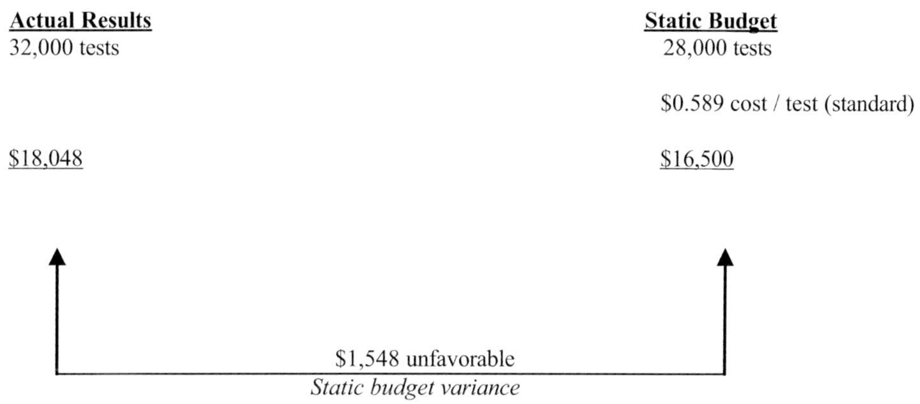

Figure 3 Direct Materials—Reagent Costs: Static Budget Variance.

budget variance is attributable to an increase in testing volume, something that may or may not be within their control, but partially offset by a favorable flexible budget variance.

Through further analysis and decomposition, though, it is possible to examine the operational costs. Although this information is not contained within the QUADRUPOL or FORESIGHT studies, we make the following assumptions to illustrate the point: The budget of 28,000 tests with a cost of $16,500 was based upon the standard of 0.125 liters of reagent per test, for a total of 3,500 liters of reagent at a cost of $4.714 per liter; we actually used 3,840 liters of reagent at an actual cost of $4.700 per liter for the period. By inserting an additional column between the actual results and the flexible budget, we can examine price and efficiency variances relating to reagent spending. Taking the actual liters and multiplying them by the standard cost per liter, we gain an added dimension to the analysis as shown in Figure 5.

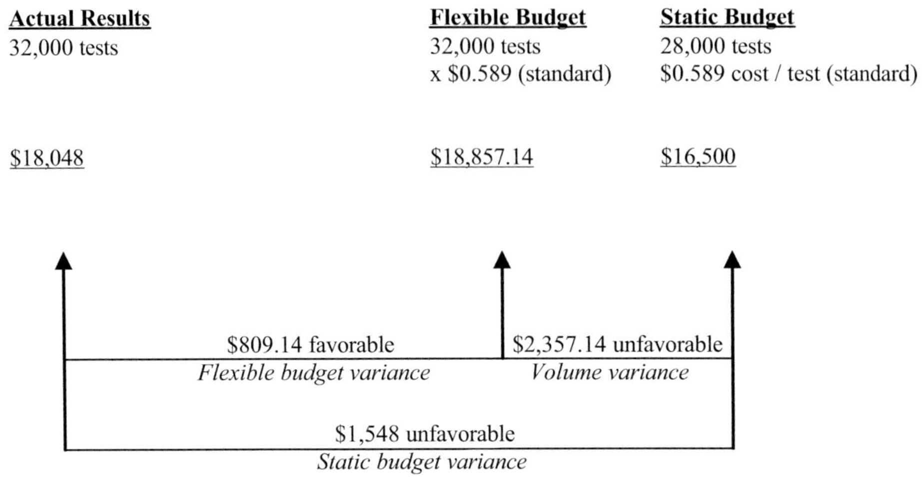

Figure 4 Direct Materials—Reagent Costs: Flexible Budget Variance.

The fourth step in the budget process involves communications of the discoveries back into the next budget cycle. The analysis of the favorable price variance shows that the laboratory spent $4.700 per liter versus the expected standard of $4.714; the favorable efficiency variance is driven by the efficient use of reagents given the increased volume of tests. Although these numbers are relatively small, the results illustrate important operational aspects of the laboratory and aid management in questions to ask and areas to focus. The better price may be due to a better negotiation, a new supplier attempting to gain a foothold at the lab, or a general price decrease in materials. The reduced consumption of 160 liters of reagent may be due to less intensive or complex testing procedures or protocols, fewer accidental mishandling of reagents, improved inventory controls, or improved testing protocols. Although these are guesses, the laboratory manager would use the information from the variance analysis to follow-up and determine root causes of both favorable and unfavorable outcomes.

As Figure 5 illustrates, the laboratory manager can go beyond the negative and simplistic statement that the laboratory overspent the budget by $1,548 to a more positive statement that, given the increased volume of tests, the laboratory actually spent $809.14 less than it should have. It may be further communicated that operationally the laboratory had a favorable price variance of $54.86 and a favorable efficiency variance of $754.28 for chemicals/reagents purchases and usage, respectfully. This analysis allows for improved communication, and just as importantly, it highlights the area or areas where the manager can apply focus to develop and share best practices or to seek improvement through the identification of inefficiencies or waste. Additionally, the information can supplement the risk assessment of operations and can accentuate areas that may merit further investigation or scrutiny by laboratory management.

The final step in the budget process is the feedback loop. The feedback loop is not a singular event, but rather is a continual process used to improve the budgetary process. As we progress period by period, we learn which indicators used in making predictions are more reliable and accurate; we gauge which standards are theoretical versus practical; we gain operational knowledge from those who help provide input into the budgeting process; and we obtain commentary from users who read and monitor laboratory performance.

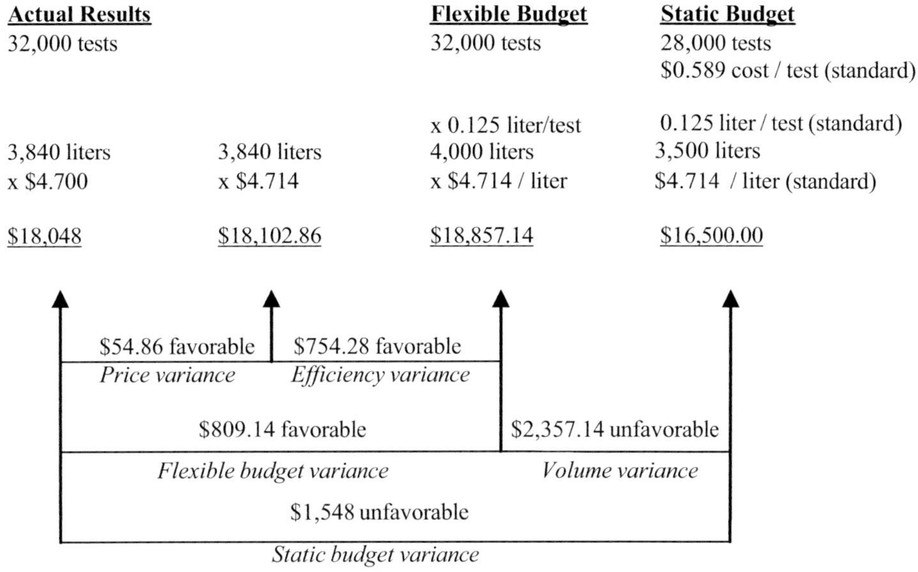

Figure 5 Direct Materials—Reagent Costs: Price & Efficiency Variance.

The feedback obtained is used to enhance and improve the planning, control, evaluation, and communication steps.

It is important to note that feedback is constructive criticism to be used for process enhancement and is not to be taken personally. The laboratory manager should view feedback as an opportunity for improvement and should look to incorporate new data and knowledge where applicable. In the example from above, the amount of reagent per test may be adjusted if it is felt that the decline in usage is permanent versus temporary.

Conclusion

Laboratories vary across jurisdictions in terms of strengths, weaknesses, capabilities, funding, mission, and strategy, but all laboratories have a universal need to monitor performance as compared to peers and within the organization. Further, laboratories need a structure and mechanism to communicate fiscal performance to mission and strategy. This can be accomplished through external benchmarking and internal budgeting. Through the construction of an applicable ROI, labs can externally benchmark to facilities with similar missions, manage and communicate their performance accordingly. Through the budgetary process, labs can improve their internal ability to plan, control, evaluate, and communicate. Utilizing both external and internal metrics optimizes the opportunity to fulfill the lab's mission and strategy.

End Note

1 The example is an illustrative amalgamation and paraphrase of the mission statement from the DoD Cyber Crime Center (http://www.dc3.mil/dcfl/dcflMission.php).

References

Dale, W.M., and W.S. Becker. 2004. A case study of forensic scientist turnover. *Forensic Science Communication* 6(3): 1–9.

Hagen, J., and I.J. Harden. 1995. Budget processes and commitment to fiscal discipline. *European Economic Review* 39:771– 779.

Houck, M., R.A. Riley, P.J. Speaker and T.S. Witt. 2009. FORESIGHT: A business approach to improving forensic science services. *Forensic Science Policy & Management: An International Journal* 1(2): 85–95.

Kahalas, H. 1977. Long range planning: An open systems view. *Long Range Planning* 10:78–82.

King, W.R., and D.I. Cleland. 1977. Information for more effective strategic planning. *Long Range Planning* 10:59–64.

Lin, W.T. 1979. Corporate planning and budgeting: an integrated approach. *Managerial Planning* 27:29–33.

Mintzberg, H. 1979. Organizational power and goals: A skeletal theory. In *Strategic Management: A New View of Business Policy and Planning,* Eds. D.E. Schendel & C.W. Hofer, p. 64–80. Boston: Little, Brown, and Company.

QUADRUPOL. 2003. European Network of Forensic Science Institutes.

Speaker, P.J. 2009a. Key performance indicators and managerial analysis for forensic laboratories. *Forensic Science Policy & Management* 1(1): 32–42.

Speaker, P.J. 2009b. The decomposition of return on investment for forensic laboratories. *Forensic Science Policy & Management* 1(2): 96–102.

Speaker, P.J., and A.S. Fleming. 2009. Monitoring financial performance: An approach for forensic crime labs. *The CPA Journal 79*(8): 60–65.

Wetherbe, J.C. 1976. A general purpose strategic planning methodology for the computing effort in higher education: development, implementation and evaluation. Unpublished PhD diss., Texas Tech. University.

Wetherbe, J.C., and J.R. Montanari. 1981. Zero based budgeting in the planning process. *Strategic Management Journal (2)*1: 1–14.

Wildavsky, A., and A. Hammond. 1965. Comprehensive versus incremental budgeting in the Department of Agriculture. *Ad-ministrative Science Quarterly, (10)*3: 321–346.

Article: Financial Management of Forensic Science Laboratories

4.3

Lessons from Project FORESIGHT 2011–2012

PAUL J. SPEAKER
College of Business and Economics, West Virginia University, Morgantown, West Virginia

Introduction

Critical to the decision making within an individual forensic science laboratory is an understanding of the efficiency and effectiveness of the laboratory. Financial management analysis in the for-profit world offers a variety of techniques to evaluate such performance, but those techniques must be adapted to the public sector. A common starting point is a comparison of the individual operation to the established standards in the industry through a review of financial ratios. Such ratios adjust for size differences and allow insight into several aspects of the operation including evaluation of efficiency, quality, risk, market nuances, and return on investment. Once an individual operation has found its comparative place among the industry, evaluation of performance over time may be tracked with the same metrics as managers begin to ascertain the effectiveness of changes for increased efficiency, quality, and return on investment. This study offers some insight into the financial performance, productivity, efficiency, and effectiveness of forensic science laboratories. Using data from the National Institute of Justice's Project FORESIGHT[1] for 2011–2012, a variety of benchmark performance data is presented in the material to follow with some analytical insight into the nature of that performance. The tabular and graphic presentations offer some insight into the current status of the forensic science industry in general and provide a basis by which individual laboratories may begin to assess their own performance with respect to both analytical efficiency and cost effectiveness.

Project FORESIGHT (Houck et al. 2009) is a cooperative effort among laboratories worldwide to connect financial and strategic management techniques to the performance of forensic laboratories. Using the individual missions of laboratories as the guidepost, project laboratories have defined a common language and common counting methods for the development of consistent metrics to measure the performance of individual laboratories and the industry as a whole for performance comparison and analysis. (Appendix A includes relevant excerpts from the FORESIGHT glossary and Appendix B contains the descriptions of the FORESIGHT Areas of Investigation.) For consistency in measurement of casework data, the data collection tool, LabRAT, offers examples to guide a laboratory's submission.[2]

 DOI: 10.4324/9780429318528-12

Project FORESIGHT data comes from voluntary participation and as such, does not represent a random sample from all laboratories. For the 2011–2012 reporting period, 81 laboratories submitted data on casework, expenditures, and personnel. All laboratories met industry quality standards via ASCLD/LAB accreditation or have been independently verified as having met ISO 17025 standards. Participating laboratories represent six continents (with all expenditure data converted to U.S. dollars), but a majority of the reporting laboratories represent U.S. jurisdictions. The laboratories represent municipal, regional, provincial, national jurisdictions as well as private laboratories.

The reported metrics represent a subset of the possible business metrics introduced in Speaker (2009a). The following section provides a univariate review by areas of investigation. These summary statistics offer a benchmark for comparison, including the detail for a disaggregation of the data (Speaker 2009b). After the summary statistics, the following section offers a bivariate view of the data with attention to economies of scale. Included in the graphical display of efficiency and cost-effectiveness are the estimated efficient relationships from which laboratories may make a finer comparison of their performance.

Concluding remarks follow the presentation of the data and stress the importance to place a holistic view when reviewing data (Houck et al. 2012). That closing section includes a brief overview of some of the other avenues of exploration and discovery that have come from the data.

Foresight Summary Statistics

The benchmark data for the 2011–2012 performance period includes laboratory submissions for several different fiscal year definitions. However, all submissions have December 31, 2011, as part of their fiscal year accounting. The majority of submissions follow a July 1, 2011, through June 30, 2012, convention. Others follow a year that begins as early as April 1, 2011, (ending March 31, 2012) while the other extreme includes laboratories with a fiscal year originating October 1, 2011, and ending September 30, 2012.

Return on Investment Metrics

Consider the summary statistics for several of the key performance indicators. The first sets of metrics offer a direct measure of a laboratory's ability to meet its mandate. The typical public sector laboratory adopts a mission statement that professes the laboratory's objective to process as much evidence as its budget allows, while following accepted scientific procedures under high standards for quality control. To convert avowed mission into a workable measure, several metrics have been proposed (Speaker 2009a), including cases processed relative to total expenditures, items examined relative to total expenditures, samples processed relative to total expenditures, and tests performed relative to total expenditures. The higher is each of these ratios for given quality, then the higher is the return on the investment from the forensic laboratory for a particular area of investigation. However, other than a comparison of one laboratory to another, the metric itself holds little common interpretation. On the other hand, the inverse of these metrics: Cost/Case, Cost/Item, Cost/ Sample, and Cost/Test, respectively, have convenient interpretations. Since the maximization of the former is equivalent to the minimization of the latter measures, then the reports to follow show the more interpretable average cost metrics.

For each metric, two estimates of central tendency, mean, and median, are reported. Because of outliers in several of the investigative areas, the most meaningful comparisons might best be made with respect to median as a representation of "typical" laboratory performance.

Consider the first of the metrics, Cost/Case (defined as Total Expenditures/Cases Processed). Table 1 highlights the summary statistics for each of the areas of investigation. The cost includes allocations for capital, wages and salary, benefits, overtime and temporary hires, chemicals, reagents, consumables, gases, travel, quality assurance and accreditation, subcontracting, service of instruments, advertisements, non-instrument repairs and maintenance, equipment leasing, utilities, telecommunications, overhead, and other expenses. A case in an investigative area refers to a request from a crime laboratory customer that includes forensic investigation in that investigative area. Note that a customer request may lead to a case in multiple investigative areas.

The latter columns in Table 1 also highlight the summary statistics for the ratio Cost/Item (defined as Total Expenditures/Items processed internally). An item is a single object for examination submitted to the laboratory. Note that one item may be investigated and

Table 1 Cost per Case and Cost per Item by Investigative Area

	Cost per Case			Cost per Item		
Investigative Area	Mean	Median	Std. Dev.	Mean	Median	Std. Dev.
Blood Alcohol	$271	$121	$360	$218	$104	$274
Crime Scene Investigation	$4,433	$5,409	$3,441	$12,878	$5,620	$18,469
Digital evidence - Audio & Video	$4,527	$4,824	$1,011	$3,767	$3,424	$1,449
DNA Casework	$1,902	$1,746	$641	$766	$689	$218
DNA Database	$71	$54	$54	$984	$52	$2,479
Document Examination	$3,965	$3,899	$2,071	$1,501	$1,219	$1,236
Drugs - Controlled Substances	$229	$187	$105	$127	$106	$87
Evidence Screening & Processing	$520	$525	$121	$260	$97	$288
Explosives	$8,542	$5,205	$6,949	$4,408	$2,801	$3,749
Fingerprints	$416	$326	$348	$191	$133	$183
Fire analysis	$2,088	$956	$1,957	$738	$444	$825
Firearms and Ballistics	$1,331	$820	$1,402	$436	$338	$321
Forensic Pathology	$3,115	$3,291	$644	$3,115	$3,291	$644
Gun Shot Residue (GSR)	$2,168	$1,215	$1,852	$1,197	$732	$1,007
Marks and Impressions	$4,349	$3,989	$3,078	$1,421	$1,086	$813
Serology/Biology	$690	$591	$370	$193	$139	$120
Toxicology ante mortem (excluding BAC)	$694	$607	$561	$486	$446	$392
Toxicology post mortem (excluding BAC)	$715	$637	$412	$373	$333	$210
Trace Evidence	$5,679	$2,843	$7,201	$3,025	$1,629	$3,581

counted in several investigation areas. For the reported metrics, items only include those items that were examined internally.

For either metric, note that several areas of investigation have noticeable differences between mean and median, suggesting some severe skewness in the distribution of performances. Likewise, many of standard deviations may appear to be quite high. In particular, blood alcohol, firearms and ballistics, and trace evidence all have a coefficient of variation (standard deviation/mean) that is greater than one. The multivariate analysis in the latter part of this study provides an explanation for these extreme values.

For some laboratories, alternative measures may hold more meaning. Table 2 offers the summary statistics for Cost/Sample (defined as Total Expenditures/ Samples processed) and Cost/Test (defined as Total Expenditures/Tests Performed). A sample is defined as an item of evidence or a portion of an item of evidence that generates a reportable result. A test is an analytical process, including but not limited to visual examination, instrumental analysis, presumptive evaluations, enhancement techniques, extractions, quantifications, microscopic techniques, and comparative examinations.

This does not include technical or administrative reviews.

As with the previous average cost metrics, the summary statistics suggest significant variation and skewness that begs explanation. This explanation will be more apparent with the decomposition of the data and the subsequent multivariate review of the data.

Table 2 Cost per Sample and Cost per Test by Investigative Area

	Cost per Sample			Cost per Test		
Investigative Area	Mean	Median	Std. Dev.	Mean	Median	Std. Dev.
Blood Alcohol	$160	$85	$175	$85	$45	$87
Crime Scene Investigation	$13,737	$940	$22,979	$7,374	$1,984	$11,094
Digital evidence - Audio & Video	$2,174	$870	$2,469	$85	$79	$28
DNA Casework	$514	$481	$222	$294	$137	$546
DNA Database	$984	$52	$2,479	$269	$45	$610
Document Examination	$1,637	$1,247	$1,481	$567	$222	$835
Drugs - Controlled Substances	$110	$81	$87	$41	$36	$25
Evidence Screening & Processing	$161	$110	$99	$46	$39	$24
Explosives	$2,821	$2,914	$485	$804	$840	$404
Fingerprints	$166	$129	$132	$62	$61	$31
Fire analysis	$743	$396	$922	$306	$191	$234
Firearms and Ballistics	$355	$277	$258	$158	$151	$131
Forensic Pathology	$2,954	$2,986	$683	$1,166	$698	$946
Gun Shot Residue (GSR)	$702	$324	$1,060	$377	$175	$384
Marks and Impressions	$1,421	$1,042	$1,203	$433	$401	$306
Serology/Biology	$127	$110	$92	$39	$29	$21
Toxicology ante mortem (excluding BAC)	$327	$332	$204	$88	$82	$53
Toxicology post mortem (excluding BAC)	$311	$333	$225	$79	$60	$54
Trace Evidence	$2,926	$1,308	$4,050	$627	$378	$554

An interpretation of the four metrics offered in Tables 1 and 2 comes from a technique known as a DuPont expansion. This technique relates the ratio metrics to the key issues for the laboratory. This includes questions regarding productivity, market conditions, analytical process, and quality. The various unit cost metrics may be interpreted using the technique highlighted in Speaker (2009b, 2010a, 2010b).

Consider the Cost/Case metric that may be decomposed (Speaker 2009b) into:

$$\frac{Cost}{Case} = \frac{Average\ Compensation \times Testsper\ Case}{Testsper\ FTE \times Personnel\ Expense\ Ratio} \tag{1}$$

From the decomposition expression for the Cost/ Case, an increase in the numerator components, Average Compensation or Testing (or Sampling) Intensity, will increase the cost per case. Similarly, a decrease in denominator component will increase the cost per case. This may occur from either a drop in productivity, as measured by cases processed per FTE, or from an increase in capital investment for future productivity but financed via a drop in personnel expenses relative to total expenses.

A similar decomposition may be made for the remaining four metrics. In particular,

$$\frac{Cost}{Item} = \frac{Average\ Compensation \times Tests\ per\ Item}{Tests\ per\ FTE \times Personnel\ Expense\ Ratio} \tag{2}$$

$$\frac{Cost}{Sample} = \frac{Average\ Compensation \times Tests\ per\ Sample\ Samples\ perF}{TE\ £\ Personnel \times Expense\ Ratio} \tag{3}$$

$$\frac{Cost}{Test} = \frac{Average\ Compensation}{Tests\ per\ FTE \times Personnel\ Expense\ Ratio} \tag{4}$$

Local Market Conditions Metric

Table 3 provides the summary statistics for the average compensation portion of these equations. As a numerator metric, then the higher is the average compensation, then the greater with be the corresponding average cost metric. Average compensation is a market measure that captures the need for the laboratory to adhere to the local conditions for personnel talent and corresponding benefits. To see the effect of market forces, a substitution of the summary statistic measures of central tendency (mean or median) offers insight into the effect of local conditions towards return on investment.

Average compensation is the total of all personnel expenditures divided by FTE, where **personnel expenditures** represent the sum of direct salaries, social expenses (employer contribution to FICA, Medicare, workers compensation, and unemployment compensation), retirement (employer contribution only towards pensions, 401K plans, etc.), personnel development and training (internal or external delivery, including travel), and occupational health service expenses (employer contribution only). A full-time equivalent (FTE) is the work input of a full-time employee working for one full year.

Table 3 Average Compensation by Investigative Area

	Average Compensation		
Investigative Area	Mean	Median	Std. Dev.
Blood Alcohol	$74,241	$76,292	$19,653
Crime Scene Investigation	$88,082	$91,150	$16,881
Digital evidence - Audio & Video	$82,709	$78,321	$10,813
DNA Casework	$95,804	$87,242	$41,919
DNA Database	$60,354	$66,988	$18,431
Document Examination	$78,142	$74,439	$19,364
Drugs - Controlled Substances	$77,120	$71,772	$16,264
Evidence Screening & Processing	$73,357	$60,732	$38,644
Explosives	$79,893	$81,758	$34,260
Fingerprints	$75,901	$72,517	$16,092
Fire analysis	$88,241	$76,038	$40,314
Firearms and Ballistics	$87,344	$82,689	$23,627
Forensic Pathology	$103,604	$102,783	$5,804
Gun Shot Residue (GSR)	$74,646	$69,843	$19,968
Marks and Impressions	$71,747	$70,943	$27,460
Serology/Biology	$74,165	$67,487	$27,022
Toxicology ante mortem (excluding BAC)	$75,675	$62,374	$25,312
Toxicology post mortem (excluding BAC)	$75,821	$66,211	$26,589
Trace Evidence	$100,599	$90,490	$62,848

Quality/Risk Metrics

From the beginning of Project FORESIGHT, laboratory directors have emphasized the need to account for quality when it comes to laboratory comparisons. As noted above, one level of quality control is captured by limiting the reported laboratories to those laboratories that have met industry quality standards via ASCLD/LAB accreditation or have been independently verified as having met ISO 17025 standards. Given this baseline, differences in laboratories with respect to quality or risk management may be perceived via the depth of the analysis before issuing a report.

Table 4 presents two metrics on sampling intensity at the case level. A higher sampling intensity indicates that a greater amount of resources are devoted to a typical case. While this may lower the uncertainty regarding the outcome of the evidence, a higher sampling intensity increases the cost of the typical case and increases the backlog percentage. The items per case metric appears in the first set of columns and samples per case appear in the last three columns. Notice that crime scene investigation, fingerprint identification, toxicology ante mortem, and toxicology post mortem all have a high degree of variability in sampling intensity across laboratories.[3]

Table 5 extends this sampling intensity to individual testing with respect to the case level and sample level and offers some suggestions regarding internal process. In Tests/Case, document examination, fingerprint identification, firearms and ballistics, and marks and impressions all demonstrate high variability. In Tests/ Sample digital evidence, document examination, firearms and ballistics and gunshot residue all have high relative variability suggesting areas for individual laboratory review of process and procedure.

Table 4 Items per Case and Samples per Case by Investigative Area

	Items per Case			Samples per Case		
Investigative Area	Mean	Median	Std. Dev.	Mean	Median	Std. Dev.
Blood Alcohol	1.26	1.03	0.38	1.69	1.51	0.80
Crime Scene Investigation	31.96	1.00	62.43	44.17	7.60	69.82
Digital evidence - Audio & Video	1.25	1.35	0.22	4.00	5.40	2.54
DNA Casework	2.63	2.35	1.00	4.08	3.98	1.52
DNA Database	0.89	1.00	0.41	0.89	1.00	0.41
Document Examination	4.35	3.46	3.28	5.08	2.66	4.59
Drugs - Controlled Substances	2.46	2.03	1.59	3.06	2.21	2.93
Evidence Screening & Processing	4.03	4.07	3.13	3.92	4.07	1.94
Explosives	2.03	1.75	0.85	4.00	3.89	2.56
Fingerprints	2.74	2.25	1.54	3.63	2.40	4.03
Fire analysis	2.99	2.54	1.41	4.09	2.88	2.90
Firearms and Ballistics	3.76	2.47	3.66	5.12	3.97	4.19
Forensic Pathology	1.00	1.00	0.00	1.00	1.00	
Gun Shot Residue (GSR)	1.96	2.22	0.76	5.15	4.58	2.85
Marks and Impressions	3.10	3.31	1.39	3.93	3.50	3.16
Serology/Biology	4.93	3.95	3.26	6.30	7.82	2.60
Toxicology ante mortem (excluding BAC)	1.71	1.35	1.19	3.16	1.65	3.55
Toxicology post mortem (excluding BAC)	2.76	2.09	2.48	4.81	2.08	6.16
Trace Evidence	2.06	2.13	0.47	3.41	2.13	3.33

Table 5 Tests per Case and Tests per Samples by Investigative Area

	Tests per Case			Tests per Sample		
Investigative Area	Mean	Median	Std. Dev.	Mean	Median	Std. Dev.
Blood Alcohol	2.98	2.19	1.73	1.80	2.00	0.44
Crime Scene Investigation*						
Digital evidence - Audio & Video	57.52	43.20	26.32	32.62	8.00	43.11
DNA Casework	15.30	10.94	12.47	3.70	3.74	2.51
DNA Database	1.35	1.06	1.24	1.89	1.10	1.44
Document Examination	30.99	11.64	43.16	21.40	3.44	45.67
Drugs - Controlled Substances	7.02	5.28	4.30	2.64	2.63	1.17
Evidence Screening & Processing	13.60	10.18	8.61	3.45	3.79	0.83
Explosives	14.07	15.31	5.90	4.31	4.02	2.14
Fingerprints	9.39	6.24	12.76	3.14	2.19	2.80
Fire analysis	7.87	6.00	3.87	2.33	2.00	1.09
Firearms and Ballistics	27.20	7.43	50.87	4.53	2.87	6.53
Forensic Pathology	3.89	5.19	2.51	3.89	5.19	2.51

(*Continued*)

Table 5 (*Continued*) Tests per Case and Tests per Samples by Investigative Area

	Tests per Case			Tests per Sample		
Investigative Area	Mean	Median	Std. Dev.	Mean	Median	Std. Dev.
Gun Shot Residue (GSR)	9.08	6.49	8.86	2.34	1.00	2.72
Marks and Impressions	16.02	10.75	16.61	3.85	3.07	2.59
Serology/Biology	20.82	21.90	10.44	3.39	2.98	1.20
Toxicology ante mortem (excluding BAC)	7.70	8.07	3.76	3.54	2.90	3.15
Toxicology post mortem (excluding BAC)	11.49	11.98	5.80	2.68	2.62	1.48
Trace Evidence	11.60	10.60	6.23	4.44	3.70	2.49

* Sample size too small to determine reliable summary statistics.

Productivity Metrics

A series of metrics on productivity by area of investigation is particularly useful, as they can be considered laboratory-wide as well as used at the individual level. Tables 6, 7, and 8 offer a glimpse at several productivity metrics. For individual laboratories, one or more of these metrics may have greater meaning depending upon historical measures of importance to the laboratory. Table 6 provides productivity summary statistics at the case and item level relative to FTE. A full-time equivalent (FTE) is the work input of a full-time employee working for one full year.

Both Cases/FTE and Items/FTE reveal some interesting trends. There appears to be a low relative variability for cases per FTE across investigative areas. For several areas of investigation, including digital evidence, DNA casework, DNA database, evidence screening and processing, forensic pathology, serology/ biology and trace evidence, the coefficient of variation is relatively low. Only crime scene investigation has a standard deviation of cases per FTE that is higher than the corresponding mean. Items per FTE, on the other hand, shows much higher relative dispersion. Crime scene investigation, document examination, drugs-controlled substances, toxicology ante mortem, and toxicology post mortem all have a standard deviation that exceeds the mean. Thus, while there is some consistency with the amount of casework handled by individuals across laboratories, there is a noticeable difference in the detail of that analysis.

When those productivity metrics are driven down to the sample and test level, that level of variability continues to be present. Table 7 presents the summary statistics for Samples/FTE and Tests/FTE and offer results consistent with items in Table 6.

However, when the productivity metrics turn to reporting, the results are similar to the case level productivity. For this metric a report is defined as a formal statement of the results of an investigation, or of any matter on which definite information is required, made by some person or body instructed or required to do so. The summary statistics in Table 8 for Reports/ FTE are similar to the Table 6 outcomes for Cases/FTE.

Analytical Process Metrics

The last ratio in the decompositions are related to the choice of analytical process. In any production process, inputs can be categorized into several different categories, including

Table 6 Cases per FTE and Items per FTE by Investigative Area

	Cases per FTE			Items per FTE		
Investigative Area	Mean	Median	Std. Dev.	Mean	Median	Std. Dev.
Blood Alcohol	857.02	830.66	580.10	1,007.75	965.78	626.81
Crime Scene Investigation	54.41	21.60	64.92	2,735.18	19.20	5,442.52
Digital evidence - Audio & Video	22.94	18.02	8.56	29.22	25.33	13.57
DNA Casework	81.96	82.46	27.77	209.84	171.17	103.00
DNA Database	2,604.33	2,523.06	493.48	2,470.79	2,395.05	1,303.69
Document Examination	32.98	26.47	25.11	197.29	79.32	371.05
Drugs - Controlled Substances	544.78	467.65	239.23	1,597.00	870.14	1,902.83
Evidence Screening & Processing	193.92	161.51	85.29	613.56	657.60	336.02
Explosives	31.90	28.15	24.19	57.35	49.63	39.95
Fingerprints	367.41	309.77	249.99	899.84	655.30	603.87
Fire analysis	102.71	98.87	68.04	273.17	265.98	161.62
Firearms and Ballistics	135.78	119.49	93.10	383.47	295.60	276.80
Forensic Pathology	46.48	43.72	6.89	46.48	43.72	6.89
Gun Shot Residue (GSR)	85.28	73.51	61.81	139.62	149.59	79.95
Marks and Impressions	33.41	24.08	26.87	83.85	81.87	43.71
Serology/Biology	167.20	164.57	77.35	644.42	559.29	329.81
Toxicology ante mortem (excluding BAC)	244.56	192.95	147.34	487.36	251.60	661.90
Toxicology post mortem (excluding BAC)	216.52	195.43	134.90	799.26	300.00	1,458.72
Trace Evidence	45.51	40.41	27.26	83.87	75.10	46.14

Table 7 Samples per FTE and Tests per FTE by Investigative Area

	Samples per FTE			Tests per FTE		
Investigative Area	Mean	Median	Std. Dev.	Mean	Median	Std. Dev.
Blood Alcohol	1,277.49	1,080.82	914.04	2,113.45	1,863.16	1,268.91
Crime Scene Investigation*	3,687.63	239.88	6,177.64			
Digital evidence - Audio & Video	98.72	99.68	79.02	1,249.12	1,418.07	444.26
DNA Casework	326.92	320.13	143.21	1,229.09	904.48	1,119.52
DNA Database	2,470.79	2,395.05	1,303.69	3,536.52	2,634.55	2,739.60
Document Examination	267.00	79.32	464.11	987.95	580.28	1,181.92
Drugs - Controlled Substances	1,822.25	1,276.01	2,394.99	3,839.15	2,748.76	3,217.22
Evidence Screening & Processing	653.64	657.60	97.32	2,258.47	2,100.00	707.15
Explosives	48.03	47.48	8.21	212.25	166.49	132.23
Fingerprints	1,080.38	655.67	981.70	2,291.67	1,935.17	2,353.78
Fire analysis	341.35	410.28	207.31	588.73	576.38	342.44
Firearms and Ballistics	421.08	334.68	227.66	1,903.75	827.58	2,813.81
Forensic Pathology	47.02	42.57	8.33	168.99	217.09	97.64
Gun Shot Residue (GSR)	382.72	385.12	254.94	1,044.03	504.35	1,873.22

(*Continued*)

Table 7 (*Continued*) Samples per FTE and Tests per FTE by Investigative Area

	Samples per FTE			Tests per FTE		
Investigative Area	Mean	Median	Std. Dev.	Mean	Median	Std. Dev.
Marks and Impressions	103.30	86.51	77.65	382.26	224.15	390.17
Serology/Biology	891.60	868.80	315.99	2,870.25	2,854.83	1,067.63
Toxicology ante mortem (excluding BAC)	882.99	386.17	1,670.15	2,049.38	1,377.95	1,771.77
Toxicology post mortem (excluding BAC)	1,600.40	379.54	3,489.75	3,209.77	2,223.02	3,846.30
Trace Evidence	117.92	87.03	87.97	384.75	410.63	203.41

* Sample size too small to determine reliable summary statistics.

Table 8 Reports per FTE by Investigative Area

	Reports per FTE		
Investigative Area	Mean	Median	Std. Dev.
Blood Alcohol	879.56	826.30	580.40
Crime Scene Investigation	62.00	21.60	82.61
Digital evidence - Audio & Video	24.77	23.70	4.33
DNA Casework	99.07	81.15	66.97
DNA Database	5,409.93	2,496.98	7,601.16
Document Examination	36.42	24.89	28.33
Drugs - Controlled Substances	609.61	510.19	228.23
Evidence Screening & Processing	157.02	157.02	27.87
Explosives	30.05	31.72	18.91
Fingerprints	395.40	342.52	264.78
Fire analysis	110.53	91.19	81.71
Firearms and Ballistics	148.83	129.81	114.78
Forensic Pathology	47.25	43.96	7.57
Gun Shot Residue (GSR)	102.99	108.24	70.79
Marks and Impressions	33.66	23.79	35.93
Serology/Biology	184.46	162.94	111.49
Toxicology ante mortem (excluding BAC)	267.04	208.10	176.55
Toxicology post mortem (excluding BAC)	209.18	137.13	145.71
Trace Evidence	45.68	35.92	32.45

capital, consumables, personnel, and overhead. We are most interested in the choice in analytical process as they become more capital-intensive versus process choices that are more labor-intensive. We would expect to see greater investment in capital as an indication of longer term investment in which only a portion of the returns would be realized in the present period. Therefore, a higher percentage of capital expenditures, while having a long-term benefit, would also show a short-term cost with higher Cost/ Case, Cost/Item, Cost/ Sample, and Cost/Test in the present period. However, that would not be viewed negatively when the benefits from the investment suggest greater long-term benefits.

Personnel expenditures represent the sum of direct salaries, social expenses (employer contribution to FICA, Medicare, Workers Compensation, and Unemployment Compensation), retirement (employer contribution only towards pensions, 401K plans, etc.), personnel development and training (internal or external delivery, including travel), and occupational health service expenses (employer contribution only). Capital expenditures include purchases of equipment, instruments, etc. with a lifetime longer than a year. These expenditures include newly purchased items as well as investments to improve the useful life of existing capital equipment. Expenses to maintain an asset at its current condition are not capitalized.

Table 9 highlights the summary statistics for the percentage of total expenditures in the form of personnel expenditures for the first three data columns and the percentage of capital expenditures for the latter three data columns. For Personnel Expenditures/Total Expenditures, a laboratory with a higher percentage will see this as an average cost reducing activity. Essentially, such a laboratory has borrowed from the future by not investing in capital, instead opting to process more cases today and thus lowering average cost.[4]

Table 9 Personnel Expenditures/Total Expenditures & Capital Expenditures/Total Expenditures by Investigative Area

	Personnel Expenditures/ Total Expenditures			Capital Expenditures/ Total Expenditures		
Investigative Area	Mean	Median	Std. Dev.	Mean	Median	Std. Dev.
Blood Alcohol	75.31%	79.17%	15.74%	5.89%	4.83%	5.15%
Crime Scene Investigation	76.62%	78.72%	15.11%	3.00%	2.52%	3.19%
Digital evidence - Audio & Video	84.16%	85.11%	2.68%	8.88%	8.24%	2.52%
DNA Casework	64.74%	65.75%	12.01%	11.10%	11.09%	4.12%
DNA Database	45.80%	54.89%	25.36%	6.93%	6.05%	5.77%
Document Examination	79.82%	82.22%	8.28%	2.62%	1.58%	2.26%
Drugs - Controlled Substances	72.66%	73.07%	8.44%	10.96%	10.49%	6.79%
Evidence Screening & Processing	83.87%	84.19%	13.44%	3.88%	4.36%	3.03%
Explosives	60.97%	63.49%	19.82%	18.46%	9.18%	18.53%
Fingerprints	79.47%	77.64%	8.96%	6.72%	5.53%	7.73%
Fire analysis	74.61%	77.49%	13.81%	5.39%	3.99%	3.77%
Firearms and Ballistics	77.43%	78.62%	9.51%	9.00%	5.48%	8.56%
Forensic Pathology	73.66%	74.44%	8.33%	5.12%	6.26%	3.30%
Gun Shot Residue (GSR)	69.98%	73.22%	12.54%	13.41%	8.14%	11.57%
Marks and Impressions	78.74%	84.91%	14.71%	4.77%	4.34%	4.29%
Serology/Biology	76.28%	76.80%	8.18%	5.63%	6.34%	3.81%
Toxicology ante mortem (excluding BAC)	63.18%	64.16%	11.40%	12.97%	9.95%	8.61%
Toxicology post mortem (excluding BAC)	62.93%	63.99%	13.21%	11.39%	10.73%	7.38%
Trace Evidence	67.75%	72.34%	16.06%	17.55%	11.41%	16.03%

The warning is that the demands for future investment will grow and the lowering of average cost is a temporary phenomenon.

The percentage of personnel expenditures are relatively stable across laboratories. Capital expenditure, on the other hand, show a great deal more variability. Some of this variability is tied to grant funding success. Table 10 highlights the percentage of expenditures for consumables. This category includes all expenditures for chemicals, reagents, gases, and any other labo-ratory consumables.

Capacity Utilization Metrics

The aforementioned metrics are all related to a direct interpretation of mission. Maximizing quality output for a given budget, and the corresponding decomposition, offer some insight into the ability of the laboratory to offer a return on investment for its constituency. In addition to maximizing output or correspondingly minimizing average cost, the justice system also expresses a keen interest in backlog and turnaround time, both issues related to capacity utilization.

Table 11 offers two measurements of turnaround time. The first measure counts turnaround time from the time of the last submission in a case until the issuance of a report. This metric follows a metric that appeared in a study of four European laboratories (European Network of Forensic Science Institutes 2003).

Table 10 Consumables Expenditures/Total Expenditures by Investigative Area

	Consumables Expenditures/Total Expenditures		
Investigative Area	Mean	Median	Std. Dev.
Blood Alcohol	8.85%	6.62%	7.87%
Crime Scene Investigation* Digital evidence - Audio & Video*			
DNA Casework	13.72%	11.48%	9.19%
DNA Database	13.42%	11.06%	11.26%
Document Examination	2.32%	1.42%	2.81%
Drugs - Controlled Substances	6.70%	6.21%	3.79%
Evidence Screening & Processing	3.89%	4.44%	2.09%
Explosives	5.12%	4.37%	3.41%
Fingerprints	3.54%	1.32%	4.37%
Fire analysis	4.68%	5.02%	2.19%
Firearms and Ballistics	2.61%	1.53%	2.49%
Forensic Pathology	3.31%	3.16%	0.69%
Gun Shot Residue (GSR)	4.04%	3.38%	3.27%
Marks and Impressions	4.76%	1.74%	4.90%
Serology/Biology	6.67%	7.50%	2.01%
Toxicology ante mortem (excluding BAC)	10.26%	10.46%	3.86%
Toxicology post mortem (excluding BAC)	8.81%	7.42%	2.35%
Trace Evidence	3.50%	2.64%	2.93%

* Sample size too small to determine reliable summary statistics.

Table 11 Turnaround Time (TAT) by Investigative Area

	TAT (measured from last submission)*			TAT (measured from first submission)**		
Investigative Area	Mean	Median	Std. Dev.	Mean	Median	Std. Dev.
Blood Alcohol	29	12	41	21	18	13
Crime Scene Investigation	39	29	36	35	27	32
Digital evidence - Audio & Video	48	48	6	76	64	40
DNA Casework	77	66	59	77	68	48
DNA Database	113	82	92	147	82	139
Document Examination	43	34	14	68	59	26
Drugs - Controlled Substances	45	38	24	49	44	32
Evidence Screening & Processing***	27	27	13			
Explosives	101	31	144	137	139	131
Fingerprints	34	35	17	44	45	18
Fire analysis	42	46	19	41	43	21
Firearms and Ballistics	52	39	57	79	62	67
Forensic Pathology***	57	57	41	28	28	
Gun Shot Residue (GSR)	44	34	39	36	34	28
Marks and Impressions	41	39	28	68	60	47
Serology/Biology	43	31	38	56	48	33
Toxicology ante mortem (excluding BAC)	33	24	27	44	36	25
Toxicology post mortem (excluding BAC)	28	24	14	42	33	22
Trace Evidence	75	68	37	86	93	35

* Turnaround time (days) with time measured from the first submission of evidence to a case.
** Turnaround time (days) with time measured from the last submission of evidence to a case.
*** Sample size too small to determine reliable summary statistics.

Since many laboratories employ a different measure of turnaround time, a second set of summary statistics appear in Table 11. For this metric, the turnaround time clock begins when the first evidence in an area of investigation is submitted to the laboratory and the time is counted until the issuance of a report. Only a few laboratories have been able to provide both metrics with approximately forty percent of the sample offering information on the first three data columns and over sixty percent offering input for the final three columns. Examination of Table 11 reveals some seeming inconsistencies. By definition, TAT—measured from last submission is less than or equal to TAT—measured from first submission. Yet, for Blood Alcohol, Crime Scene Investigation, Fire Analysis, Gunshot Residue, and Forensic Pathology, the mean and/or median TAT relationship violates this ordinal relationship. However, these anomalies are easily explained. Since few laboratories report both TAT measures, and some laboratories are unable to report either TAT measure, the reduced sample size for each metric may suffer from being drawn from a small number of observations. This is certainly true for Crime Scene Investigation, Fire Analysis, Gunshot Residue, and Forensic Pathology. For Blood Alcohol, the relationship between the median TAT's appears as expected, but the median TAT's do not. The extraordinarily large standard deviation associated with TAT—measured from last submission is suggestive of some very large TAT submissions which have skewed the distribution and led to a higher mean.

The final table provides some detail on the percentage of cases older than thirty days. While thirty days is acknowledged to be a rather arbitrary measure of backlog, it does offer some measurement consistency across laboratories.

Table 12 offers the summary statistics for the number of cases over 30 days old as a percentage of the volume of casework for an entire year by each area of investigation. The backlog situation is heavily skewed in several areas of investigation, including DNA casework, document examination, firearms and ballistics, forensic pathology, marks and impressions, and trace evidence. In several areas of investigation the median is very low, suggesting a fairly efficient throughput in the typical laboratory.

Kobus et al (2011) and Newman et al. (2011) offer some lessons regarding the use of the above metrics into the strategic planning and corresponding change in the laboratory. Comparison of an individual laboratory's metrics to the norms of the industry offer a notion of where to look for situations in which the laboratory excels and those areas in which it may be deficient. The resulting strategic changes should lead to a series of allocations across areas in the laboratory budget (Speaker and Fleming 2010).

Foresight: Multivariate Analysis

While the metrics highlighted in Tables 1–12 offer some insight into the comparative performance of individual laboratories, they are limited to a univariate viewpoint. Such a view is not as problematic in the forprofit world where the invisible hand of market forces

Table 12 Backlogged Cases (30 days C) as a percentage of Total Caseload by Investigative Area

	Backlog Cases/Total Cases		
Investigative Area	Mean	Median	Std. Dev.
Blood Alcohol	2.62%	0.50%	5.34%
Crime Scene Investigation	9.53%	0.27%	16.24%
Digital evidence - Audio & Video	9.55%	9.55%	0.64%
DNA Casework	18.99%	9.34%	22.41%
DNA Database	26.92%	25.76%	20.51%
Document Examination	40.19%	18.48%	57.79%
Drugs - Controlled Substances	6.84%	5.75%	5.45%
Evidence Screening & Processing	33.44%	33.44%	11.23%
Explosives	31.07%	20.00%	21.24%
Fingerprints	13.52%	5.74%	19.39%
Fire analysis	6.18%	7.50%	6.82%
Firearms and Ballistics	32.43%	16.46%	34.45%
Forensic Pathology	54.06%	39.38%	52.85%
Gun Shot Residue (GSR)	10.45%	4.32%	15.67%
Marks and Impressions	51.98%	43.75%	74.05%
Serology/Biology	6.71%	8.20%	4.21%
Toxicology ante mortem (excluding BAC)	6.74%	5.83%	4.96%
Toxicology post mortem (excluding BAC)	11.17%	9.74%	6.75%
Trace Evidence	53.19%	25.87%	91.74%

guides organizations to the "right size" via competitive forces. Without such market forces for public institutions, a jurisdictional-based laboratory size retains some inefficiencies with respect to cost effectiveness (Maguire et al. 2012b). In particular, the Cost/Case ratio, and similar metrics, must be reconsidered with respect to the volume of activity in the jurisdiction.

This multivariate analysis was developed in response to inquiries into the best model for the delivery of forensic science. Bedford (2011) offers an argument into the value of the market forces in competitive markets and the efficiencies that arise from a profit motive. Maguire et al (2012b) following a study of Canada's provision of forensic science services (Maguire et al. 2012a), suggest that there is a natural sense of best that is less model-driven, and more size-driven by economies of scale. Following these developments, it may be the case that the success attributed to market forces in Bedford (2011) is shared with other economies (Speaker 2013).

This analysis follows from the law of diminishing marginal returns (LDMR) in economics that leads to the conclusion that there is an ideal size in every industry. The notion of "ideal size" is a reflection of perfect economies of scale or the level of output at which average costs are minimized. The corresponding depiction of average total cost is a U-shaped curve as average total cost is related to volume of production. For each of the areas of investigation, a second-degree polynomial regression was estimated where:

$$\frac{Cost}{Case} = \alpha + \beta_1 Cases + \beta_2 Cases^2 \tag{5}$$

The results of those regressions using (5) are presented below in equations (6)–(20) for each area of investigation[5] and the figures that follow provide a depiction of the data and the estimated curve. For better comparison that the univariate summary statistics in Table 1, a laboratory may put its case level into the estimated equation and obtain the mean response for that level of casework as a comparison to their actual performance.

Blood Alcohol

The estimate of equation (5) for blood alcohol analysis is:

$$Average\ Cost\ (Blood\ Alcohol) = 247.424 - 0:02573Cases + 0.0000011467Cases^2 \tag{6}$$

Figure 1 illustrates the data from the FORESIGHT submissions with the estimate in (6) drawn as a solid curve. At the minimum point on the curve is an estimate of perfect economies of scale and for blood alcohol analysis this occurs at approximately 11,218 cases. This area of investigation includes analysis of blood or breath samples to detect the presence of and quantify the amount of alcohol. The quadratic equation estimated in (6) offers a fairly simple relationship between casework and average cost. More sophisticated econometric techniques might offer more detail at the low casework levels to explain more of the variation, but above a 5,000 caseload, the estimated equation offers valuable comparative information for laboratories.

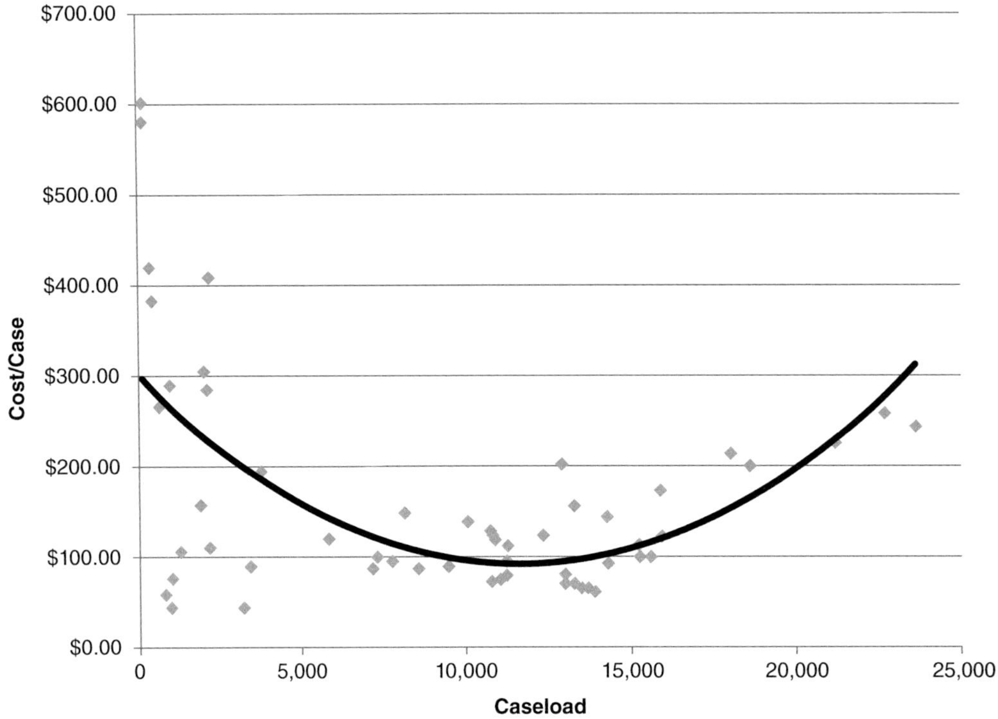

Figure 1 Blood alcohol analysis economies of scale.

DNA Casework

The estimate of equation (5) for DNA casework analysis is:

$$\textit{AverageCost}\,(\textit{DNAcasework}) = 2{,}700.513 - 0:35531\textit{Cases} + 0.000032\textit{Cases}^2 \tag{7}$$

Figure 2 illustrates the data from the FORESIGHT submissions with the estimate in (7) drawn as a solid curve. At the minimum point on the curve is an estimate of perfect economies of scale and for DNA casework analysis this occurs at approximately 5,550 cases. The simple model estimated for DNA casework in (7) provides a fairly good fit to the data. More sophisticated econometric analysis may be found elsewhere (Maguire et al. 2012a), but the essence of the relationship is captured in the quadratic estimate in (7).

DNA Database

The estimate of equation (5) for DNA database analysis is:

$$\textit{Average Cost}\ (\textit{DN Adatabase} = 140.486 - 0.00544\textit{Cases} + 0.000000064326\textit{Cases}^2 \tag{8}$$

Figure 3 illustrates the data from the FORESIGHT submissions with the estimate in (8) drawn as a solid curve. At the minimum point on the curve is an estimate of perfect economies of scale and for DNA database this occurs at approximately 42,301 cases. As seen in Figure 3, the sample of laboratories analyzing and entering DNA results into the database is a smaller subset of the submitting laboratories and very few laboratories are operating at

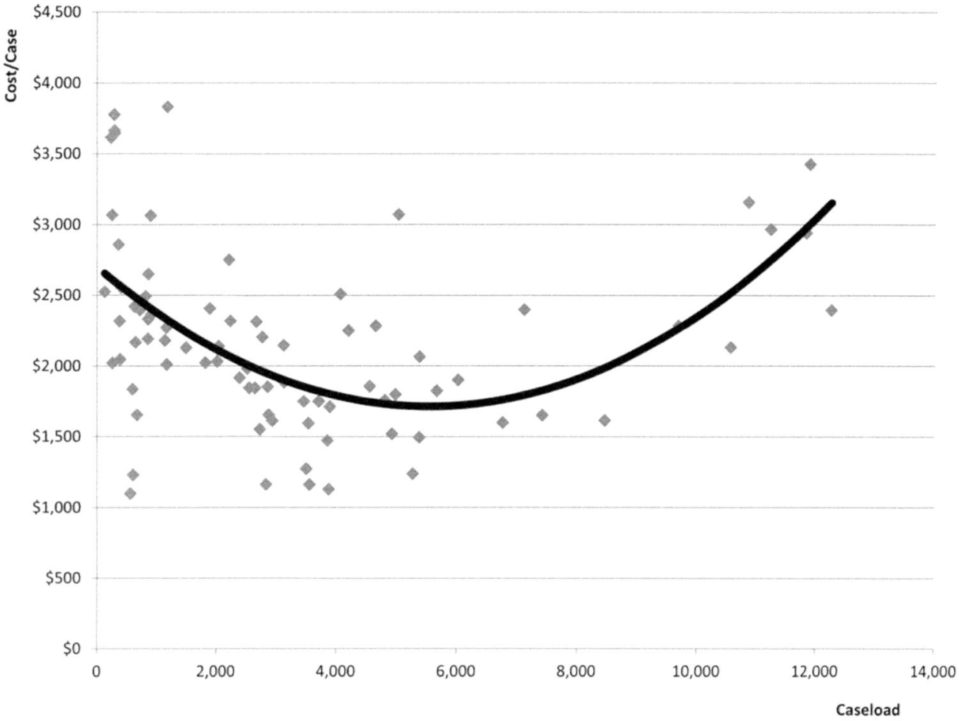

Figure 2 DNA casework analysis economies of scale.

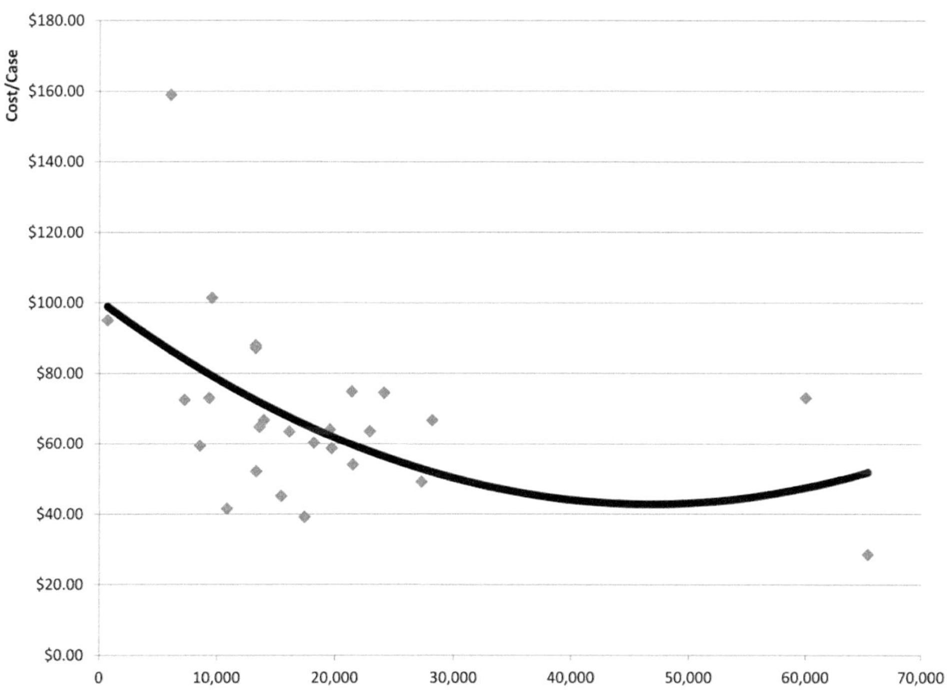

Figure 3 DNA database analysis economies of scale.

a level large enough to take advantage of economies of scale. As more laboratories provide data in the future, a better comparative picture should emerge on the relationship between average cost and caseload.

Document Examination

The estimate of equation (5) for document examination is:

$$Average\ Cost\ (document\ examination) = 5.399.327\ \text{-}\ 21.4757 Cases \quad (9)$$

Figure 4 illustrates the data from the FORESIGHT submissions with the estimate in (9) drawn as a solid curve. At the minimum point on the curve is an estimate of perfect economies of scale and for document examination this occurs at approximately 283 cases. Document examination includes the analysis of legal, counterfeit, and questioned documents, including handwriting analysis.

Drugs—Controlled Substances

The estimate of equation (5) for drugs—controlled substances is:

$$Average\ Cost\ (Drugs) = 380.95\ \text{-}\ 0.01725 Cases + 0.000000366 Cases^2 \quad (10)$$

Figure 5 illustrates the data from the FORESIGHT submissions with the estimate in (10) drawn as a solid curve. At the minimum point on the curve is an estimate of perfect

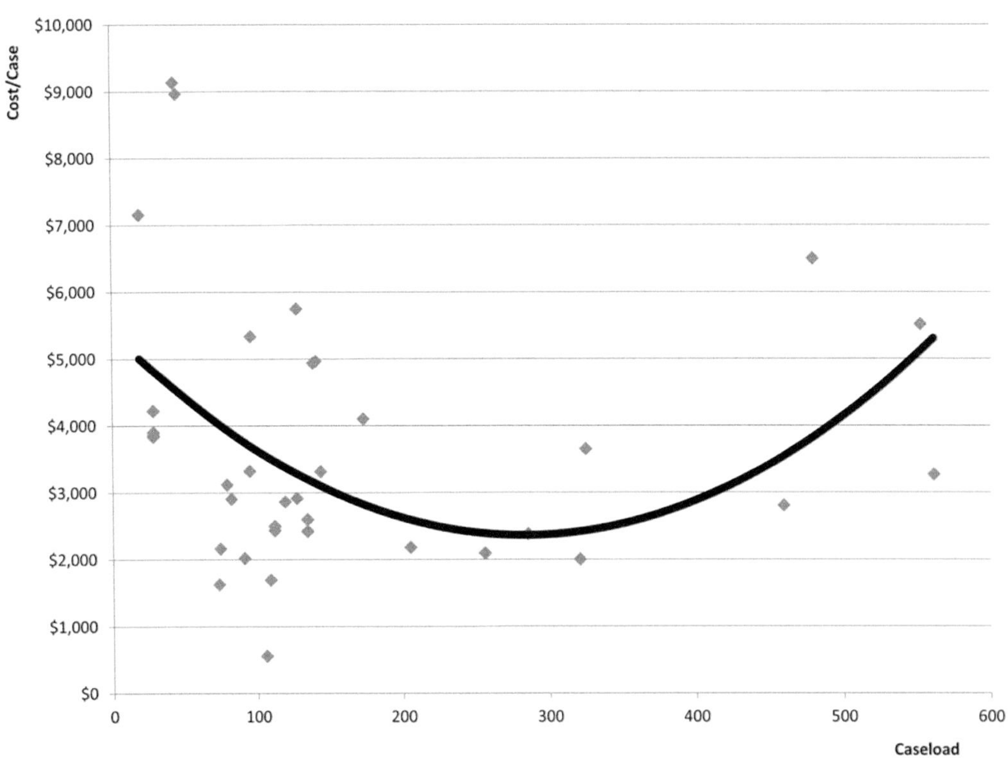

Figure 4 Document examination (including handwriting) analysis economies of scale.

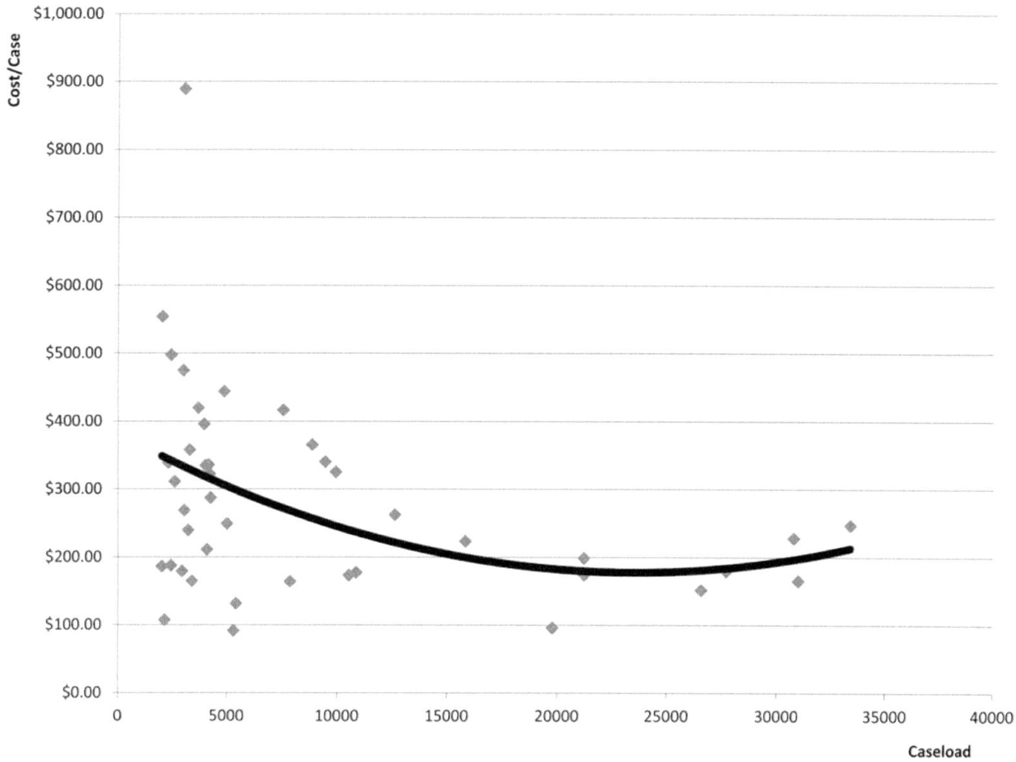

Figure 5 Drugs-controlled substance analysis economies of scale.

economies of scale and for drugs/controlled substance analysis this occurs in approximately 23,547 cases. Drugs/controlled substances include analysis of solid dosage licit and illicit drugs, including precursor materials.

Explosives

The estimate of equation (5) for explosives is:

$$Average\ Cost = 11.714.063 \ - \ 161.2313Cases + 0.99573Cases^2 \quad (11)$$

Figure 6 illustrates the data from the FORESIGHT submissions with the estimate in (11) drawn as a solid curve. At the minimum point on the curve is an estimate of perfect economies of scale and for explosives analysis; this occurs at approximately 81 cases. The explosives area of investigation includes analysis of energetic materials in pre- and post-blast incidents.

Fingerprint Identification

The estimate of equation (5) for fingerprint identification is:

$$Average\ Cost = 690.727 \ - 0.1548Cases + 0:000011Cases^2 \quad (12)$$

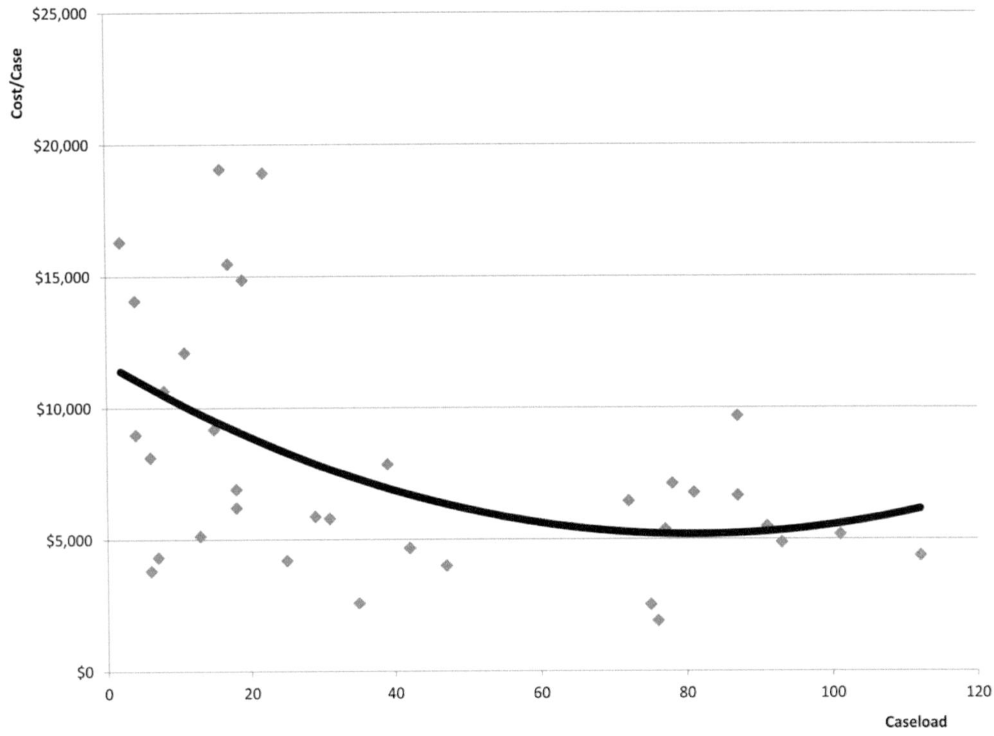

Figure 6 Explosives analysis economies of scale.

Fingerprint identification involves the development and analysis of friction ridge patterns. Figure 7 illustrates the data from the FORESIGHT submissions with the estimate in (12) drawn as a solid curve. At the minimum point on the curve is an estimate of perfect economies of scale and for fingerprint identification; this occurs at approximately 7,006 cases.

Fire Analysis

The estimate of equation (5) for fire analysis is:

$$Average\ Cost = 4{,}184.652 - 19{,}9925 Cases + 0.028006 Cases^2 \tag{13}$$

This area of investigation involves the analysis of materials from suspicious fires to include ignitable liquid residue analysis. Figure 8 illustrates the data from the FORESIGHT submissions with the estimate in (13) drawn as a solid curve. At the minimum point on the curve is an estimate of perfect economies of scale at 357 cases.

Firearms and Ballistics

The estimate of equation (5) for firearms and ballistics analysis is:

$$Average\ Cost = 2{,}067.331 - 0.958 Cases + 0.00013 Cases^2 \tag{14}$$

Firearms and ballistics involves the analysis of firearms and ammunition, to include distance determinations, shooting reconstructions, integrated ballistics identification system

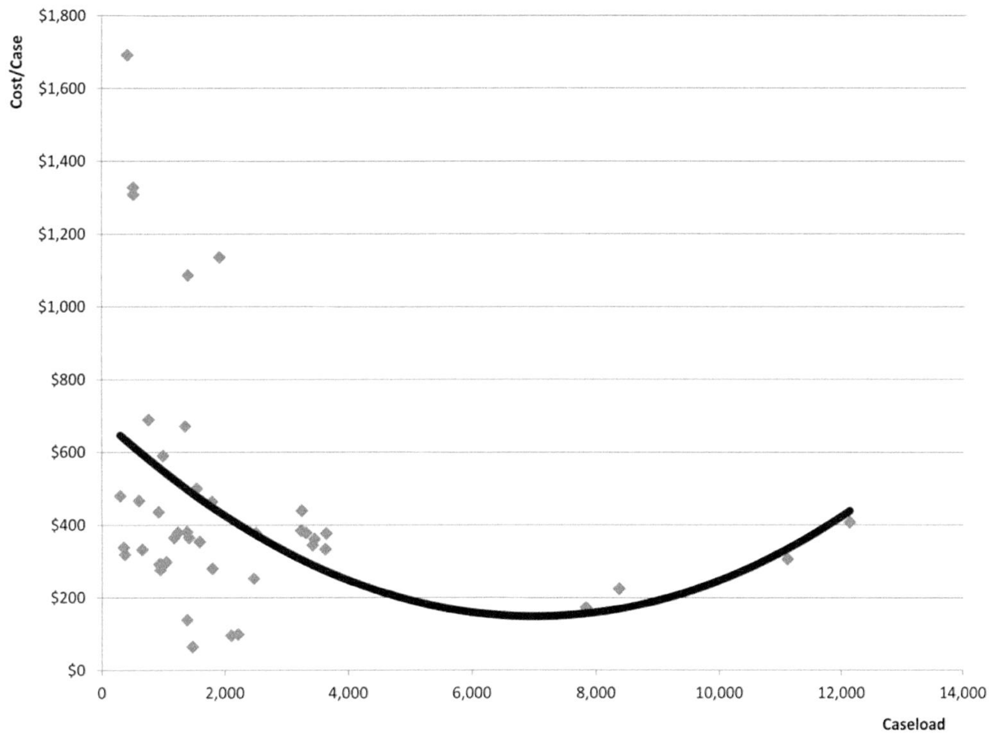

Figure 7 Fingerprint identification analysis economies of scale.

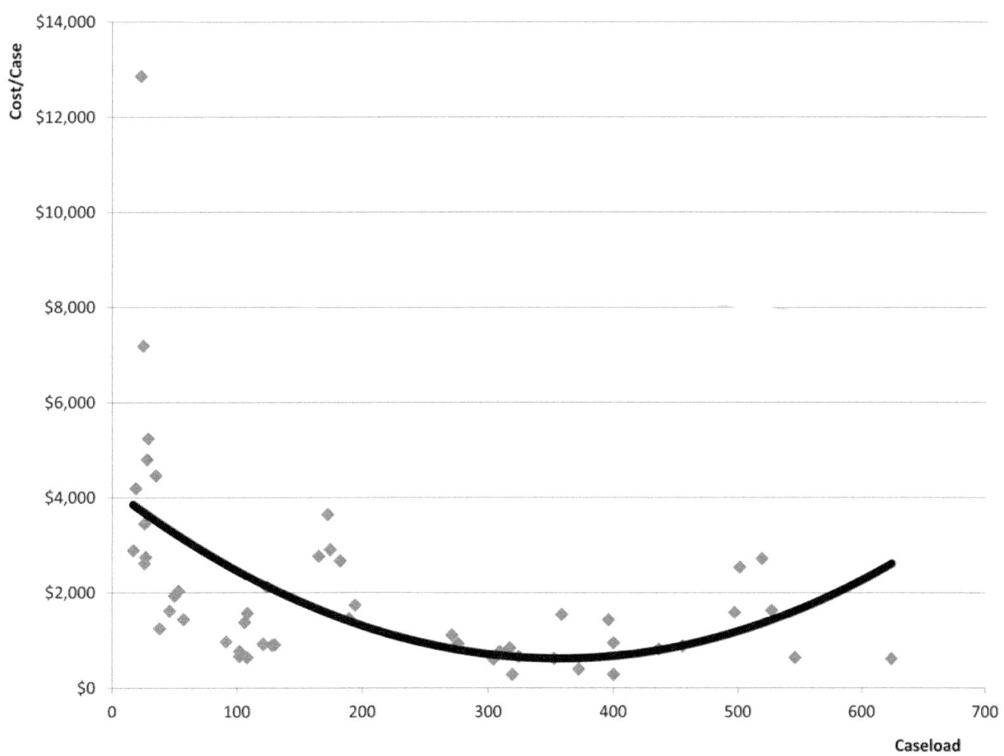

Figure 8 Fire analysis economies of scale

(IBIS), and tool marks. Figure 9 illustrates the data from the FORESIGHT submissions with the estimate in (14) drawn as a solid curve. At the minimum point on the curve is an estimate of perfect economies of scale at 3,680 cases per year.

Gun Shot Residue

The estimate of equation (5) for the analysis of gunshot residue is:

$$Average\ Cost = 3{,}804.241 - 17.420 Cases + 0.02165 Cases^2 \quad (15)$$

Gun shot residue (GSR) is the analysis of primer residues from discharged firearms (not distance determinations). Figure 10 illustrates the data from the FORESIGHT submission with the estimate illustrated in equation (15) drawn as a solid curve. The estimation suggests that perfect economies of scale are realized at an estimate of 402 cases annually.

Marks and Impressions

The estimate of equation (5) for the analysis of marks and impressions was not evaluated as specified, given the relationship observed between the average cost per case and the caseload. Visual inspection of the data in Figure 11 suggests that the expected U-shaped curve is not fully revealed. Rather the relationship appears to only reveal part of the

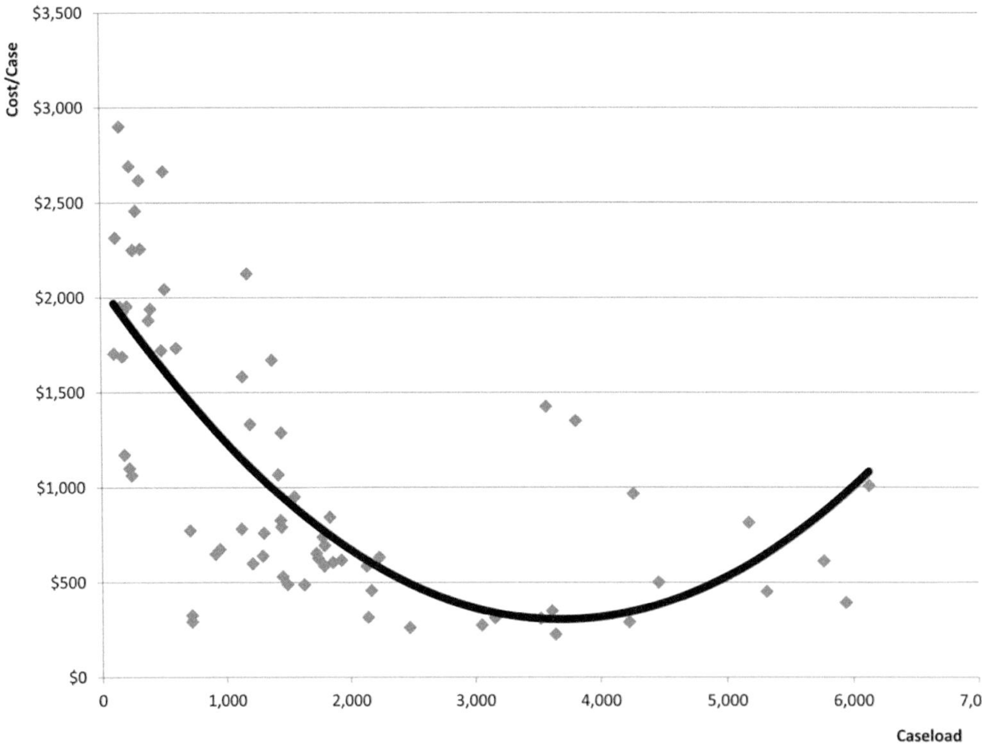

Figure 9 Firearms & ballistics analysis economies of scale.

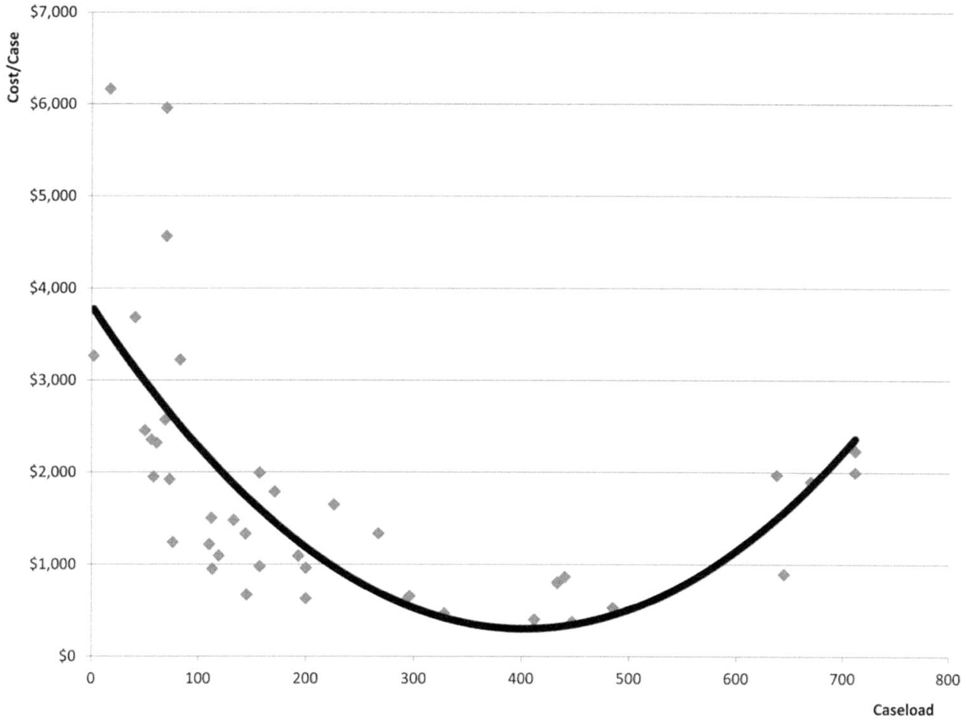

Figure 10 Gun shot residue analysis economies of scale.

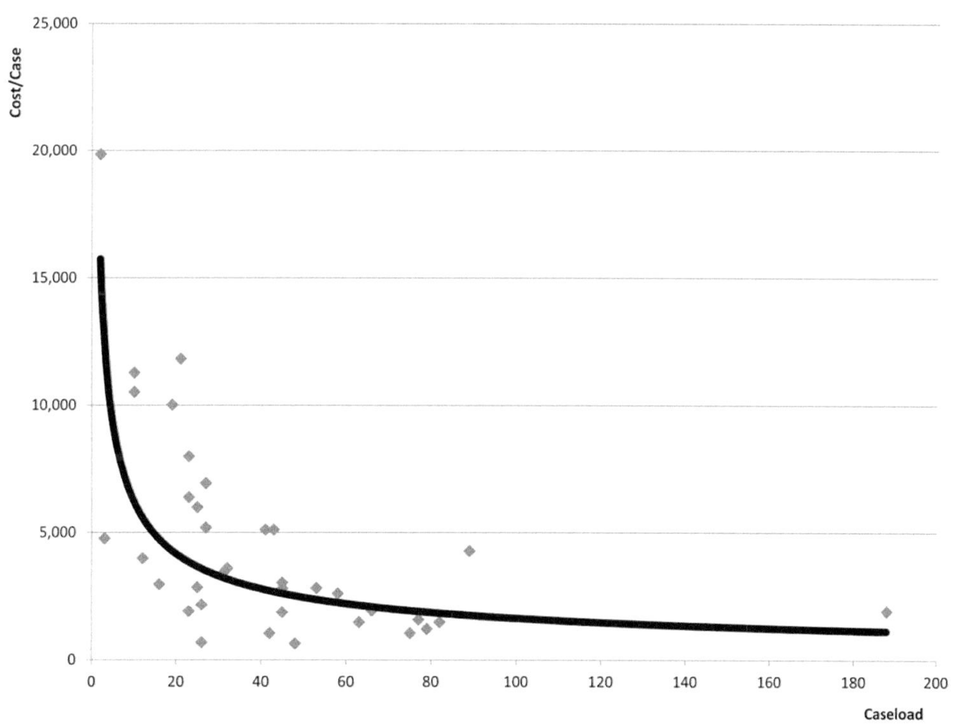

Figure 11 Marks and impressions analysis economies of scale.

downward-sloped portion of that expected U-shaped curve. To account for this, a power function is estimated:

$$Average\ Cost = 10.063Cases^{-0.57679} \tag{16}$$

Marks and impressions is the analysis of physical patterns received and retained through the interaction of objects of various hardness, including shoeprints and tire tracks. As illustrated in Figure 11, perfect economies of scale are undefined since the caseload does not appear to have reached the minimum for the laboratories examined.

Serology/Biology

The estimate of equation (5) for serology and biology analysis is:

$$Average\ Cost = 993{:}986 - 0.23738Cases + 0.0000224Cases^2 \tag{17}$$

Biology/serology includes the detection, collection, and non-DNA analysis of biological fluids. Figure 12 illustrates the data from the FORESIGHT submissions with the estimate in (17) drawn as a solid curve. At the minimum point on the curve is an estimate of perfect economies of scale at 5,300 cases per year.

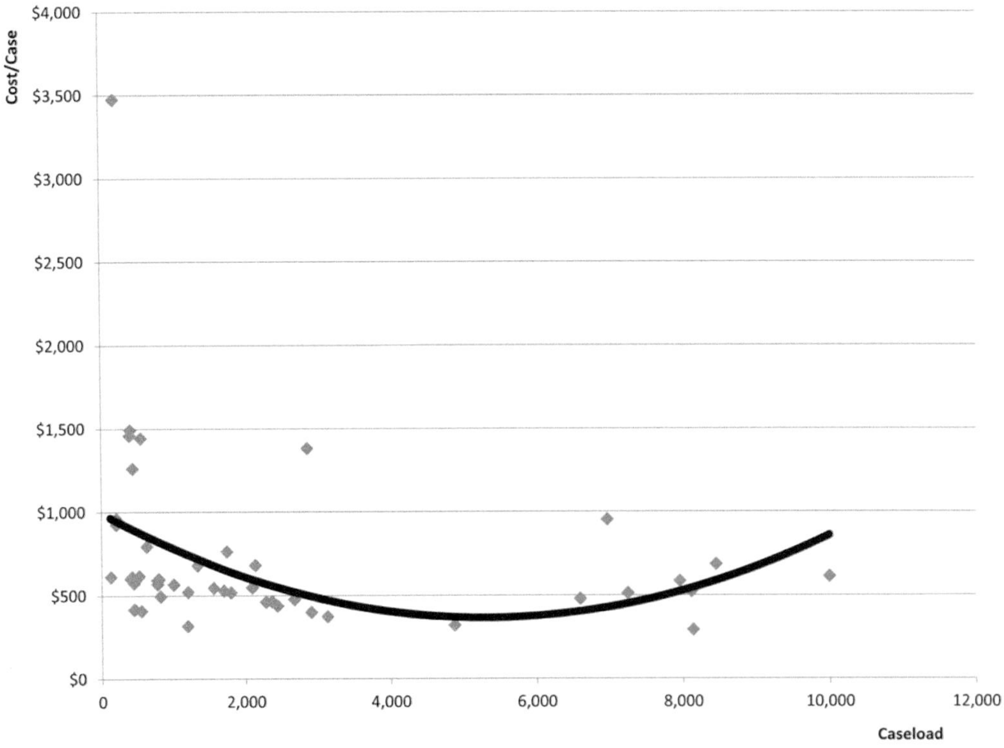

Figure 12 Serology and biology analysis economies of scale.

Toxicology Ante Mortem

The estimate of equation (5) for the analysis of toxicology ante mortem is:

$$Average\ Cost = 1{,}288.9914 - 0.37679 Cases - 0.00003487 Cases^2 \quad (18)$$

Toxicology ante mortem involves the chemical analysis of body fluids and tissues to determine if a drug or poison is present in a living individual, to include blood alcohol analysis (BAC). Toxicologists are then able to determine if a drug (including alcohol) or poison is present in a living individual and at what concentration. Figure 13 illustrates the data from the FORESIGHT submissions with the estimate in (18) drawn as a solid curve. At the minimum point on the curve is an estimate of perfect economies of scale estimate at 5,402 cases per year.

Toxicology Post Mortem

The estimate of equation (5) for the analysis of toxicology post mortem is:

$$Average\ Cost = 1{,}451.3074 - 0.47299 Cases + 0:00004815 Cases^2 \quad (19)$$

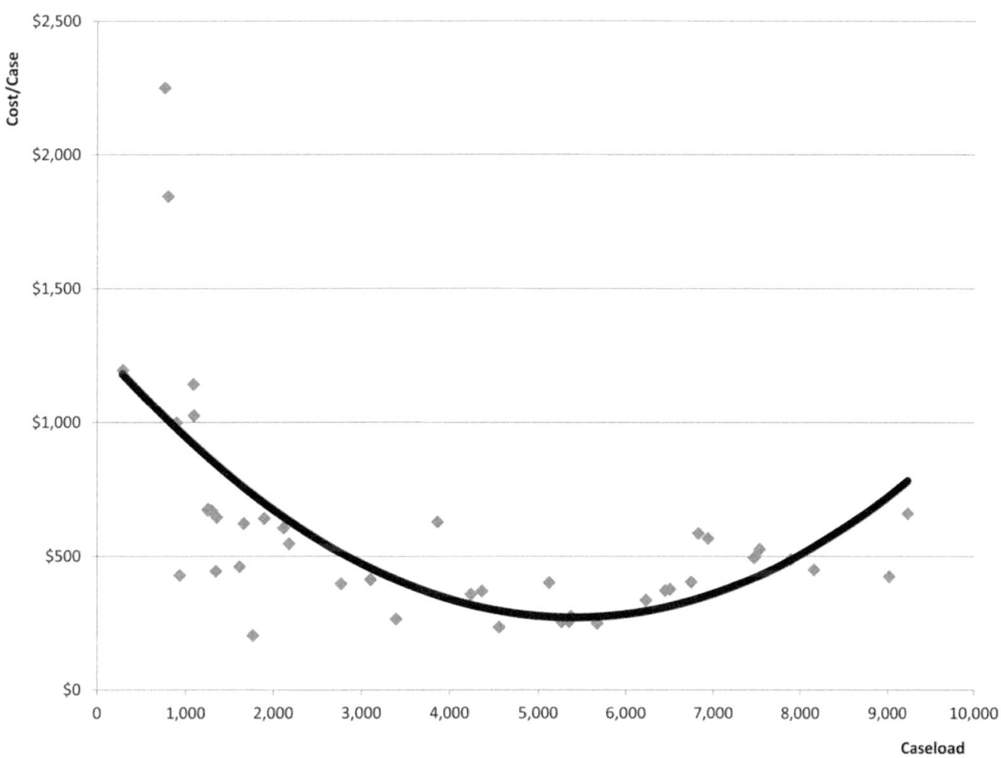

Figure 13 Toxicology ante mortemante-mortem analysis economies of scale.

Toxicology post mortem involves the chemical analysis of body fluids and tissues to determine if a drug or poison is present in a deceased individual. Toxicologists are then able to determine if a drug (including alcohol) or poison is present in a deceased individual and at what concentration. Figure 14 illustrates the data from the FORESIGHT submissions with the estimate in (19) drawn as a solid curve. At the minimum point on the curve is an estimate of perfect economies of scale estimate at 4,912 cases per year.

Trace Evidence

The estimate of equation (5) for the analysis of trace evidence is:

$$Average\ Cost = 7{,}949.999 - 18.9443Cases + 0.012983Cases^2 \tag{20}$$

Trace evidence includes the analysis of materials that, because of their size or texture, transfer from one location to another and persist there for some period of time. Microscopy, either directly or as an adjunct to another instrument, is involved. Includes analysis of hairs and fibers and paint and glass. Figure 15 illustrates the data from the FORESIGHT submissions with the estimate in (20) drawn as a solid curve. At the minimum point on the curve is an estimate of perfect economies of scale estimate at 730 cases per year.

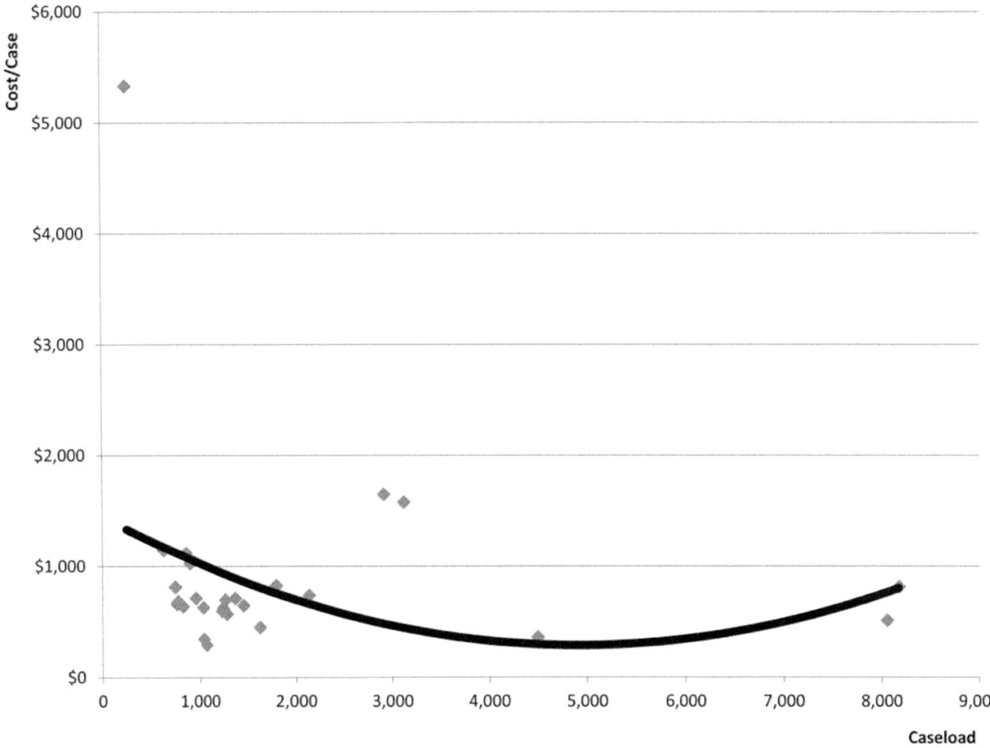

Figure 14 Toxicology post- mortem analysis economies of scale.

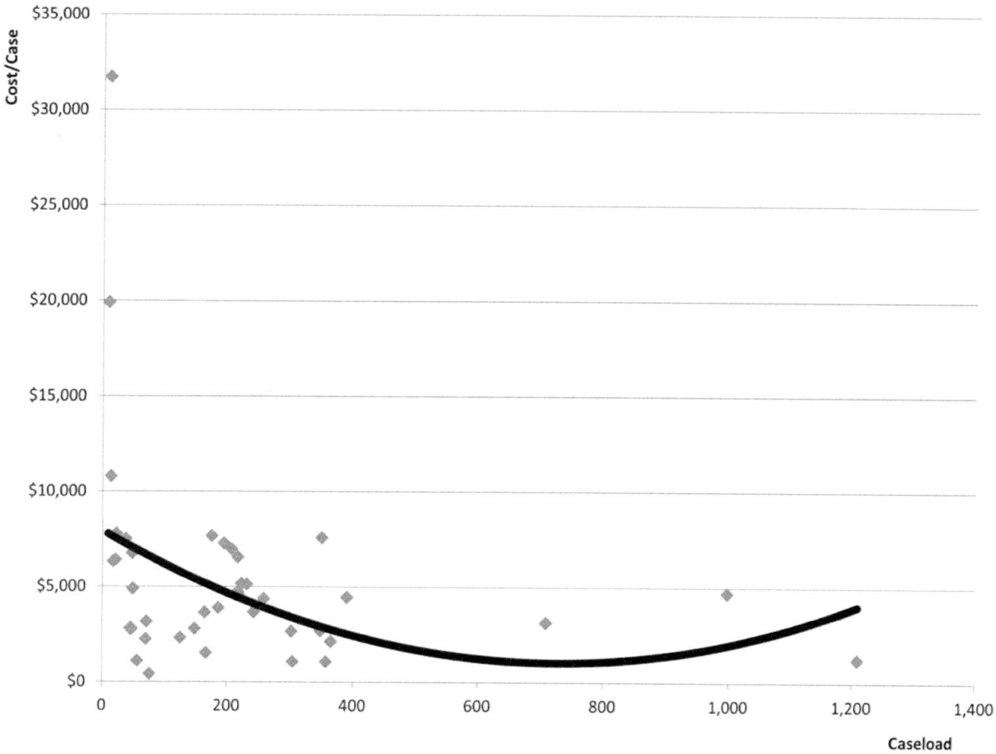

Figure 15 Trace evidence (including paint & glass and hairs & fibers) analysis economies of scale.

Concluding Remarks

Project FORESIGHT offers a detailed glimpse into the operation of forensic science laboratories. The project has analysed data annually from the 2005–2006 fiscal years to the present and will continue into the foreseeable future. While data submissions started small, they have grown in the recent years and the 2011–2012 reporting period offers some large sample properties from the data submissions. It is these insights that have been highlighted in the previous sections. We remind the reader that Project FORESIGHT data comes from voluntary participation and as such, does not represent a random sample from all laboratories. For the 2011–2012 reporting period, 81 laboratories submitted data on casework, expenditures, and personnel. As noted, all laboratories met industry quality standards and the participating laboratories represent six continents with a majority of the reporting laboratories representing U.S. jurisdictions. The laboratories represent municipal, regional, provincial, national jurisdictions as well as private laboratories. It is anticipated for the 2013–2014 reporting period that the number of laboratories submitting data will far exceed one hundred laboratories.

The summary statistics offer a good starting point for any laboratory to analyze their internal operations. Such a self-study can be easily accomplished using the tables, equations, and figures above. Note, however, that any laboratory may choose to participate in Project FORESIGHT by submitting data using the LabRAT tool (available at http://www.

be.wvu.edu/forensic/foresight.htm). Following submission, participating laboratories receive immediate access to industry comparables and time series analysis of performance to match against strategic initiatives.

Beyond the positioning of an individual laboratory using the analysis above, Houck et al. (2012) highlights how the financial management is only a part of the performance analysis, but an integral part. Expanding analysis into a balanced scorecard model enables a continual cycle of continuous improvement. Newman et al. (2011) provide a best practices approach to strategic improvement with the FORESIGHT data helping to guide the direction for the laboratory. McAndrew and Speaker (2012) extend the analysis to industry trends that may expand the laboratory's ability to meet its mission. Combining FORESIGHT results with independent evaluations of the external return on investment offers direction for future work. That pathway has been illuminated with the impact from DNA Database CITATION Dol13 /l 1033 (Doleac, 2013) and applied by Lodhi et al. (2014). As more laboratories participate, so will the knowledge shared with all laboratories.

NOTES

1 NIJ award #1004801AR (West Virginia University Forensic Science Initiative 2014).

2 The LabRAT data collection tool is available via http://www.be.wvu.edu/forensic/foresight.htm.

3 As one reviewer noted, the large variability in Crime Scene Investigation can be affected by rate-limiting process steps at the laboratory level. FORESIGHT does not track such restrictions and comparison across laboratories may be difficult. As noted in Houck et al. (2012), a balanced scorecard approach to the evaluation of any metrics is important for a contextual understanding of the measures.

4 Across areas of investigation there are considerable differences in the median percentages allocated for personnel. Comparisons across areas of investigation are difficult to make and are not suggested. Some areas are naturally capital-intensive and others are naturally personnel-intensive. As such, comparisons should generally be limited to other laboratories with an area of investigation or of the laboratory over time in a particular area of investigation.

5 In a few areas of investigation there was insufficient data to provide a reliable estimate and those areas have been omitted.

REFERENCES

Bedford, K. 2011. Forensic science service provider models—Is there a 'best' option? *Australian Journal of Forensic Sciences* 43(2–3): 147– 156.

Doleac, J. 2013, January. *The Effects of DNA Databases on Crime*. Charlottesville, Virginia, USA: Frank Batten School of Leadership and Public Policy, University of Virginia Faculty Working Papers.

European Network of Forensic Science Institutes. 2003. *QUADRUPOL*. ISIN II programme of the EU. The Hague, The Netherlands: ENFSI.

Houck, M. M., Riley, P. J. Speaker, and T. S. Witt. 2009. FORESIGHT: A Business Approach to Improving Forensic Science Services. *Forensic Sciece Policy and Management* 1(2): 85–95.

Houck, M. M., R. A. Riley, P. J. Speaker, andA. S. Fleming. 2012. Balanced scorecard approach. *Science and Justice* 52(4): 209–216

Kobus, H., M. M. Houck, P. J. Speaker, and T. S. Witt. 2011. Managing performance in the forensic sciences: expectations in light of limited budgets. *Forensic Science Policy and Management 2* (1): 36–43.

Lodhi, K. M., R. L. Grier, and P. J. Speaker. 2014. A novel approach to forensic molecular biology education and training: it's impact on the criminal justice system. *Australian Journal of Forensic Sciences* DOI: 10.1080/00450618.2014.925974, 1–12.
Maguire, C. N., M. M. Houck, P. J. Speaker, and R. Williams. 2012a. *A feasible and sustainable model for forensic service delivery in Canada*. Ottawa, Canada: Her Majesty the Queen in Right of Canada, Public Safety Canada.
Maguire, C., M. Houck, R. Williams, and P. Speaker. 2012b. Efficiency and the cost effective delivery of forensic science servicess: insourcing, out-sourcing, and privatization. *Forensic Science Policy and Management* 3(2): 62–69.
McAndrew, W. P., and P. J. Speaker. 2012. Expanding budgets via strategic use of leasing. *Forensic Science Policy and Management* 3(4): 169–179.
Newman, J., D. Dawley, and P. J. Speaker. 2011. Strategic management of forensic laboratory resources: From Proect FORESIGHT metrics to the development of action plans.*Forensic Science Policy and Management* 2(4): 164–174.
Speaker, P. J. 2009a. Key performance indicators and managerial analysis for forensic laboratories. *Forensic Science Policy and Management*1 (1): 32–42.
Speaker, P. J. 2009b. The decomposition of return on investment for forensic laboratories.*Forensic Science Policy and Management* 1(2): 96–102.
Speaker, P. J. 2010a. A new twist on old tricks: dupont expansions for public entities. *Journal of the Academy of Business Education, Proceedings*, 1–7.
Speaker, P. J. 2010b. The KPI of CSI: A Business Application in the Public Sector. *Journal of the Academy of Business Education, Proceedings*, 1–12.
Speaker, P. J. 2013. Forensic science service provider models: data-driven support for better delivery options. *Australian Journal of Forensic Sciences*, DOI:10.1080/00450618.2013.773076.
Speaker, P. J., and A. S. Fleming. 2010. Benchmarking and Budgeting Techniques for Improved Forensic Laboratory Management. *Forensic Science Policy and Management* 1(4):199–208.
West Virginia University Forensic Science Initiative. 2014. *West Virginia University's Forensic and Investigative Science Outreach Final Report 2009-DI-BX-K012, Award 1004801AR*. Washington, DC: National Institute of Justice.

Appendix A—Foresight Glossary

Assistant/Analyst An individual carrying out general casework examinations or analytical tests under the instruction of a reporting scientist or reporting analyst and who is able to provide information to assist with the interpretation of the tests.

Backlog Open cases that are older than 30 days.

Capital Expenditure Purchases of equipment, instruments, etc. with a lifetime longer than a year. These expenditures include newly purchased items as well as investments to improve the useful life of existing capital equipment. Expenses to maintain an asset at is current condition is not capitalized. Case—institute case: A request from a crime lab "customer" that includes forensic investigations in one or more investigative areas.

Case—area case A request for examination in one forensic investigation area. An area case is a subset of an institute case. Case (as reported in the LabRat form): Cases reported in LabRat are "area cases."

Casework All laboratory activities involved in examination of cases.

Casework time Total FTE's for operational personnel in an investigation area (in hours) subtracted by the hours of research and development, education and training, and support and service given to external partners.

Floor area Total of all floor area including office, laboratory, and other.

Full-time equivalent (FTE) The work input of a full-time employee working for one full year.

Investigation area Area limited by item type and methods as they are listed in the definitions of investigative areas tab.

Item A single object for examination submitted to the laboratory. Note: one item may be investigated and counted in several investigation areas.

Laboratory area Floor area used for forensic investigation, including sample and consumable storage rooms.

Non-reporting manager An individual whose primary responsibilities are in managing and administering a laboratory or a unit thereof and who is not taking part in casework.

Office area Floor area of offices (square feet).

Operational personnel Personnel in operational units providing casework, research and development (R and D), education and training (E and T) and external support services. Non-reporting unit heads are included.

Other floor area Floor area of space not belonging to laboratories or offices, i.e. corridors, lunch corners, meeting rooms, etc. (square feet).

Personnel expense Sum of direct salaries, social expenses (employer contribution to FICA, Medicare, workers comp, and unemployment comp), retirement (employer contribution only towards pensions, 401K plans, etc.), personnel development and training (internal or external delivery, including travel), and occupational health service expenses (employer contribution only).

Report A formal statement of the results of an investigation, or of any matter on which definite information is required, made by some person or body instructed or required to do so.

Sample An item of evidence or a portion of an item of evidence that generates a reportable result.

Support personnel Forensic laboratory staff providing various internal support services. Management and administration personnel not belonging to the operational units are included.

Test An analytical process, including but not limited to visual examination, instrumental analysis, presumptive evaluations, enhancement techniques, extractions, quantifications, microscopic techniques, and comparative examinations. This does not include technical or administrative reviews.

Turn-around time (TAT) The number of days from a request for examination in an investigative area until issuance of a report. (Note that an area case may have multiple requests and each new request has a separate turn-around time.)

Workload Total time spent on all work related to job, including overtime.

Appendix B—Foresight areas of Investigation

Biology/Serology The detection, collection, and non-DNA analysis of biological fluids.

Blood Alcohol The analysis of blood or breath samples to detect the presence of and quantify the amount of alcohol.

Crime Scene Investigation The collection, analysis, and processing of locations for evidence relating to a criminal incident.

Digital evidence—Computer, Audio and Video The analysis of multimedia audio, video, and still image materials, such as surveillance recordings and video enhancement. Includes the analysis of computers, computerized consumer goods, and associated hardware for data retrieval and sourcing.

DNA Casework Analysis of biological evidence for DNA in criminal cases.

DNA Database Analysis and entry of DNA samples from individuals for database purposes.

Document Examination The analysis of legal, counterfeit, and questioned documents, including handwriting analysis.

Drugs—Controlled Substances The analysis of solid dosage licit and illicit drugs, including pre-cursor materials.

Evidence Screening and Processing The detection, collection, and processing of physical evidence in the laboratory for potential additional analysis.

Explosives The analysis of energetic materials in pre- and post-blast incidents.

Fingerprint Identification The development and analysis of friction ridge patterns.

Fire analysis The analysis of materials from suspicious fires to include ignitable liquid residue analysis.

Firearms and Ballistics The analysis of firearms and ammunition, to include distance determinations, shooting reconstructions, NIBIN, and tool marks.

Forensic Pathology Forensic pathology is a branch of medicine that deals with the determination of the cause and manner of death in cases in which death occurred under suspicious or unknown circumstances.

Gun Shot Residue (GSR) The analysis of primer residues from discharged firearms (not distance determinations).

Marks and Impressions The analysis of physical patterns received and retained through the interaction of objects of various hardness, including shoeprints and tire tracks.

Toxicology, ante-mortem Toxicology involves the chemical analysis of body fluids and tissues to determine if a drug or poison is present in a living individual, to include blood alcohol analysis (BAC). Toxicologists are then able to determine how much and what effect, if any, the substance might have had on the person.

Toxicology, post-mortem Toxicology involves the chemical analysis of body fluids and tissues to determine if a drug or poison is present in a deceased individual. Toxicologists are then able to determine how much and what effect, if any, the substance might have had on the person.

Trace Evidence The analysis of materials that, because of their size or texture, transfer from one location to another and persist there for some period of time. Microscopy, either directly or as an adjunct to another instrument, is involved. Includes analysis of hairs and fibers and paint and glass.

Performance Is Everything, but What Is It?

5

All managerial actions must result in solid performance at a reasonable cost. They are rated on these factors: Output and cost. Because of this, managers may focus on short-term outcomes that are easily measured, like cost, turnaround time, and output rather than longer term, qualitative issues, like quality and performance.

The short view and the long view are both necessary to produce the desired outcome, that is, be effective. Performance is a flabby word; it just means accomplishing an action, task, or function. But that's the word that gets used instead of effective, producing a *desired outcome*. Generally, rating the performance of a manager or a process just means, is the desired outcome being produced. Because costs are one of the few economic measures relevant to FSPs, **efficiency, the extent to which time and effort are used to produce the desired outcome**, is also important. Efficiency is often confused with effectiveness but they are related. We'll sort that out in this section.

Public Sector Costs Are Its "Profits"

In a for-profit organization, performance is measured by the bottom line of the income statement, profit. Profit is an easy, single metric to gauge success. Organizations without a profit mandate (like governments), as the name implies, have money as an input (budgets) but not as an output. **They need to measure their success in other ways.** Despite the public concern and administrative angst over FSPs' performance and capacity, these terms have gone undefined as industry standards. Despite, or maybe because of, a lack of explicit understanding about the forensic industry, FSPs are held accountable by one or more agencies for their performance (whatever that means) based on non-standard, ill-defined, or non-existent criteria. Any successes or improvements the FSP makes may go unrecognized. The FSP may not have the information it needs to advance its mission and goals. **How, then, can FSPs define and measure their performance?**

As discussed previously, you can think of governmental FSPs as organizations that produce intelligence, reports, and testimony as their output. The staff are scientists and technicians who use their education, training, and knowledge to convert evidence and data into intelligence, reports, and testimony. The staff specialize in these "transactions" and, therefore, simplify them for the benefit of the criminal justice system. Investigators and attorneys don't need to run around, finding a variety of experts to do the specific tests required for a case. Instead, they go to the FSP as a "one stop shop."

Why This Matters

If the costs (not just financial but also ease, timeliness, access, and other factors) of using the governmental FSP are less than going to a variety of other (external) providers, then

DOI: 10.4324/9780429318528-13

the FSP has a value to its customers. If, however, the costs of using the FSP are greater than the costs of private FSPs, then there may be trouble. **Nearly all FSPs are governmental and are housed in law enforcement agencies**. This placement controls the FSP's relationships with their parent agency (definitely hierarchical) and other related agencies (maybe hierarchical, like prosecutors, and maybe lateral, such as medical examiners). If a FSP is too expensive (time, money, whatever), the customer (like the police) may opt to look elsewhere; if there is a service request that the FSP can't provide, they may choose to outsource that service to keep the customer happy. If it's the latter, outsourcing adds additional costs to the FSP's tally.

Personnel Are Central to Success

Think of the following set of logical statements:

- Services are dependent upon expertise.
- FSP professionals embody expertise.
- Therefore, services depend on FSP professionals.

Seems simple, doesn't it? Unlike other forms of labor, work that is dependent on knowledge, like FSP casework, is not solely *experience*-based, but rather is *learning*-based. For example, you can't just sit down to a spreadsheet and do somebody's taxes for them; it takes education in accounting, tax law, and financials to be able to do it properly. Knowledge work like this uniformly has high entry costs, like education, training, and credentialing. You can't go from work on a factory assembly line (skilled though it may be) and simply move into knowledge work or services work. Beyond the entry costs, major changes in attitudes, values, and beliefs are required. The term knowledge worker was coined by Peter Drucker in 1959 to describe the then-rising group of workers whose jobs required extensive education, the application of theoretical and analytical knowledge, and continuous learning. **This means FSP professionals are knowledge workers**. What FSP personnel can and cannot do, therefore, is directly linked to what services your FSP offers and, to a large degree, your **performance**.

Why Are Knowledge Workers Key to Performance?

The advance of science has taught us one thing about knowledge: Specialization and not generalization, is what makes it useful in the hands of a knowledge worker. The more specialized knowledge is, the more useful it becomes. Think of DNA analysis: From basic genetics, to combinatorial DNA, to analysis, to population genetics, to mixture interpretation, each of these digs deeper into the broader topic of "DNA analysis." Two things come along with specialization of knowledge workers:

- First, **knowledge workers operate in teams**. Teams balance out the necessary specializations for knowledge to be applied properly, like DNA, firearms, fingerprints, and trace evidence. Forensic scientists, the various investigators, laboratory colleagues, and other related organizations constitute one or more participatory teams.

- Second, **knowledge workers have to have access to an organization**. The organization provides the basic continuity that allows the knowledge workers' efforts to be converted into *performance*. That's the key point mentioned earlier. The continuity of the organization translates to a "memory" of not only methods and procedures but a culture that reflects the purpose of the FSP (mission) and the way its members work together and with others (values).

Who is a knowledge worker? Traditionally, they are defined by six criteria:

1 They must know what the task is and how it relates to the overall goals of the organization.
2 Knowledge workers have to manage themselves and be responsible for their own productivity.
3 Continuous improvement and innovation are inherent in their efforts; they have agency over their changes.
4 Continuous learning and teaching are necessary.
5 Quantity of production is at least as important as the quality of production.
6 The organization must view knowledge workers as "assets" or "investments" and not "costs"; the knowledge workers must want to work for their organization above all other opportunities.

Are FSP personnel knowledge workers? ***Yes***, definitely. And here's the reason we've spent all this time talking about knowledge workers: **Managing them is far more complex than typical industrial or manufacturing workers.** Collaboration and professionalism have to be the main emphasis (teams and organizations, right?).

Performance Is Effectiveness (and Maybe Efficiency)

Not to be pedantic but a few definitions are in order. First, **effectiveness is the capability of producing an outcome, frequently a specific desired effect**. You have a headache, you take an aspirin and the headache goes away: It has produced the desired effect. You have a sports car and it gets you from Point A to Point B: It is effective as transportation. Note that effectiveness can be a vague concept or non-quantitative assessment; in the aspirin example, what does "go away" mean? No headache? Less headache? And, if the "less headache," *how much less*? Actually producing the desired effect is called **efficacy**, like saying 325 mg of aspirin is efficacious for relieving a headache. **Efficacy is purely achieving the effect, regardless of the resources expended**, such as time or money. Whether you bought an inexpensive generic aspirin from a large pharmacy or paid way too much for a name brand aspirin at a convenience shop in an airport, if they both relieve your headache, both are efficacious (being effective). If you drive your sports car to the grocery store and your used ten-year-old minivan to the grocery store, both get you there and home. Effective.

The difference in the prices of the aspirins or the vehicles leads to the concept of **efficiency, which is being efficacious in the most economical way**. The least amount of input produces a minimum of, if not more than, the desired output. **Thus, what is**

effective (aspirin) is not necessarily efficacious (the dose may be too small), and what is efficacious (a sports car gets you to the grocery store) may not be necessarily efficient (sports cars are expensive!).

Effectiveness and Performance of FSPs

The effectiveness of an organization is a more complicated, dynamic concept than that of aspirin or sports cars. **An FSP's effectiveness combines the effectiveness of all its individual processes *and* those external values that may not relate to money, people, or time, such as social responsibility, professional ethics, and stewardship**. Communication and leadership are deeply intertwined with organizational effectiveness. Unless the organization's vision and goals can be clearly stated to everybody—inside and outside the FSP—the FSP will not be effective.

Efficiency is a useful tool to judge the effectiveness of either individuals or organizations. Efficiency can be expressed as **a specific or quantitative effort to yield a specific outcome**, such as a vehicle's miles-per-gallon of gasoline rating. Higher values of output (miles) per amount of input (gas) relate to greater efficiency; less is more, so to speak. Often, efficiency **is expressed as a ratio of input to maximal possible output**, like miles-per-gallon (miles: gallon) which allows for comparison between other like sources of output (other vehicles, for example) or to an idealized goal (100 miles-per-gallon or an average rating for a company's fleet of cars). Thus, efficiency loosely corresponds to the ratio,

$$e = O/I$$

where e is efficiency, O is the amount produced (output), and I is the amount of resources consumed in the process (input). Efficiency may be stated as a single number (ratio) but it reflects the process of a system and the assumptions that underlie it. **This is one of the reasons why FORESIGHT uses ratios**. Think back to the problematic fingerprint unit: *The ratios of cases/FTE and cost/case are measures of effectiveness; when compared to FORESIGHT averages or other FSP values, they become measures of efficiency*—that is, how much good are you doing with your resources.

Let's Get Down to Brass Tacks

Simply put, "The effectiveness of forensic science is directly related to the quality of the scientific analysis and the timeliness with which that analysis is provided, while efficiency is associated with attempts to minimize costs without negatively impacting quality." That's from Hilton Kobus and co-authors (including us) talking about how to manage performance with limited resources (the article is at the end of this section). Most FSPs have greater demands for service than they can respond to; this results in case backlogs and slow turnaround times. Submitters become unhappy, management above the FSP leaders share that unhappiness, and the organization may face a number of results, not all of which are good. **FSPs can't raise prices like a for-profit can, so they must minimize costs**, be they time, money, or otherwise.

Here's an example. Your FSP has a service mandate forced on it, like all arrestees must have DNA samples taken and run in a narrow window of time or sexual assault kits must be reported out within a limited number of days or similar. Your FSP must react to that mandate. Your FSP is charged with providing quality analysis of evidence in a timely manner. Given unlimited resources, effectiveness of services boils down to timeliness. However, because we live in this reality, your FSP resources are insufficient to satisfy the demand for the new mandated services. Effectiveness faces a double-edged sword: To improve quality on your current cases, you assign more resources (both time and money) ***but*** this investment comes at the cost of *not assigning* resources on cases not yet received. Broadly speaking, this is an **opportunity cost, which is the value of the next best alternative**; that is, choosing to take a particular action (like focusing on current cases) implies foregoing some other action (like focusing on future cases) and that forgone action is the true measure of "cost." Your FSP must balance a tradeoff between resources allocated to current cases and future cases.

How to Balance the Tradeoff

> *Sometimes it seems as if there are more solutions than problems. On closer scrutiny, it turns out that many of today's problems are a result of yesterday's solutions.*

Thomas Sowell, an American author and economist, was right on both fronts when he said that. There are always more solutions than problems; otherwise, no problems would ever get solved. On the other hand, not all of the solutions are that great and may, in the short or long term, cause other problems. Review the section on systems thinking and ask yourself, Why would a solution to one problem cause others? The answer is, of course, problems range from the simple to the complicated to the complex. You can't always see all the elements that influence a system, so squeezing the balloon *here* can make it smaller, yes, but will also force the air to make that section *there* larger. With that in mind, consider a few suggestions to help balance between potential solutions for an inefficient process.

Automation

Routine laboratory testing, like in clinical laboratories or FSPs, lends itself to automation. Time and money (or time *is* money) relate to the perceived evidentiary value of the examination or analysis.

Again, with the opportunity costs! Your choice to spend time and money on a particular exam implies that you have decided not to use alternative options for that time and money. If you can minimize either cost (time, money), say, by automating the examinations that are labor intensive, you can improve their use over those that are labor intensive and of low value: You shift your emphasis from the lower right quadrant and maximize the upper left quadrant. Doing so should improve efficiency and cost effectiveness but remember that some processes do not lend themselves to automation and will remain high input activities. These high-labor/high-value examinations fall into the upper right quadrant. The key to efficient use of resources committed to the upper right quadrant category cases, therefore, is minimizing those in the lower right quadrant. This emphasis may happen on a case-by-case basis as the value of evidence shifts due to circumstances, such as when textile fiber evidence takes center stage.

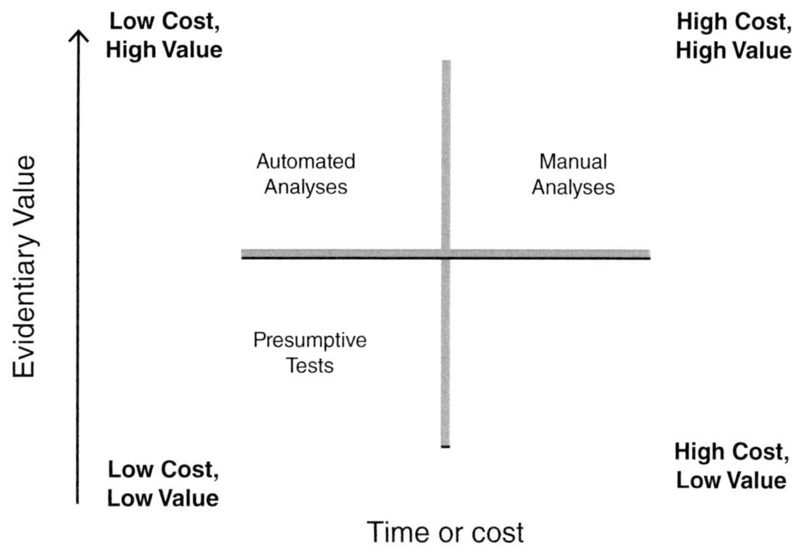

Figure 5.1 The perceived relationships between evidentiary value and time or cost of analysis (from Kobus et al. (2011)).

Triage

The low-input/low-value examinations in the lower left quadrant do serve a purpose, however. Rapid presumptive tests, such as color tests for blood and drugs, are a good example. Although not positive identification of a substance, they can suggest if it is worth moving on to further examinations. They are an aid in decision making, a type of triage. While these types of tests have low evidentiary value, they may make useful contributions to laboratory decision making and the (re)allocation of limited resources. The goal is to use valid sampling to decide on which samples to analyze, that is, focusing on a reduced number for analysis.

Process Improvement

The perceived value of a forensic exam is a kind of goal, a desired outcome. Moving any one exam from the high cost quadrant to the low-cost quadrant takes time, research, testing, and evaluation. This is what process improvement is all about and why you are reading this book. Data like that from FORESIGHT provides a starting point for the plan-do-check-act cycle described by Shewhart:

- **Plan**: Establish objectives necessary to achieve goals and intended deliverables and outcomes.
- **Do:** Implement the processes.
- **Check**: Measure the results of the new process and compare against previous and expected data.
- **Act**: Analyze any differences and look for causal connections.
- **Repeat**.

The Shewhart cycle is a scientific approach to quality where a hypothesis is developed, tested, and refined. The key word is "cycle": Quality is a process, not an event (read Deming to have this drilled into your head), and that's why the last (or is it the first?) step is "repeat."

Knowledge Boundaries

Many forensic cases involve multiple items of evidence. This means that choices must be made about which samples need to be analyzed and have to be based on a combination of factors, like evidentiary value, turnaround time, resolution of the method (precision), and the goals of the investigation. The optimal starting point is a conversation between investigators and the FSP professionals. Being knowledge workers, the FSP professionals should be consulted about what is being requested and the significance of the items submitted. The testing should be done only on relevant items with the mutual understanding of the costs and the realities of what is being asked. As a noted UK police author says:

> *What is required by all forensic disciplines is a clear objective and an understanding of how it is to be achieved. This requires a dialogue between the investigators and the forensic practitioner, thus enabling the latter to apply their specialist knowledge to the problem and identify the most appropriate techniques to achieve the objective. It is not necessary that the investigator understands the technique, so long as they set a clear objective and know what type of result can be expected.*
>
> (*Stelfox 2009, p. 134*)

Consider the situation where a woman is dead and her ex-husband is a prime suspect. Requesting DNA analysis might make investigative sense but only if they have not been in recent contact prior to the death. Knowing, for example, that they routinely had intimate interactions although living apart and divorced may limit the value of the analysis or even rule it out. Context trumps evidence every time. But the FSP professionals, the specialists, should be the ones to decide, that is, after all, their job.

Go Deeper

Collins, J., 2005. *Good to great and the social sectors*. New York: Harper Collins.

Deming, W.E., 1986. *Out of the crisis*. Boston, MA: MIT Center for Advanced Engineering Study.

Drucker, P., 2006. *The effective executive: The definitive guide to getting the right things done*. New York: Collins.

Kobus, H., Houck, M., Speaker, P., Riley, R. and Witt, T., 2011. Managing performance in the forensic sciences: Expectations in light of limited budgets. *Forensic Science Policy & Management: An International Journal*, 2(1), pp.36–43. (See full article in Chapter 4)

Kurimski, L.M., Speaker, P.J. and Bassler, J.R., 2017. Project FORESIGHT and return on investment: Forensic science laboratories and public health laboratories. *Forensic Science Policy & Management: An International Journal*, 8(1–2), pp.1–12.

Maguire, C., Houck, M.M., Williams, R. and Speaker, P.J., 2012. Efficiency and the cost-effective delivery of forensic science services: Insourcing, outsourcing, and privatization. *Forensic Science Policy & Management: An International Journal*, 3(2), pp.62–69.

McAndrew, W.P., 2018. National versus local production: Finding the balance between fiscal federalism and economies of scale. *Public Finance Review*, 46(6), pp.926–948.

Mintzberg, H., 2005. *Managers not MBAs*. San Francisco, CA: Berrett-Koehler.
Shewhart, W., 1939. *Statistical method from the viewpoint of quality control*. New York: Dover.
Speaker, P.J., 2015. Financial management of forensic science laboratories: Lessons from project FORESIGHT 2011–2012. *Forensic Science Policy & Management: An International Journal*, 6(1–2), pp.7–29. (See full article in Chapter 4)
Speaker, P.J., 2017. Process improvement and the efficient frontier: Forecasting the limits to strategic change across crime laboratory areas of investigation. *Forensic Science Policy & Management: An International Journal*, 8(3–4), pp.109–127.
Stelfox, P., 2009. *Criminal investigation*. Portland, OR: Willan Publishing.

Article: Project FORESIGHT and Return on Investment
Forensic Science Laboratories and Public Health Laboratories

5.1

LORELEI MORELAND KURIMSKI[1], PAUL J. SPEAKER[2], AND JOHN R. BASSLER[3]

[1]*State Hygienic Laboratory, University of Iowa, Coralville, IA;*
[2]*College of Business & Economics, West Virginia University, Morgantown, WV;*
[3]*School of Public Health, West Virginia University, Morgantown, WV*

Introduction

The emergence of the European QUADRUPOL project (European Network of Forensic Science Institutes 2003) and the subsequent Project FORESIGHT (Houck et al. 2009) were the response to the forensic science's need for the development of business tools for forensic laboratory managers. Forensic laboratories experienced a growing demand for services that had far outpaced the resources allocated. These projects have led to the development of managerial tools for optimizing efficiency and the cost-effective delivery of forensic science services (Kobus et al. 2011).

These developments have attracted the attention of other public sector operations. Following guidelines for independence of the crime laboratory from policing agencies as highlighted in the 2009 National Research Council of the National Academies of Science Report (NAS 2009), the district government of Washington, DC created the Consolidated Forensic Laboratory (CFL) that included the Forensic Science Laboratory, Crime Scene Sciences, and the Public Health Laboratory (PHL). The CFL Management's adoption of Project FORESIGHT metrics as an integral decisionmaking tool for the forensic science laboratory sparked the interest of managers in the PHL portion of the operations.[1] The CFL leadership invited the Association of Public Health Laboratories (APHL) to join in the discussion of business metrics for the PHL, which led to a multi-year project to develop similar metrics for public health laboratories across the U.S.

The PHLs share many of the characteristics of the forensic science laboratory making the conversion of the Project FORESIGHT data collection easier than it might have been with other public sector activities. Both sectors are dominated by the work of scientists with many of the applications of similar techniques from toxicological examination to applications of DNA technologies. Similar to the forensic sciences, public health laboratorians

DOI: 10.4324/9780429318528-14

work in a culture that requires standard operating procedures. The accompanying data collection tool for the PHLs needed to be easy to complete while capturing essential information, yet providing immediate feedback. The data collection tool's immediate feedback would include output to demonstrate PHL value to decision makers, and be presented in the language of those decision makers; that is, using some standard business metrics that were familiar to those decision makers.

Project FORESIGHT data has enabled research into several areas for forensic laboratories. From the development of basic business ratios (Speaker 2009a) to managerial interpretation of those ratios (Speaker 2009b) and subsequent strategic change (Newman et al. 2011), the gains in efficiency can be dramatic (van Asten 2014), but are predictable (Maguire et al. 2012). The reaction to the widening gap between available resources has prompted questions regarding the optimal model for the delivery of forensic laboratory services (Bedford 2011; Tjin-A-Tsoi 2013). The availability of industry level data from QUADRUPOL and FORESIGHT enables such questions to be examined from a data-driven basis (Speaker 2013).

Observing these developments with forensic science laboratories, APHL introduced a project to create data collection tools for their membership, and to offer a foundation for research into the issues and challenges facing these laboratories. In the sections to follow, we outline the progression of events where the forensic science laboratory experience has been extended to the PHLs. The PHL project has, in turn, developed some metrics that may prove valuable to the forensic laboratory. The PHL project went beyond the internal inspection of the business experience of the laboratories and connected the internal efficiencies to the public outcomes via the development of return on investment (ROI) metrics. Careful observation of the PHL project provides some future direction for extension of the use of business metrics for forensic science laboratories.

The measurement of a return on investment may come in different forms. In a financial realm, the return on investment is a measurement of the net gain in dollars relative to the dollars invested. For the PHLs, the measurement of the net gain is a representation of the avoidance of treatment expenses for the dollars expended or a measure of lives saved or improvement in the quality of life from avoidance of illness or disease. Extending such measures for the forensic laboratory, the measures could represent savings to society from faster detection and the savings from additional crimes avoided or lives saved or improved.

While the development of managerial tools for forensic science laboratories has progressed vigorously over the past decade, the contribution of forensic sciences to the justice system has been difficult to measure (Peterson et al. 2010). The lessons from the PHLs offer a direction for forensic laboratory future research to demonstrate the return to the public welfare that results from laboratory contributions to the justice system. The PHL project demonstrates that the generalities of the overall societal gains from PHL testing can be connected to the internal efficiencies of individual laboratories to show the return on investment from each laboratory. Following this direction should enable forensic science laboratories to connect the justice system societal gains to the contributions by individual laboratories and use that connection to serve as the basis for public funding to be expanded to meet the growing demands for forensic laboratory services.

The public health laboratory metrics project

Project FORESIGHT began with a series of sessions where laboratory representatives across North America outlined their biggest management issues, which led business faculty to link those issues to relevant data for participants to collect regarding casework, budgets, and personnel allocation. To frame the direction for the project, the FORESIGHT team had to first link laboratory missions with a common language to permit industrywide measurement. It took nearly two years of discussion before any data could be collected. The PHL metrics project was able to proceed at a much more rapid pace than the forensic science laboratories experienced, largely due to two factors: the existence of a data collection repository through the APHL; and the documented experience via Project FORESIGHT and other developments in the business of forensics (Houck et al. 2015).

Initial data review

Before devising a data collection tool to survey PHLs on their laboratory, personnel, and expenditure data, a review of existing APHL data collection was undertaken. APHL had previously conducted a survey of its membership with information regarding testing, expenditures, and personnel levels. Using the APHL *Core 2011* survey data, base level metrics were developed along the lines of the metrics used in Project FORESIGHT (Speaker 2015). Key among those measures was the construction of an output metric showing the average total cost of testing which allowed laboratories to make a comparison of their average total cost of testing to other laboratories with a similar number of tests performed. With forensic science laboratories, the data showed substantial economies of scale, where the average cost of testing fell until an optimal level of testing was performed. A first look at the PHL data revealed a similar sense of economies of scale with a dramatic fit of the data to the theoretical relationship between the average total cost of testing as it relates to the volume of testing. Figure 1 highlights the data by plotting average total cost per test against the test level for each reporting laboratory. A glance at the data appears to demonstrate the expected relationship as theorized in economics; namely, when average total cost is mapped against the scale of operations, a U-shaped curve emerges. Economies of scale are realized as the level of testing is increased. While we would normally expect that at some point diseconomies of scale may be realized (i.e., the rising portion of the U-shaped average total cost curve), it does not appear that the testing level has been reached where diseconomies have been experienced by the laboratories in the APHL's *Core 2011* survey.

The reference to efficiency and cost effectiveness, can be seen through the U-shaped average total cost curve. Efficiency is a reference to the lowest cost production for a given level of output. All points along the U-shaped average total cost curve represent a point of efficiency. If a laboratory operates at a point above the average total cost curve, it is inefficient. That is, the laboratory should be able to produce that same level of output for a lower average cost. Cost effectiveness, on the other hand, is a reference to a specific point on that average total cost curve where perfect economies of scale have been achieved. The laboratory is efficient and operates at the lowest average cost for any level of output.

As had been done with the forensic science laboratory data, the relationship between the average total cost and the level of testing was estimated using a non-linear regression technique. From an examination of Figure 1, it does not appear that any of the laboratories

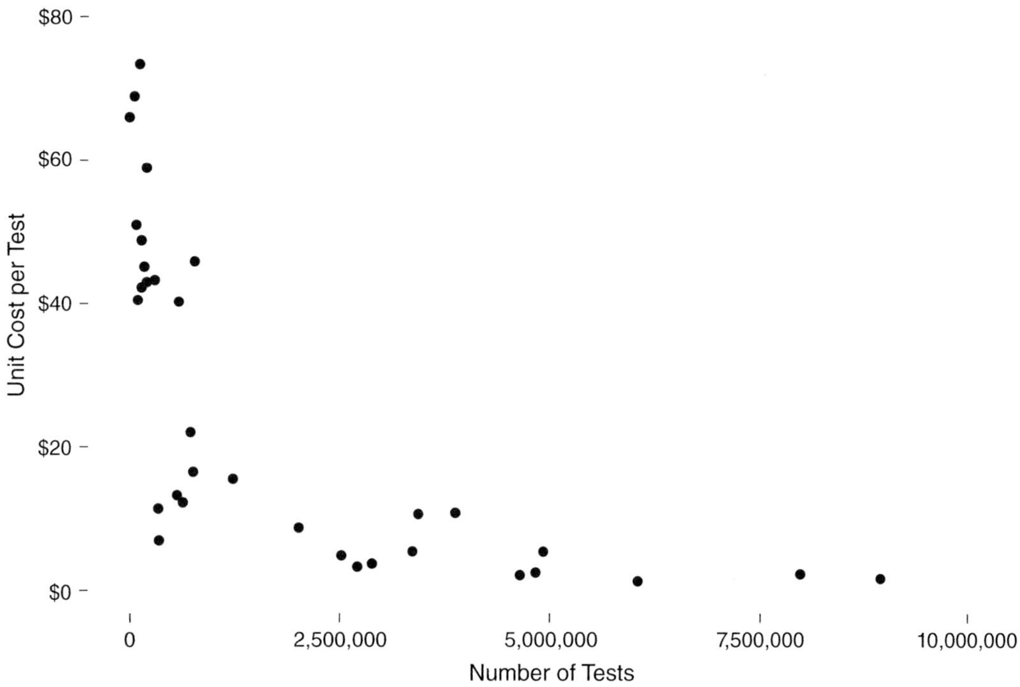

Figure 1 Cost per test vs. test volume.

had achieved a level of testing beyond the minimum average cost. The data suggest an image that is reminiscent of the downward sloped portion of the expected U-shaped curve. The functional form of a power function offers a potential nonlinear relationship between the average total cost and the level of testing that may be estimated. That is,

$$Cost\,/\,Test = \alpha\ Tests^{\beta}$$

Plotting histograms of the number of Tests and the Cost/Test, there is a right skewness in each of the variables; after a log transformation of each variable, when each histogram is replotted, the resulting distribution is much closer to a normal distribution. This observation is strengthened after calculating p-values using the Shapiro-Wilk test for each log transformed variable, Cost/ Test and Tests, with p-values of 0.6727 and 0.0904, respectively, which suggests that we cannot reject the null hypothesis that these samples are taken from a normal population. Given this information, a log-log regression model is estimated to preserve the properties of a linear model, while providing an interpretation of the nonlinear relationship between the Cost/Test and Tests.

Taking natural logarithms of each side we run the regression:

$$LN(\text{Cost}\,/\,Test) = LN(\alpha) + \beta \times LN(Tests)$$

Figure 2 illustrates the selected log-log model with corresponding 95% confidence interval of the plotted regression line:

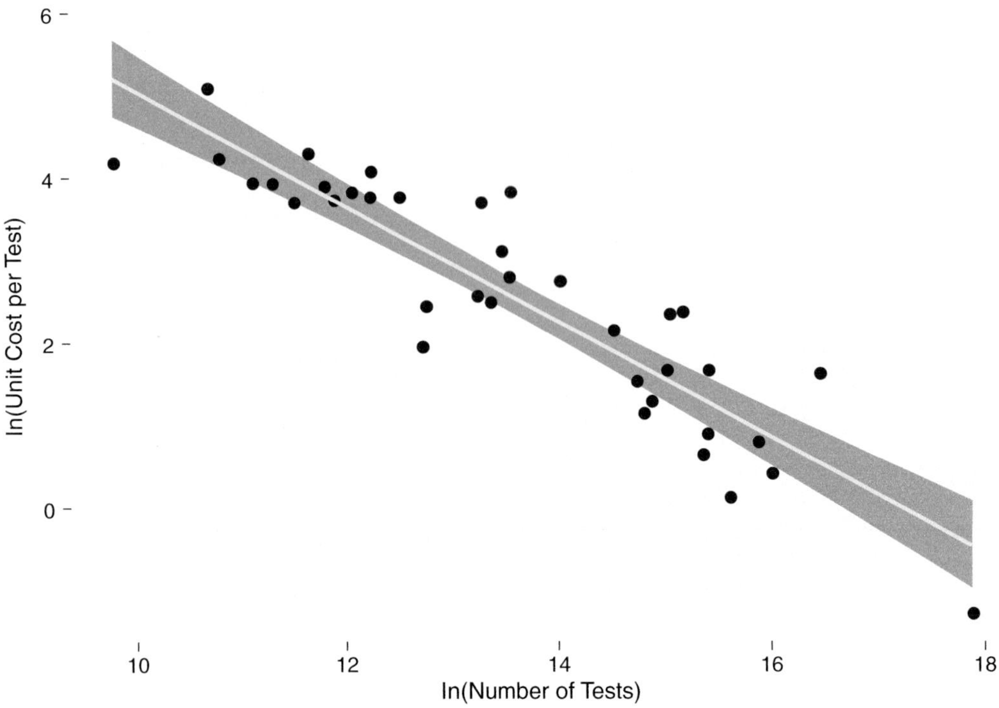

Figure 2 Natural log of cost per test vs. Natural log of test volume.

$$LN(\text{Cost} / Test) = 12.04660 - 0.69797 \times LN(Tests)\ (0.7475)\ (-0.0546)^{***}$$

***significant at the 0.01 level;
adjusted R-squared D 0.8187

Transforming these estimates of the log-log transformation back to the original relationship between Costs/Test and Tests yields.

$$\text{Cost} / Test = 170{,}519 \times Tests^{-0:6979}$$

The strength of this initial relationship was encouraging in that PHLs appear to provide a very good fit to the expected relationship formed by economic theory. This early finding was used to support the recruiting of laboratories to participate in the initial development of data collection tools. A few existing PHL committees were tapped for initial participation in the metrics workgroup; the Laboratory Systems & Standards committee offered a systems perspective and the Knowledge Management committee provided both state and local laboratory perspectives. This newly formed metrics workgroup followed a similar path to that forged by the original Project FORESIGHT members. However, the guidance permitted from the earlier project permitted a much faster pace for the PHL workgroup and within the first nine months, the workgroup had reached general consensus on language with the adoption of several key definitions (Appendix A) and test counting conventions.

As with the forensic science laboratories, the connection between mission and metrics had to be addressed. To develop appropriate business metrics for the PHLs, an examination of the mission of each laboratory had to be considered and the results synthesized to establish a common optimization problem faced by each laboratory. Fortunately, such an examination had already been performed through the APHL and the common missions inherent in a listing of core activities.

Public health laboratory core activities

In the Project FORESIGHT review of the mission and mandates of forensic science laboratories, most mission statements centered on casework activity with some missions also including a research function. However, the breakdown of activity for forensic science laboratories left little time beyond casework and casework related activities. Data collected for each year of Project FORESIGHT revealed a very small percentage of forensic laboratory time devoted to research or other non-casework activity. For the PHLs, however, the range of activities is much broader with time allocated across a list of core functions (see Appendix B for a detailed description of each core function) with testing comprising an average of approximately 70% of laboratory activity.

PHL core functions

1. Disease Prevention, Control, and Surveillance
2. Environmental Health and Protection
3. Public Health Preparedness and Response
4. Integrated Data Management
5. Reference and Specialized Testing
6. Food Safety
7. Laboratory Improvement and Regulation
8. Policy Development
9. Public Health Related Research
10. Training and Education
11. Partnerships and Communication

To isolate the expenditures attributable to each activity of the laboratory, the project design required each laboratory to provide a breakdown of effort across the laboratory to meet each of these core functions. From these breakdowns of activity, laboratory expenditures could then be allocated accordingly.

Pilot testing

While recognizing the breadth of PHL mandates, the PHL workgroup decided to concentrate trial efforts in a few areas of hot button issues and easily recognizable efforts of the laboratories. These included program level testing for the areas of foodborne illness, influenza, newborn screening, safe drinking water, and tuberculosis. These program areas have a national impact and are common to most PHLs, and had a direct public interest with a great deal of relevant scientific literature on impact. Meeting on the heels of some critical public health crises (Ebola outbreak and Zika virus), PHLs effort towards emergency preparedness was also included in the data collection. During the course of the 2015–2016 fiscal year, the pilot study participants began to collect the data, refine definitions and counting techniques, and submitted data that permitted individual output standards to be connected to the body of literature on the benefits received from each area of testing. As had occurred with forensic science laboratories, multiple meetings were held to provide details for categories with examples, glossary definitions, updates, consistent test measurement, and collection methodology. These clarifications were discussed and revisited on multiple occasions.

As with the forensic science laboratories, the collected data concentrated on the information needed to directly address the consensus missions across laboratories. For testing activity, this led to metrics that would permit evaluation of a mandate to provide high quality testing for the budgets allocated to meet this mission. The data collection tool was designed to include revenue levels and sources (national, state, and local funding; federal and state grants; and for-fee services), and to separate expenditures into personnel, capital, consumables, and other expenditure categories. Additional data captured the distribution of c3. ore activities, full-time equivalent personnel distribution, and testing volumes across various laboratory activities.

Output

The forensic science data collection tool, LabRAT, included data entry pages for casework and laboratory financials which were linked to an output page for immediate feedback. That feedback page provided immediately usable metrics that followed the analysis highlighted in the literature (Houck et al. 2015). The project team directed the PHL data collection tool to offer some of the same immediate feedback, including output with average total cost metrics as well as the introduction of return on investment metrics.

Many PHL programs are designed for early detection of potential public health problems. The medical and health economics literature offer a large number of studies that measure the direct individual and indirect social costs from adverse medical conditions. From these studies, health economics calculate costbenefit comparisons that connect the benefits to society from preventive and early detection via costs avoided when preventive care is able to eliminate or reduce the need for health care treatments of medical conditions.

Unfortunately, the project design in the medical and health economics literature was not created for interpretation for the specific cost structure of an individual laboratory. Because their service areas are generally associated with some political jurisdiction, PHLs, as with forensic science laboratories, operate at different average total costs, often reflecting an underachievement of economies of scale. As a result, these different average cost structures do not permit a generalized return on investment that is directly applicable to each individual laboratory. The estimation of the ROI for each individual laboratory requires conversion from the average total cost in published studies to the individual laboratory's experience as reflected in the individual laboratory's cost structure.

The output sections of the data collection tool included indications of the average total cost for the laboratory and the conversion to the corresponding return on investment in each program evaluated. The nature of the returns were presented in multiple forms, a dollar return per dollar invested, a health economics metric marking the improved lifestyle (termed a Quality Adjusted Life Year (QALY)), and a combined metric noting dollar treatment expenditures avoided plus QALY gains.

Quality adjusted life years

A Quality Adjusted Life Year (QALY) is a measure widely used in matters of health economics. It is a metric that is frequently adopted as a means to quantify cost-benefit decisions in healthcare. QALY's are the product of life expectancy and a quality metric associated with any health condition and these are compared to the average quality of a person without the associated condition. The value of a particular treatment as measured in QALY's is the difference between the quality of an average person's life and the expected QALY associated with the condition in question.

A person in perfect health would have a QALY valued at 1.00. For severe conditions, the QALY may be negative. Note that the average person has a QALY of 0.87 for a typical year (Scharff et al. 2015). The common use of the QALY is to then value the QALY expected gain at $50,000, $100,000, or up to $200,000 for the benefit comparison to the cost of treating the condition, or making a capital expenditure, etc.

Measuring social benefit

Fortunately, there is an extensive literature that provides detail regarding the likelihood of the adverse events tested in public health laboratories, as well as the distribution of costs incurred for treatment and the loss of function associated with the adverse condition. A measurement of the social benefit from detection and/or treatment relies upon the estimated value of costs avoided, pain and suffering, and any associated deterioration in the quality of life from the adverse condition. Several medical and other scientific studies were reviewed for detail on cost of treatment, value of statistical life, QALYs, etc. In particular, the benefit measurement included a review of economic studies, particularly in health economics, for use of detailed cost-benefit analyses. Thus,

$$Social\ Benefit = Tests \times p\ (positive\ test) \times E\ (cost\ of\ treatment\ avoided),$$

where the probability of a test subject yielding a definitive positive indication of the adverse condition, *p(positive test)* and the expected cost of treatment, *E(cost of treatment avoided)* are each taken from medical/health/ economic studies. In addition to the costs avoided, other social benefits may include values for the pain and suffering avoided, both directly to the person with the diagnosed condition and indirectly to others, such as the anguish to parents from a previously undiagnosed condition. Those direct pain and suffering effects are generally measured in QALYs and provided a dollar value per QALY that ranges from $50,000 to as much as $200,000.

While many health care professionals have a claim that they have provided a contribution to the benefit or costs avoided (e.g., laboratory, physician, hospitals, over-the-counter treatment, environmental engineers), at the margin, many may rightly take credit for the public benefits. The benefits claimed by the individual contributors do not sum to the total social benefit, but at the margin, all may claim the social benefits. That is, the ROI is claimed by many who each make critical contributions for the social benefit. The size of the ROI across laboratories differs according to the internal productivity, processes, and market conditions in which they operate.

Return on investment methodology

From the medical and health economics literature, details were extracted on costs of treatment, QALYs lost, and statistical analysis of incidence. The total social benefits were then approximated from these prior studies for use in the determination of part of the numerator of ROI. As for the denominator, the total costs for testing were collected from the same studies and then amended from the detail collected for the pilot laboratories. These measures of the total cost also affect the numerator, where the term represents the net benefit to society; that is, net of the costs.

The individual laboratory's total cost determination for a particular activity is the key contribution of the pilot study. Consider the average cost of a test for a particular laboratory compared to the volume of testing in that laboratory. To illustrate the conversion from published studies to a determination of the return on investment to an individual laboratory i (ROIi), assume that the average benefit per test is independent of the laboratory doing the testing. That is, if ATBL represents the average social benefit per test across all laboratories, we assume that ATBi D ATBL. The average return on investment across all laboratories relates the total social benefit, TB, to the total expenditures for testing, TC:

$$ROI_L = (TB - TC) / TC$$

Divide numerator and denominator by the total number of tests N by all laboratories yields:

$$ROI_L = \left(\frac{TB}{N} - \frac{TC}{N}\right) \Big/ \frac{TC}{N} = \left(ATB_L - ATC_L\right) / ATC_L$$

where ATCL is the average total cost across all laboratories.

Solving for the average total benefit, ATB_L, we obtain

$$ATB_L = ATC_L \times (1 + ROI_L)$$

For laboratory i:

$$ROI_i = {(ATB_i - ATC_i)} \Big/ {ATC_i} \cdot = {ATB_L - ATC_i} \Big/ {ATC_i}$$

where ATC_i is the average total cost for laboratory i. Substituting for ATBL from above, we obtain

$$ROI_i = {(ATC_L \times (1 + ROI_L) - ATC_i)} \Big/ {ATC_i} = \left({ATC_L} \big/ {ATC_i}\right) \times (1 + ROI_L) - 1:$$

Several of the health economics studies have attempted to capture the cost of particular testing by counting hours devoted to a single test and measuring materials cost employed in the test. While such a micro level examination offers some information, it does not capture the full cost to the laboratory, including capital requirements, overhead, and incidental expenditures. Taking a macro view of the laboratory and breaking down the fully-loaded costs, including facilities, infrastructure, training and development, education, preparedness, etc., then a more complete picture of the costs may be determined. The use of the core functions breakdown to help allocate the infrastructure costs is an explicit recognition of the complex mission of the public health laboratory.

Return on investment

In a for-profit environment, ROI represents the dollar returns minus the total expenditures, relative to the total expenditures. While typically presented as a percentage, it may also be presented as a dollar of benefit per dollar spent. To put this magnitude into perspective, consider the ROI for the thirty firms in the Dow Jones Industrial Average, where published ROIs range from near zero to a high of 28.23% (Google.com 2016). A comparison of the returns to these leading companies offers some perspective to the public health laboratories and their societal returns.

Extending the concept of return on investment to the public sector, the ROI numerator, net benefits, may be measured in dollars saved, quality adjusted life years (QALY), or some combination that accounts for costs avoided for treatment as well as pain and suffering. The alternative measures may be extracted from the existing health, medical, and health economics literature and then applied to the detailed expenditure structure of the PHL's.

ROI example

Consider the role of the PHLs with respect to food safety. The Centers for Disease Control and Prevention (CDC) program, PulseNet USA, provides a nationwide network to speed up the identification and subsequent intervention regarding foodborne illness.

> "PulseNet is a national laboratory network that connects foodborne illness cases to detect outbreaks. PulseNet uses DNA fingerprinting, or patterns of bacteria making people sick, to detect thousands of local and multistate outbreaks. Since the network began in 1996, PulseNet has improved our food safety systems through identifying outbreaks early. This allows investigators to find the source, alert the public sooner, and identify gaps in our food safety systems that would not otherwise be recognized." (Centers for Disease Control and Prevention 2016)

The Individual laboratory ROI was suggested by the following details in the medical and health economics literature (Scharff 2012). Total national net benefit from PulseNet was evaluated using data from 1994–2009 (Scharff et al. 2015). Those benefits represent costs avoided because of early detection. Multiple models were employed to measure the social benefits including direct benefits from product recall and the indirect effects from averted harm. The measurement of these benefits ranged from $491–654 million, with the lower bound capturing the benefits from costs avoided and the upper bound reflects the inclusion of both costs avoided and QALYs which were valued at $100,000 per QALY.

The determination of the ROI also requires measurement of the total costs of the investment. For the PulseNet project, the nationwide costs of the program totaled $7.3 million. For the total program, the Return on Investment in terms of cost avoided exceeds

6,600% or stated in terms of dollar returns per dollar invested, the PulseNet program returns $67.26 per dollar invested. And when the returns capture both costs avoided and quality of life from early detection, the ROI exceeds 8,850%, a return of $89.59 per dollar invested in the program.

While those program totals for social benefits and total program costs provide a sense of the ROI for the entire program, the localized laboratory returns required adjustment for laboratory productivity and associated economies of scale. The economies of scale adjusted returns can be quite dramatic. As a laboratory increases its testing level towards the optimum sized testing facility, average total costs decline, while social benefits per test remain constant.

To estimate the per laboratory performance, we examined each pilot laboratory's level of effort with the Core Function "Food Safety," which includes the laboratory effort in PulseNet. Total laboratory expenditures were allocated to the food safety category according to its corresponding level of effort. That expenditure was then divided by the number of reported PulseNet tests as a proxy for the average cost of testing. The overall PulseNet project ROI was reallocated to the individual laboratory as described in the ROI methodology section. The result was a range of individual laboratory ROI.

Figure 3 demonstrates the relationship between the program wide average return on investment and the return on investment to the individual laboratories. The humped shaped curve in the upper graph is the result of a quadratic regression of the data points relating the individual laboratory return on investment (the ROIi) measured on the vertical axis to the level of testing on the horizontal axis. The flat line in the center of the upper graph illustrates the average ROI (ROIL). Directly below in Figure 4 are the associated average total costs per laboratory (ATCi), which are also fitted with a quadratic

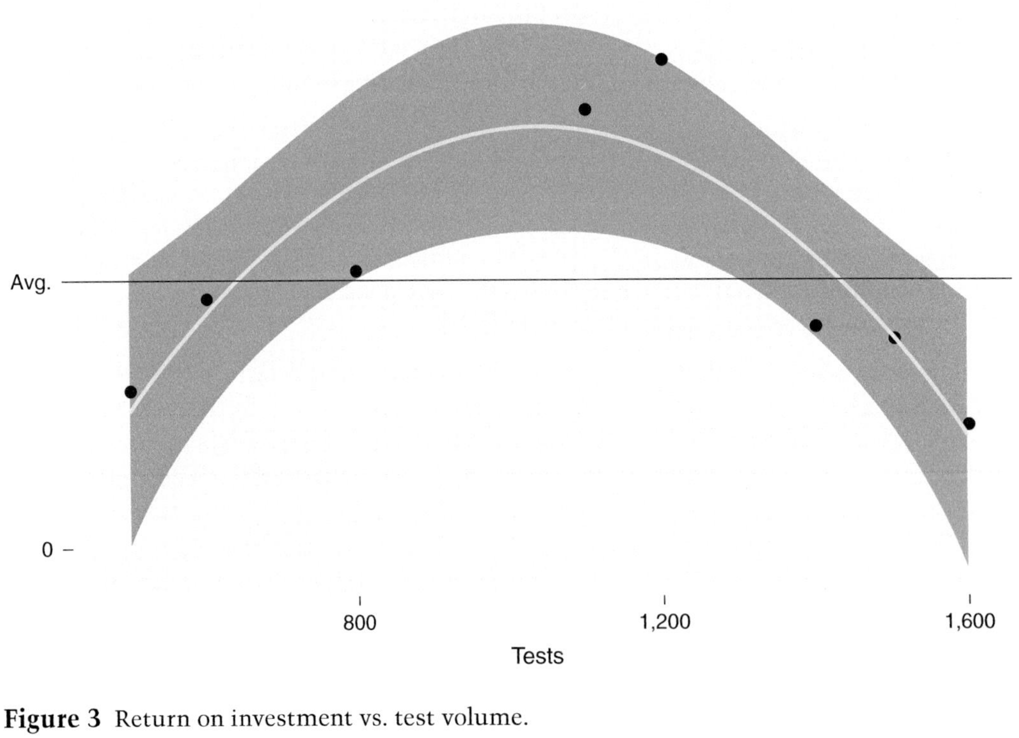

Figure 3 Return on investment vs. test volume.

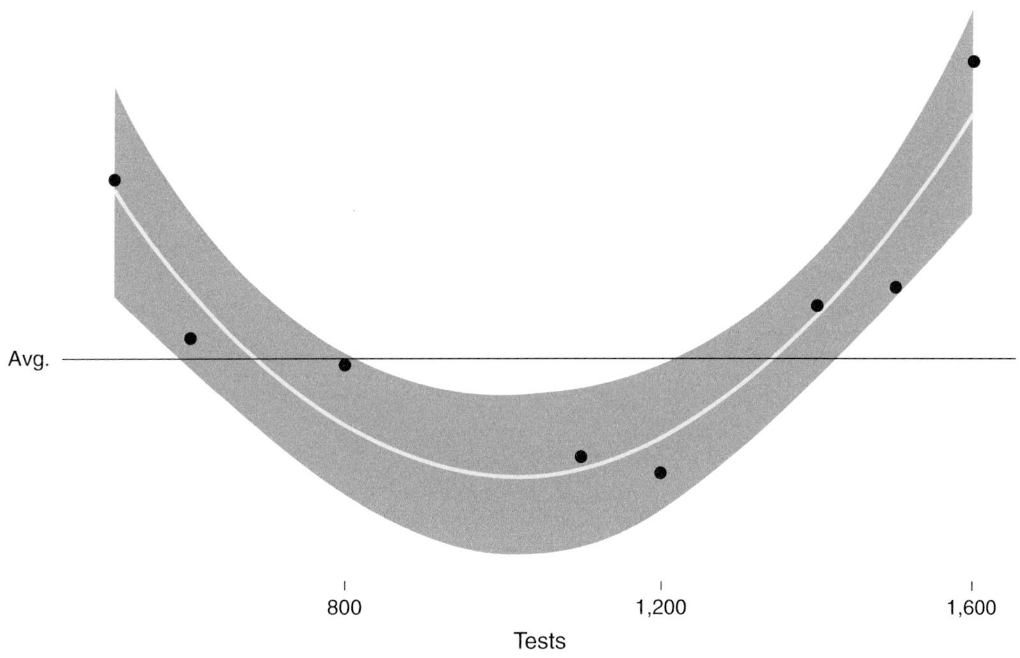

Figure 4 Average total cost vs. test volume.

regression to represent the U-shaped curve showing economies of scale. Again, the mean average total cost (ATCL) is plotted as a flat line for the overall program across all laboratories.

While the APHL pilot study results only offer a first pass look at the relationship between the individual laboratory performance and the resulting societal outcomes, the study offers valuable direction as the project moves forward. Certainly, a more detailed breakdown of activity is needed to more accurately account for expenditures and future data collections will offer that detail. Those future data collections include more detail on the assignment of full-time equivalent employees, a complete allocation of testing activity across all programs within the laboratory, and continued improvement in the description of processes for consistent measurement across all participants. However, the current proxy measures also provide direction for research into the contributions of other entities, including forensic science laboratories. The direction of the PHL study to connect operational and financial efficiency to the resultant gains to society should be a direction followed in other public sector studies.

Concluding observations

Moving forward, the direction of the PHL project offers suggestions for continued business research for forensic laboratories, especially the determination of ROI for forensic laboratories. For the 19 areas of investigation in Project FORESIGHT, can the associated literature provide some connections to the benefits from casework testing? If so, this offers a leap forward for the justice system and for managers of forensic science laboratories. For the system as a whole, a societal return on investment from various efforts can help to guide the allocation of public funds towards optimizing the social benefits. That optimization of fund allocation cuts across all public sectors. It permits a more informed discussion of what can be expected for each dollar invested.

A move toward measures of the contribution of forensic science laboratories to the justice system has not been a simple process. However, the expansion of the FORESIGHT database enables the measurement of the fully loaded costs of forensic science services down to the individual investigative areas. When coupled with economic, social, and criminology studies on outcomes, the return on investment in selected forensic areas of investigation may be developed. While it is not expected that all areas of investigation will be easily addressed, other areas should yield some quickly obtained, useful measures of the societal returns from forensic laboratories.

For laboratory managers, a greater sense of the societal return on investment allows more informed decision-making with the allocation of resources within the laboratory. While developments to-date from QUADRUPOL and FORESIGHT offer insight into the cost structure towards efficiency and cost effectiveness, a move to connect the societal benefits would offer a more complete picture to influence the allocation of scarce resources. Armed with the returns information, laboratory leadership can better address funding bodies to inform of the societal contribution of areas of investigation in support of mounting requests for services.

Funding

This publication was supported by the Cooperative Agreement Number U60OE000103, funded by the Centers for Disease Control and Prevention through the Association of Public Health Laboratories. Its contents are the sole responsibility of the authors and do not necessarily represent the official views of the Centers for Disease Control and Prevention, the Department of Health and Human Services, or the Association of Public Health Laboratories.

Note

The CFL declined participation in Project FORESIGHT after the 2015 fiscal year.

References

Bedford, K. 2011. Forensic science service provider models – Is there a 'best' option? *Australian Journal of Forensic Sciences* 43(2–3): 147–156.

Centers for Disease Control and Prevention. 2016. PulseNet, About PulseNet. September 29. Centers for Disease Control and Prevention. https://www.cdc.gov/pulsenet/about/index. html

European Network of Forensic Science Institutes. 2003. QUADRUPOL. OISIN II programme of the EU.

Google.com. 2016. Apple Inc. September 29. https://www.goo gle.com/finance?qDNASDAQ%3AAAPL&eiD5GDtVmYJMG-mAGezIqoCQ

Houck, M. M., W.P. McAndrew, M. Porter, and B. Davies. 2015. A review of forensic science management literature. *Forensic Science Review* 27(1): 53–68.
Houck, M. M., R.A. Riley, P.J. Speaker, and T.S. Witt. 2009. FORESIGHT: a business approach to improving forensic science services. *Forensic Sciece Policy & Management: An International Journal* 1(2): 85–95.
Kobus, H., M. M. Houck, P.J. Speaker, and T.S. Witt. 2011. Managing performance in the forensic sciences: expectations in light of limited budgets. *Forensic Science Policy & Management: An International Journal* 2(1): 36–43.
Maguire, C., M. Houck, R. Williams, and P. Speaker. 2012. Efficiency and the cost effective delivery of forensic science servicess: in-sourcing, out-sourcing, and privatization. *Forensic Science Policy & Management: An International Journal*, 3(2): 62–69.
National Research Council of the National Academies of Science (NAS). 2009. *Strengthening Forensic Science in the United States: A Path Forward.* Washington, D.C.: The National Academic Press.
Newman, J., D. Dawley, and P.J. Speaker. 2011. Strategic management of forensic laboratory resources: from proect FORESIGHT metrics to the development of action plans. *Forensic Science Policy & Management: An International Journal* 2(4): 164–174.
Peterson, J., I. Sommers, D. Baskin, and D. Johnson D. (2010). *The Role and Impact of Forensic Evidence in the Criminal Justice Process.* Washington, DC: U.S. Department of Justice. Scharff, R. L. 2012. Economic burden from health losses due to foodborne illness in the United States. *Journal of Food Protection* 75(1): 123–131.
Scharff, R. L., J. Besser, D.J. Sharp, T.F. Jones, P. Gerner-Smidt, and C.W. Hedberg. 2015. An economic evaluation of pulsenet a network for foodborne disease surveillance. *American Journal of Preventive Medicine* 50(5): S66–S73.
Speaker, P. J. 2009a. Key performance indicators and managerial analysis for forensic laboratories. *Forensic Science Policy & Management: An International Journal* 1(1): 32–42.
Speaker, P. J. 2009b. The decomposition of return on investment for forensic laboratories. *Forensic Science Policy & Management: An International Journal* 1(2): 96–102.
Speaker, P. J. 2013. Forensic science service provider models: data-driven support for better delivery options. *Australian Journal of Forensic Sciences* 45(4): 398–406.
Speaker, P. J. 2015. Financial management of forensic science laboratories: lessons from project FORESIGHT 2011–2012. *Forensic Science Policy & Management: An International Journal* 6(1–2): 7–29.
Tjin-A-Tsoi, T. 2013. *Trends, Challenges and Strategy in the Forensic Science Sector.* Gaithersburg, MD: National Institute of Standards and Technology (NIST).
van Asten, A. C. 2014. On the added value of forensic science and grand innovation challenges for the forensic community. *Science and Justice* 54: 170–179.

Appendix A: Key definitions

FORENSIC SCIENCE POLICY & MANAGEMENT 11

Glossary

Administrative & Support Staff Administration includes the laboratory director, associate directors and other key personnel whose primary job function is related to executive leadership and organizational decision-making. Support staff includes laboratory personnel whose primary job functions are not directly related to performing the test and whose salary is contributed to overhead. Examples of job functions include central services, accessioning/sample receiving, specimen transport, finance/billing, customer service, outreach, education/training, facilities, safety, emergency preparedness, HR, public relations, quality assurance, and IT.

Capital Expenditure Purchases of equipment, instruments, etc., over $1,000 with a lifetime longer than one year.

Consumables Chemicals, reagents, gases, and other laboratory supplies.

Employee Benefits Employer contribution to FICA, Medicare, Workers Comp, and Unemployment Comp, retirement (employer contribution only towards pensions, 403B or 401K plans, etc.), and occupational health service expenses (employer contribution only).

Equipment Leasing Payments for rent or lease contracts for capital equipment, instruments, etc. with a lifetime longer than one year (the lease contract may include maintenance as part of the annual payment).

Fee-for-Service (FFS) Payment of public health laboratory services are based on unit price such as test fees. Includes FFS contracts and direct invoicing.

Grants/Contracts Includes all federal, in-state and out-of-state, and local grants and contracts. Includes federal pass-through funding, cooperative agreements, and memorandums and understanding. Budgets are itemized by such categories as personnel, equipment, and supplies. Does not include any contracts that are for Fee-For-Service.

Internal TAT The time measured from the point of receipt of a sample or specimen within the laboratory until the generation of a final result.

Other Expenses Includes any laboratory expenditure not included in the other subcategories.

Other Revenue Includes any revenue source not included in State Appropriations, Grants/Contracts, or Fee-for-Service.

Program Includes all tests specific to any organized public health action or activity. Defined by specific criteria or legislation and partially or fully supported through federal, state, or local funding.

Service of Instruments Includes maintenance contracts for upkeep of capital equipment and unplanned expenditures for equipment/instrument repair. State Appropriations Funding dedicated by the state legislature for the purposes of public health laboratory services.

Technical Staff Excludes any support personnel including those who perform sample accessioning, kit preparation, client services, facilities, safety, quality assurance, administrative functions, education/outreach, information technology, billing and finances, and grants/ contracts.

Test A standard laboratory procedure or method

Transport Time The time measured from the point of collection of a sample or specimen until received by the laboratory.

Wages & Salaries Includes personnel expenditures paid directly to the employee including overtime and temporary.

Appendix B: Core functions of public health laboratories

Core Function
Disease Prevention, Control, and Surveillance
Provide accurate and precise analytical data in a timely manner in support of the:

- Prevention and control of infectious, communicable, genetic and chronic diseases, and environmental exposure. This may include testing for emerging and re-emerging microbial agents, immune status, antibiotic resistance, screening for inherited neonatal metabolic disorders, environmental toxins, and heavy metals such as blood lead.
- Recognition of outbreaks and other events of public health significance, by the identification and characterization of the causative agents of disease and their origin.
- Population-based surveillance for conditions of public health importance and to guide programmatic decisions.
- Early detection of congenital disorders in newborns leading to timely diagnosis and treatment.
- Monitoring of low incidence and/or high risk diseases, such as antibiotic-resistant tuberculosis, influenza, botulism, and rabies.
- Investigation and control of communicable or environmental diseases when testing is not available in the private sector.

Environmental Health and Protection
Collaborate with partners to coordinate and ensure scientific analysis of environmental and human samples to identify, quantify, and monitor potential threats to health by:

- Testing for toxic chemical, radiological, and microbiological contaminants in air, water, soil, and hazardous waste.
- Conducting biomonitoring of human specimens in the assessment of toxic chemical exposure.
- Testing of environmental samples in support of federal and state regulations, aiding in the compliance with those regulations.
- Industrial hygiene/occupational health testing to assist in efforts to protect indoor air quality and worker health, such as routine analysis of asbestos, lead, pesticides, and radon.
- Participating in the Chemical Laboratory Response Network (LRN-C) and the Environmental Response Laboratory Network (ERLN).

Public Health Preparedness and Response
Fulfill a key partnership role in local, state and national disaster preparedness and response by:

- Functioning as a Laboratory Response Network (LRN) Reference laboratory for biological agents and as an LRN Chemical Laboratory at a level designated by CDC.

- Assuring the triaging of environmental samples for the rapid identification of threat agents (chemical, biological, radiological and nuclear – CBRN); and food samples as a part of the Food Emergency Response Network (FERN).
- Planning for and ensuring that surge capacity is available during a public health emergency.
- Having a Continuity of Operations Plan in the event of a disruption of laboratory services.
- Participating in the Environmental Response Laboratory Network (ERLN).

Integrated Data Management
Serve as the conduit for scientific data and information in support of public health programs through the:

- Capturing of laboratory data essential for public health analysis and decision making, including detecting trends and sentinel events.
- Use of standardized data formats.
- Influencing public health policy.
- Linkage with CDC and other national and international surveillance databases.
- Collaboration with state and national laboratory systems.
- Continuous improvement of laboratory data systems

Reference and Specialized Testing
Serve as centers of excellence using their expertise, reference and resources in the areas of biological, chemical and radiologic issues of public health importance to:

- Support the diagnosis of and surveillance for unusual and emerging pathogens.
- Confirm atypical laboratory test results.
- Verify results of other laboratories' tests.
- Provide reference services to laboratories that may not have the capability to fully identify disease agents of public health importance.
- Provide diagnostic testing for diseases of public health importance directly to providers when testing is not readily available.
- Test for diseases of public health importance that are too rare and unusual for other laboratories to maintain capacity

Food Safety
Collaborate in the detection, monitoring and response to food safety issues by:

- Testing samples from persons, food and beverages implicated in food-borne illness outbreaks to detect and identify potential food-borne pathogens.
- Characterizing isolates and participating in national strain characterization databases, such as PulseNet, to inform epidemiologic investigations.
- Analyzing food specimens to detect, identify and quantify toxic contaminants such as pesticide residues, heavy metals and volatile organic compounds.
- Monitoring for radioactive contamination.
- Participating in the Food Emergency Response Network (FERN).

Laboratory Improvement and Regulation

Provide leadership for laboratory improvement in areas of public health importance by:

- Promoting quality improvement programs for partner laboratories through activities such as training, consultation and proficiency testing.
- Developing and overseeing statewide laboratory improvement programs to ensure the reliability of laboratory data used for environmental monitoring and communicable disease surveillance and control.
- Promoting safe laboratory practice through education, training and consultation.
- Assessing and improving the State Public Health Laboratory System by implementing the Laboratory System Improvement Program (L-SIP).
- Guiding the creation of and supporting enforcement of regulations and laws that contribute to laboratory improvement.

Policy Development

Play a role in the development of state and federal health policy by:

- Generating scientific evidence that informs public health practice and law.
- Monitoring the impact of public health laboratory practice on health outcomes.
- Serving as centers of expertise, reference and resources in the areas of biological, chemical and radiologic issues of public health importance.
- Participating in the development and evaluation of standards related to the operation and performance of laboratories involved in public health testing.
- Advocating for the use of sound reasoning in the application of laboratory science and system infrastructure sustainment.
- Engaging in strategic planning at local, state and national levels.

Public Health Related Research

Engage in research to improve and expand the scientific and policy bases of public health laboratory practice and assure their optimal application in support of the public health system by:

- Developing, evaluating and implementing new technologies and methodologies.
- Partnering with other public health disciplines.
- Collaborating with academic institutions to carry out clinical and translational science.
- Conducting public health systems and service research.
- Working with the private sector to foster scientific innovation.

Training and Education

Facilitate access to training and education by:

- Sponsoring training opportunities to improve scientific and technical skills within the public health laboratory system.
- Supporting management and leadership development opportunities.
- Participating in the training of both domestic and international scientists.

- Partnering with academia to provide experiential learning opportunities.
- Providing continuing education in the area of laboratory practice.

Partnerships and Communication
Support their respective state public health laboratory systems by:

- Highlighting the importance of laboratory contributions in support of public health.
- Maintaining a strong communication plan that links all system partners.
- Utilizing information technology for robust connectivity;
- Engaging traditional and non-traditional partners.
- Coordinating activities through the use of a laboratory program advisor, (i.e., laboratory system coordinator).
- Linking the SPH Laboratory System to appropriate national surveillance networks.

Source: (Association of Public Health Laboratories, 2014)

Article: Efficiency and the Cost-Effective Delivery of Forensic Science Services

5.2

Insourcing, Outsourcing, and Privatization

CHRIS MAGUIRE,[1] MAX M. HOUCK,[2] ROBIN WILLIAMS,[1] AND PAUL J. SPEAKER[3]

[1]*Northumbria University Centre for Forensic Science, Newcastle-upon-Tyne, United Kingdom*
[2]*Consolidated Forensic Laboratories, Washington, DC, USA*
[3]*College of Business and Economics, West Virginia University, Morgantown, West Virginia, USA*

Introduction

The 2012 closure of the Forensic Science Service (FSS) in the United Kingdom sent a shock wave through forensic science laboratories around the world (Dougan 2012). If the continued existence and operation of one the world's great forensic science innovators was not safe from government cuts in a severe economic climate, then no institution could consider itself immune from the austerity measures under consideration across the globe (Lawless 2011). Such austerity measures may have unintended consequences upon the ability to carry out the mission previously championed by the public sector laboratory (Houck 2011).

Rather than argue whether the FSS should have anticipated the changed economic conditions that put it out of business, we consider some of the realities that exist for all forensic laboratories and suggest some actions that other laboratories might take before being faced with such an undesirable fait accompli. The alternative strategies are rooted in the fundamental laws of economics. Economics has been branded the dismal science (Carlyle 1849), but unlike other social sciences it does share some characteristics with the hard sciences. Economics is the only social science able to confidently propose laws of behavior, and these laws pertain to private and public sectors alike. Notable among the laws of economics are the law of demand and the law of diminishing marginal returns. The former law deals with the desires of a customer for goods and services, given the limitations of their ability to purchase from their income and wealth and given their willingness to purchase as defined by their tastes and preferences. The latter, the law of diminishing marginal returns, guides the realities of the productivity of resources and the affiliated costs as goods or services are made available for customers. Failure to understand the implications of these laws could make a laboratory victim to the harsh realities of economic life (Kobus et al. 2011).

 DOI: 10.4324/9780429318528-15

To monitor its performance in the economic world, a laboratory must have something to measure that it connects in a meaningful way to its decision making. Ideally, the laboratory will have metrics that connect its performance to its goals and strategies. Such metrics should allow comparison to the laboratory's internal performance over time and permit external comparison in a given time period to other laboratories that provide the same forensic science services. These comparisons should extend to both the public and private provision of these services with recognition of the different motives of each type of organization.

Once armed with the knowledge of its position in the economic world, a laboratory opens itself up to more meaningful evaluations of the cost-effectiveness and efficiency of its operations. Such evaluations offer better opportunities to develop strategies to enhance efficiency and achieve more cost-effective use of its resources (Speaker & Fleming 2010). These strategic decisions include opportunities to expand, contract, or eliminate the provision of existing services and evaluate opportunities to expand operations into new investigative areas. The application of quality forensic science coupled with equally strong economic science offers an opportunity for laboratories to continually monitor their operational strategies and to preempt or at least mitigate the austerity measures of their funding bodies.

This article begins with an initial discussion of tools for the measurement and monitoring of performance, both scientific output and corresponding accounting costs. This includes a look at the data available across the forensic science industry that might be used for comparison. Next, a brief presentation of two laws in economics and their implications for every laboratory is offered. These laws govern the provision of every good and service across all industries. An application of the economic laws through a demonstration of the metrics for a large sample of forensic laboratories in the provision of one investigative area, DNA casework, is provided.[1] For DNA casework, the data suggests that a large number of laboratories are conducting analysis at too small of a scale to be cost-effective. While they may be highly efficient in the conduct of their internal tasks, crime rates and subsequent submissions within the jurisdiction are insufficient to permit the most cost-effective delivery of services. The strategic alternatives that a laboratory might consider when faced with the realities of the data are then examined. As a point of reference, one laboratory from the sample was selected for a demonstration of how a laboratory might react to the data and its position in the economy.

Measuring and Monitoring Laboratory Performance

To identify appropriate metrics, it is first necessary to connect goals with measurable outcomes that directly or indirectly provide evidence of the degree of success in meeting those goals. For forensic laboratories, there are a host of viable options available from measurement in the laboratory for successful determination of scientific tests to traditional business metrics on productivity, efficiency, risk, and return. Many of these business metrics involve ratios that permit performance comparison for an entity across time and comparison with respect to others in the industry. Speaker (2009a) provides a description of some of these alternatives for forensic science laboratories.

However, for successful comparison between laboratories, it is essential that each laboratory consistently defines measurement; otherwise, comparisons will be meaningless. Forensic science currently lacks a universally accepted nomenclature from which this consistent measure may be derived. Fortunately, recent efforts have been launched to overcome this lack of a common language between laboratories (Houck et al. 2009). From their Project FORESIGHT, forensic scientists from laboratories across North America developed and defined a process for measurement that includes consistent metrics on casework, expenditures, and personnel allocation across a range of areas of forensic science analysis and investigation. FORESIGHT data from 64 forensic laboratories worldwide has now been collected and is being analyzed.

Some metrics may be directly linked to the mission, goals, and strategies of the laboratory. For the private laboratory, a metric such as return on equity (defined as profit/shareholder contribution) provides a direct link to the profit motive of the business operating the laboratory. For the public laboratory, alternative metrics for the return on investment are available. Consistent with maximizing the public laboratory's return on investment is the minimization of the average cost of successfully processing cases. And while it is convenient to relate goals to one metric, it is also true that no single metric tells the whole story. There are well-known processes available to look at a series of metrics as they are related to the return on investment through a decomposition of the return on investment ratio into a series of performance metrics that help to tell the entire story (Speaker 2009b).

For the analysis presented in the following sections, the metrics developed from the FORESIGHT Project will be used. Additionally, examples use the data from submissions using the LabRAT tool.[2] Note that the LabRAT tool includes the automatic calculation of the metrics for return on investment and the decomposition metrics that highlight market conditions, productivity, local economic conditions, and proxy measures for risk.

There is a linkage between these performance metrics and the development of strategies to increase cost-effectiveness and efficiency. Newman, Dawley, and Speaker (2012) provide a demonstration of how one laboratory, Ontario's Centre of Forensic Sciences, managed to implement a strategic planning cycle in which the metrics were regularly monitored and coordinated with project goals. Houck et al. (2012) offer a broader perspective in which cost-efficiency is evaluated alongside other laboratory goals. This allows the assessment of the impact across broader missions in a balanced scorecard approach.

Among the metrics common to these studies is the direct measure of the cost minimization goal. For any investigative area, the metric COST/CASE (defined as total expenditures/cases processed) offers a ratio that adjusts to the scale of operations and is therefore useful for comparisons across time and across laboratories. The total expenditure component is extensive in its coverage of costs. It includes all personnel expenditures from direct wages and salaries to benefits, all capital expenditures (allocated across time following Internal Revenue guidelines for laboratory equipment), consumables, utilities, and all forms of overhead.

In the analysis below, another aspect of the COST/CASE metric is examined as it relates to caseload and cost theory. Economic theory has demonstrated that there is range of output that is optimal for a good or service provider in any industry. In the next section, that aspect of economic theory is introduced and it is demonstrated how this relates to forensic science laboratories is introduced.

Laws of Economics

The demands for forensic science support through police investigations, the subsequent development of a prosecution case, and the adversarial criminal justice system often outstrips the availability of such support from forensic science laboratories, leading to delays, backlogs, and adverse judicial comment (Associated Press 2011; Michigan State Police Forensic Science Division 2011; Ministry of Community Safety and Correctional Services 2009; Nelson 2011; National Institute of Justice 2003).

There are two laws in economics that are particularly important for understanding what is meant by efficiency for forensic science laboratories and how that sense of efficiency may differ from cost-effectiveness. The first of these economic laws, the law of demand, governs the behavior of the customer for any good or service, including forensic science services. The second, the law of diminishing marginal returns (LDMR), governs the production of goods and services and has direct implications for the cost of forensic science services. Understanding the implications of both of these laws is critical for the management of a successful laboratory. The old adage has particular relevance here: Ignorance of the law is no excuse.

The law of demand is a direct consequence of what is termed the "economic problem" (Kobus et al. 2011). That is, society has unlimited wants, but only limited resources to meet those wants. As a result, economic agents must find some mechanism to ration those scarce resources and attempt to put them to their highest valued use. In a capitalist economy, that rationing mechanism is price, where the relative price of one good versus another reflects the relative value to consumers, given the scarcity of the good or service in question. The law of demand indicates how consumer preferences for a good or service will be expressed with respect to the price, given fixed levels of income, wealth, tastes and preferences, and the prices of other goods and services. The law of demand tells us that as price is increased, eventually the quantity demanded of that good or service falls. Figure 1 shows a typical downward-sloped demand schedule, D_0, along which we

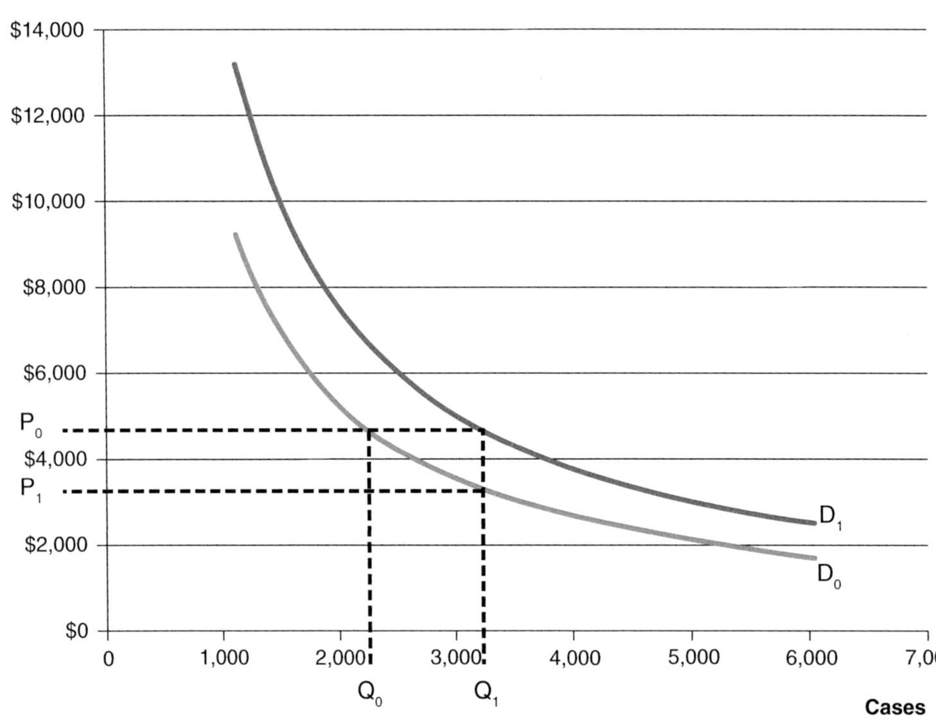

Figure 1 Demand schedule for cases (color figure available online).

highlight two prices: P_0 and P_1. If the initial price is P_0 and the corresponding quantity of cases demanded at that prices is Q_0, then given the budget restrictions of consumers of casework (e.g., police and prosecutors), a higher quantity demanded, such as Q_1, will be the result only when the price of casework falls to a level such as P_1. The alternative way in which a higher quantity demanded might emerge is through the relaxation of the assumptions regarding the factors held constant in the construction of the demand schedule. That is, if there were an increase in income or wealth, a change in attitude (tastes and preferences) toward a particular investigative area, or a change in the price of alternative services, then demand could change at every price level. Consider the second demand curve D_1 in Figure 1. Suppose there had been an increase in the budget for this investigative area (e.g., receipt of a grant), then the quantity demanded is higher at every price level. In that situation, at the initial price P0 the higher quantity demanded Q_1 is the result of a shift in demand from D_0 to D_1.

For the public sector forensic laboratory, the law of demand has important implications (Gronberg & Hwang 1992). Public sector budgets, once established, limit the laboratory, police, or prosecutor to a fixed amount of spending within any fiscal period. Any increase in demand (a rightward shift of the demand schedule in Figure 1) can only be met via a reduction in the demand for some other good or service. The economic cost of increasing demand can be stated in terms of the drop in demand for some other service. For example, the cost of one DNA case might be expressed as a reflection of its actual accounting cost of $1,900, or it might be recognized by its economic cost—say four fingerprint identification cases. The fixed budget forces decision makers to choose between services. The second law, LDMR, indicates the level of output expected as inputs are increased, holding all other conditions constant. For example, imagine the blood alcohol analysis section of a laboratory in which a fixed amount of space has been dedicated and that space has been outfitted with a fixed number of workstations, equipment, chemicals, and other supplies. As the number of scientists/analysts in that section is increased, output can be expected to increase as each additional analyst is added to the section. Output may well increase at an increasing rate initially; however, a point will be reached when the hiring of an additional person increases output, but the gain is less than was found with the previous hire. This behavior holds for all industries and for all inputs.

The implications of the LDMR for cost are important to consider. If new inputs (capital, consumables, personnel, etc.) are less productive, then at some point it will become more costly to provide additional services. Figure 2 illustrates this for a total cost curve as the quantity of services is increased. Notice that the total cost curve has a backward-bending S-shape. In the early levels of production, the greatest productivity gains are realized and thus the cost of higher levels of output rises at a slower rate. Beyond the point of diminishing marginal returns, costs begin to rise at a higher rate. The point of tangency of this S-shaped curve with a line drawn from the origin (i.e., the [0,0] point in Figure 2) represents the lowest average cost attainable.[3]

The LDMR affects the production of all goods and services and that includes all of the activities of the forensic science laboratory. So, what are the implications for the laboratory with respect to efficiency and cost-effectiveness? Consider a series of lines from the origin and note where they intersect the total cost curve—this will depict the average total cost curve, as shown in Figure 3. The average total cost curve (ATC) shows the efficient response to the provision of any possible quantity of output. Each and every point along the curve in Figure 3 is efficient. For example, if the caseload was level Q_0, then the corresponding

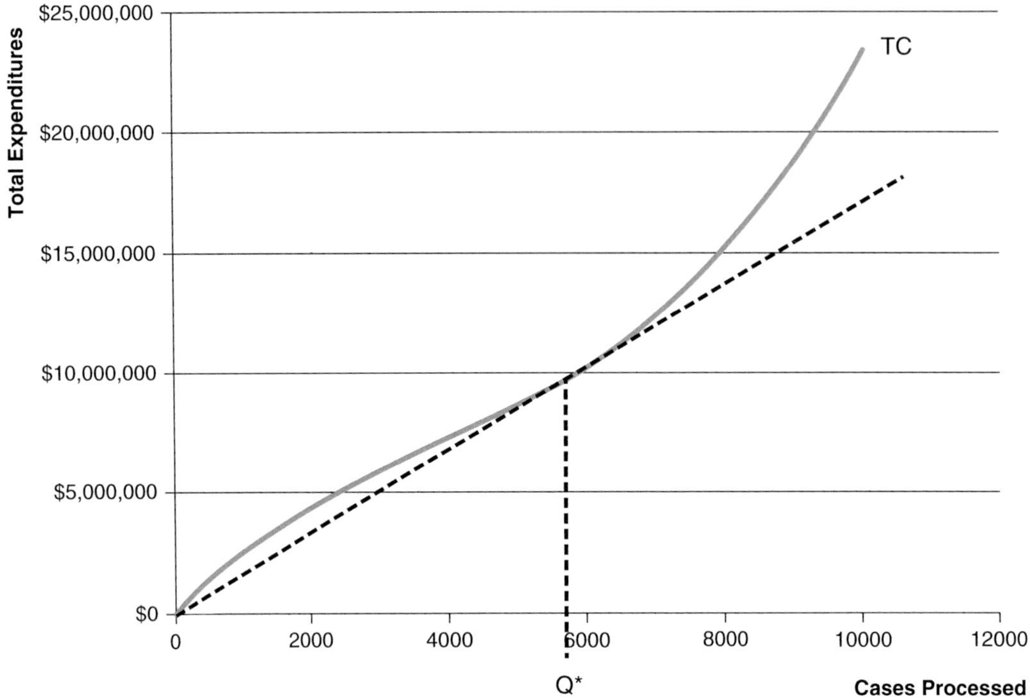

Figure 2 Total expenditures versus cases processed with minimum average total cost (color figure available online).

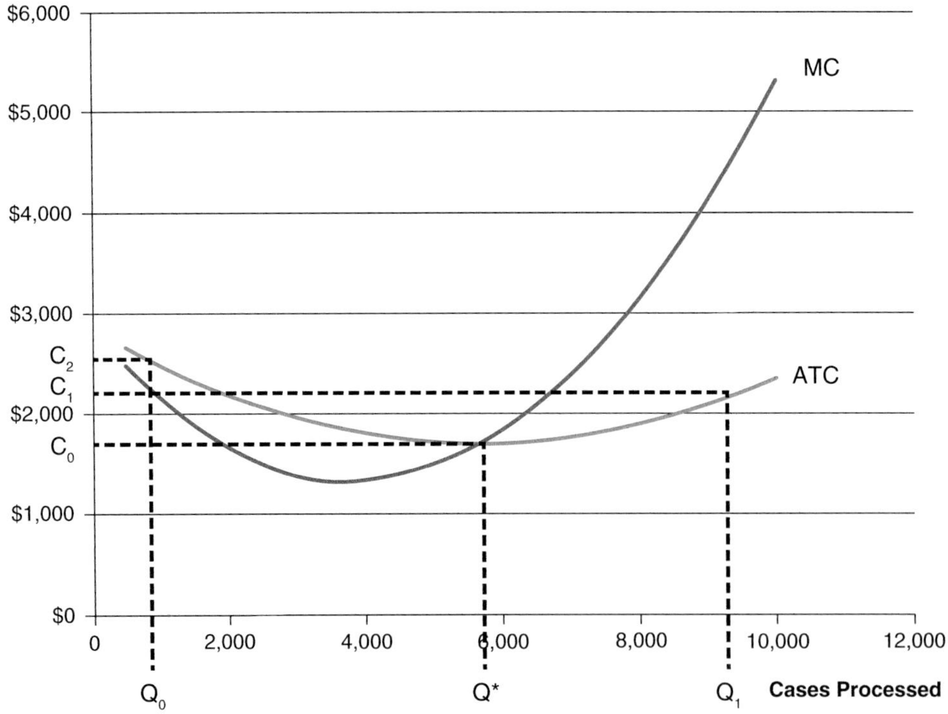

Figure 3 Average total cost and marginal cost versus cases processed and perfect economies of scale (color figure available online).

COST/CASE, C_0, represents the most cost-effective way to process those Q_0 cases. Likewise, at caseload Q_1, the corresponding COST/CASE, C_1, is the efficient response.

The second curve illustrated in Figure 3 is the marginal cost curve (MC). This represents the additional cost from the production of one more unit (i.e., processing one more case). It also is a product of the total cost curve in Figure 2 where the value of the marginal cost is a reflection of the slope of the TC curve at any level of case processing. Notice that when the MC is less than the ATC then the ATC is falling, and when the MC is above the ATC then the ATC is rising. It is when the additional cost, MC, is exactly equal to the average total cost, ATC, that the minimum ATC is achieved.

Now consider the minimum COST/CASE, C_0. It represents an efficient response to the processing of Q^* cases, but as the minimum average total cost it also represents the most cost-efficient means of production and yields the "best" response to the economic problem. At Q^* cases, perfect economies of scale are achieved. Along the downward-sloped portion of Figure 3, higher case processing provides economies of scale. Beyond Q^*, higher case processing results in diseconomies of scale as the laboratory volume has exceeded its ideal size.

Economic Law into Practice

These two economic laws when taken together provide an indication of the nature of market solutions to the economic problem. On the one hand, the demand for any good or service may be represented by a demand schedule in which there is an inverse relationship between price and the quantity demanded for a given level of income or wealth. This relationship holds for all goods and services. For a fixed budget over any given period of time, such as that of a public laboratory for a fiscal year, this implies that decisions to conduct tests in a particular area of investigation will have the simultaneous effect of reducing the ability to conduct tests in another area of investigation for either the same case or for other cases. Couple the behavior of the demand schedule with the law of diminishing marginal returns and the implication is that the cost of additional services eventually reaches a point where costs will increase as more services are provided.

In private sector markets, the implication of the foregoing analysis is that all surviving businesses in an industry tend to produce a level of output near the quantity associated with perfect economies of scale, $Q*$. Any business failing to produce at that low cost level will find that mar-ket prices will be insufficient to generate a profit level for survival as competing businesses force them to either right-size their operating level or leave the industry altogether. Imagine any product or service and there will be countless examples of businesses tending toward a similar size for that industry.

The public sector, however, does not share this characteristic. Rather than economic forces determining the size of services, other factors, such as population or the crime rate in the case of the forensic laboratory, are the influencing factor on size. While the primary size-determining factors for public sector operations allow a political entity to exercise great control, they do not necessarily lead to a cost-effective approach to the provision of services (McAndrew 2012). Lacking the pressure from competitors to find more cost-effective solutions, the public laboratory may continue to operate at an efficient level. However, it runs the danger of forced change in an economic crisis, and this may have undesirable consequences, as seen with the FSS (Dougan 2012).

If a laboratory is cognizant of the cost-effective level of output—the caseload which is consistent with perfect economies of scale—then the most efficient level of public service might be provided, one that is both efficient and minimizes the average total cost. This minimum cost level is not necessarily the same for public and private laboratories. A public laboratory does have an advantage over any private laboratory that might compete to provide the same service because the public laboratory is not burdened with the economic cost of a rate of return on investment for its owners (Speaker & Fleming 2009).

Cost Curves for DNA Casework

To illustrate the alternatives for a laboratory, consider the performance of laboratories in DNA casework using a sample of laboratories. The data represent 42 voluntary submissions of self-reported data on casework, expenditures, and personnel allocation using the LabRAT tool for their performance in fiscal 2010.[4] The sample includes some submissions to Project FORESIGHT from public sector jurisdictions that operate a single laboratory facility and it includes some submissions from single-facility laboratories that do not participate in the FORESIGHT project. However, all laboratories have been certified by ASCLD/LAB or are ISO/IEC 17025 certified. Laboratories in the sample represent metropolitan, regional, provincial, and national laboratories.

Figure 4 maps the individual laboratories for cases processed and the corresponding average total cost for the provision of the DNA casework services.[5] Visual inspection of the

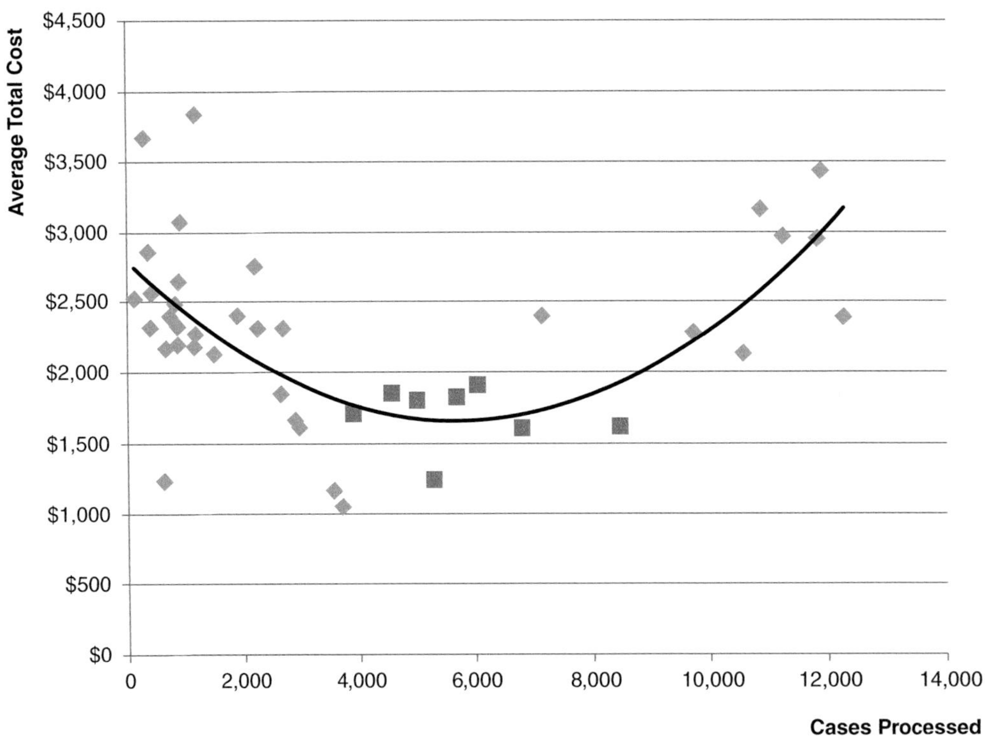

Figure 4 Average total cost versus cases processed in DNA casework (color figure available online).

data points is suggestive of the U-shaped curve for average total cost presented in Figure 3 and projected by economic theory (Witt & Speaker 2012). The curve presented in Figure 4 is representative of a curve fitted to the data using ordinary least squares of a quadratic function (i.e., a U-shaped second degree polynomial) relating the average total cost to cases processed. The second-degree polynomial regression permits the formation of the best unbiased estimator of the relationship between caseload and average total cost while observing the assumptions of the regression model.

Notice that a large portion of the sample involves laboratories operating at a fairly low caseload level when compared to the caseload of laboratories operating near the minimum average total cost level of the fitted curve (the observations near the perfect economies of scale are highlighted by squares, while the other observations are noted by diamond-shaped points). The sample suggests that cost-minimizing analysis of DNA casework occurs somewhere in the range of roughly 4,000 to 8,000 cases per year (with a point estimate of 5,727 cases). Nearly two-thirds of the sample laboratories are operating at a level of casework below that range, and approximately 15% of the sample is operating above the upper limit of that caseload range.

What are the implications of these laboratory data when reviewed with respect to economic theory? First, consider this question as if the market for DNA casework was a private sector market. In that case, private laboratories would have to compete on price, and only those laboratories that are able to generate enough business to lower their average total cost to levels near the perfect economies of scale would survive. Those surviving businesses would have experienced an average total cost curve that was similar to the one pictured in Figure 4.[6] Any laboratories that insisted on a level of operation outside the range of perfect economies of scale would find that they would be priced out of the market. They would be forced to operate at a loss, eventually realize that long-term profit could not be earned and shut down. This would be true regardless of the size of the laboratory's caseload; high caseload facilities might adapt by either opening a second facility or by lowering their case volume and associated average total cost of servicing that lower volume, but they would have to react in some fashion or be forced into financial distress (i.e., bankruptcy).

Second, consider the scenario when forensic science services are delivered through the public sector. In that case, the freedom of entry and exit into the market faces significant barriers, and as a result of these barriers to entry the pressure to compete on price is not a direct consequence of market activity. The allocation of scarce re-sources is based upon other criteria where the body politic makes the decision for the general population (McAndrew 2012). In times of plentiful public coffers, the decision by a jurisdiction to maintain a forensic laboratory might not face much opposition. However, in times of economic stress, when public budgets face greater scrutiny, the ability to maintain laboratories that operate at the low-volume end of the caseload spectrum in Figure 4 comes into question.

It is these stressful economic conditions, or other political commitments to austerity, in which change is forced upon the public sector. That stress is exacerbated when coupled with an increased perception of what forensic science may be able to answer (Houck 2006). The results of desperate circumstances can have tragic consequences for the provision of these services as has been experienced in the United Kingdom with the Forensic Science Service (Dougan 2012).

However, public sector agencies can act to avoid the direct consequences of these pressures. Recognition of the significance of economic laws and the forces contained within them can empower jurisdictions of any size to take a proactive stance in the determination of their own destinies. In the situation of the public forensic laboratory, there are a variety of choices. These include the continuation of the status quo, insourcing, outsourcing, and abandonment of services. The expansion of services through insourcing could include cross-jurisdictional work in existing expertise or expansion of the types of services within an investigative area. For example, several geographically adjacent laboratories may combine resources to address a desired service that no single laboratory has the capacity to sustain—questioned documents, for example. These laboratories might parcel out tasks by types of case (misdemeanors vs. felonies), examinations (handwriting vs. ink chemistry), or unit of time (month to month, quarter to quarter, etc.). Thus, the same or a greater level of service could be achieved through a minimal realignment of existing resources to balance supply with demand.

Similarly, outsourcing of services could experience similar gains as a laboratory releases its workload to another public or private laboratory in exchange for a lower cost for these services. Outsourcing is generally assumed to mean the private sector, but it is not inevitable. Much depends upon the potential gains from privatization (Schmidt 1996). These gains are lessened or eliminated by a stance that is made in response to, rather than in advance of, economic realities. The potential for crossjurisdictional agreements is only limited by the existing infrastructure of political laws, agency policies, and the creativity of the interested parties to achieve their desired goals. Typically, the most difficult limitations to overcome are structural and not administrative. For example, one jurisdiction having a vehicle to make and receive payments with another jurisdiction can be a vital link to a successful collaboration of resources; without such a financial channel, the jurisdictions have to devise other means (such as memoranda of understanding or barter) to exchange resources.

A transition to an alternative model will not come without other costs, and these costs must also be weighed in the decision. For example, consider a small county laboratory operating at the low end of the caseload spectrum in Figure 4. Even if the small laboratory is operating efficiently (along the average total cost curve), the corresponding cost level may far exceed the cost level associated with perfect economies of scale. Suppose that one alternative for the county laboratory is to outsource its DNA casework to a larger state laboratory that currently performs at a significantly lower average total cost that is somewhere in the range where there are still some economies of scale to enjoy from a higher caseload. Both laboratories could experience a lower average cost from a combination of services into a single laboratory. However, unless the dominant state laboratory is able to offer the lower cost service with a fair queuing scheme, one that treats the county laboratory cases on equal footing with the host laboratory cases, then the potential gains to both laboratories could be lost. The county laboratory would have to weigh the gains from the average cost reduction against the backlog cost of greater wait time. At the same time, the larger state facility must evaluate the gains from insourcing cases from other laboratories and recognize the economic reality that the provider gains as a direct result of serving the needs of its customer: the better the treatment of the county customer, the greater the gains to be enjoyed by customer and provider alike. If they cannot come to some fair terms for the sharing of facilities, then each laboratory will continue to face higher costs and continue to be subject to the threat of private laboratory replacement of services.

Conclusions

The public sector is not immune from economic reality and that is true at both the macroeconomic and microeconomic levels. Macroeconomic forces, as seen through global recession, have forced dramatic changes in the modes of operation of many. In some cases, the macroeconomic forces have led to draconian policies, including the closure of public facilities in favor of privatization of many services.

Attention to the nature of economic forces at the laboratory level—the microeconomic level—offers a means for public sector laboratories to fend off unwanted change deriving from economic circumstances. Laboratories must first understand foundational economic laws in order to exploit them for their own benefit. And, as shown in the example of DNA casework, many laboratories currently are operating at levels that are far from a cost-effective delivery of services. To remedy the situation, laboratories need to understand the proactive alternatives that couple the efficient delivery of services with a cost-effective level of activity. That cost-effective level could involve cross-jurisdictional delivery of services through agreements on insourcing or outsourcing cases. It could also include outsourcing some casework to the private sector or even abandonment of some services to private providers.

Endnotes

1 The data come from a voluntary sample of laboratories using the LabRAT tool from Project FORESIGHT (Houck et al. 2009).
2 The LabRat tool and corresponding definitions are available at http://www.be.wvu.edu/forensics/LaRAT.xlsx.
3 A ray from the origin that passes through the total cost curve represents the average cost for that case level. The point of tangency represents the only ray from the origin that touches a single point, the minimum average cost. For any other steeper ray, it intersects the total cost curve at two different levels of case processing. The slope of each ray is the measure of the average cost.
4 The specific dates for the fiscal years differed across laboratories, with some laboratories reporting a fiscal year of January 1, 2010, through December 31, 2010; other laboratories reporting a fiscal year of April 1, 2010, through March 31, 2011; and a third portion of laboratories reporting a fiscal year of July 1, 2010, through June 30, 2011. The data are expressed in 2010 U.S. dollars for consistency across the measures.
5 We follow the Project FORESIGHT definition for this area of investigation, where the DNA casework investigative area is defined as "analysis of biological evidence for DNA in criminal cases" (Houck et al. 2009, p. 90).
6 The difference between the cost curve for the public sector laboratory and the for-profit laboratory comes from one portion of the cost, the profit portion required by the for-profit firm. A return on investment that is sufficient for the risk undertaken is part of the economic cost structure that would be embedded in a cost curve that is a bit higher at all levels of service.

References

Associated Press. 2011. Budget cuts cause delays in Alabama forensic tests. *Alabama Live*, December 27. Accessed 24 May, 2012, https://www.al.com/wire/2011/12/budget_cuts_cause_delays_in_al.html.

Carlyle, T. 1849. Occasional discourse on the negro question. *Fraser's Magazine for Town and Country* 60(2): 672–673. Dougan, J. 2012. What future for forensics? *RSC News*: 10–11.

Gronberg, T.J., and H. Hwang. 1992. On the privatization of excludable public goods. *Southern Economic Journal* 58(1): 904–921.

Houck, M.M. 2006. CSI: Reality. *Scientific American*: 84–89.

Houck, M.M. 2011. Editorial: Forensic science: Whose job is it anyway? *Forensic Science Policy & Management* 2(2): i–ii.

Houck, M.M., R.A. Riley, P.J. Speaker, and T.S. Witt. 2009. FORESIGHT: A business approach to improving forensic science services. *Forensic Science Policy & Management* 1(1): 85–95.

Houck, M.M., P.J. Speaker, A.S. Fleming, and R.A. Riley. 2012. The balanced scorecard: Sustainable performance assessment for forensic laboratories. *Science & Justice*: doi:10.1016/j.scijus.2012.05.006.

Kobus, H., M. Houck, P. Speaker, R. Riley, and T. Witt. 2011. Managing performance in the forensic sciences: Expectations in light of limited budgets. *Forensic Science Policy & Management* 2(1): 36–43.

Lawless, C.J. 2011. Policing markets: The contested shaping of neo-liberal forensic science. *British Journal of Criminology* 51:671–689.

Maguire, C.N. 2012. Development of a forensic market in England and Wales and the impact of U.K. government organisation in other jurisdictions. Northumbria University Centre for Forensic Science working paper.

McAndrew, W. 2012. Privatization in forensic science. West Virginia University Center for Forensic Business working paper. Michigan State Police Forensic Science Division. 2011. Forensic science for Michigan: Backlog prevention. Accessed 24 May, 2012, http://www.michigan.gov/msp/0,4643,7-123-60141 60468 60473—,00.html.

Ministry of Community Safety and Correctional Services. 2009. *Chapter 4, Section 4.02: Centre of Forensic Sciences: Follow-up on VFM section 3.02, 2007 annual report*. Access 24 May, 2012, https://www.al.com/wire/2011/12/budget_cuts_cause_delays_in_al.html.

National Institute of Justice (NIJ). 2003. Report to the attorney general on delays in forensic DNA analysis. *NIJ Special Report*. Washington, DC: NIJ.

Nelson, M. 2011. *Making sense of DNA backlogs, 2010—Myths vs. reality. NIJ Special Report*. Washington, DC: National Institute of Justice.

Newman, J., D. Dawley, and P.J. Speaker. 2012. Strategic management of forensic laboratory resources: From project FORESIGHT metrics to the development of action plans. *Forensic Science Policy & Management* 2:164–174.

Schmidt, K.M. 1996. The costs and benefits of privatization: An incomplete contracts approach. *Journal of Law, Economics, & Organization* 12(1): 1–24.

Speaker, P.J. 2009a. Key performance indicators and managerial analysis for forensic laboratories. *Forensic Science Policy & Management* 1(1): 32–42.

Speaker, P.J. 2009b. The decomposition of return on investment for forensic laboratories. *Forensic Science Policy & Management* 1(2): 96–102.

Speaker, P.J., and A.S. Fleming. 2010. Benchmarking and budgeting techniques for improved forensic laboratory management. *Forensic Science Policy & Management* 1(4): 199–208.

Speaker, P.J., and A.S. Fleming. 2009. Monitoring financial performance: An approach for forensic crime labs. *The CPA Journal* 79(8): 60–65.

Witt, T.S., and P.J. Speaker. 2012. The power of information. *Forensic Magazine*: 1–5.

Article: Process Improvement and the Efficient Frontier

5.3

Forecasting the Limits to Strategic Change across Crime Laboratory Areas of Investigation

PAUL J. SPEAKER
Finance Department, WVU College of Business and Economics, College of B&E, Morgantown, WV

Introduction

The Netherlands Forensic Institute (NFI) embraced a Lean Six Sigma change process that led to dramatic increases in capacity, while simultaneously reducing turnaround time (TAT) to a fraction of their historical experience (van Asten 2014; Tjin-A-Tsoi 2013). The adoption of a Lean Six Sigma strategy by NFI was aligned with the mandates of the National Academies of Science report (NAS 2009). Pressure upon the NFI, as with forensic laboratories worldwide, was evidenced by greater demands for services without corresponding increases in resources to meet those demands. As a response, they took a customer-centric approach towards process improvement as a means to respond to the widening gap between customer demands and available resources

The account of the NFI process improvement (van Asten 2014) includes the prospects for further gains via innovation in forensic science research and offers some grand challenges for further improvements (some challenges which have already been addressed elsewhere (Castillo-Peinado and de Castro 2016). In this article, we offer observations and analysis to complement the observations of the NFI experience through insights into the economic foundations that provide direction into the forecasting of results from such process improvement. Economics is a social science guided by laws; and rudimentary knowledge of the workings of the affiliated laws enables laboratory management to better prepare for the resulting change from strategic initiatives. From those economic foundations, econometric modeling of performance across laboratories in the forensic sciences leads to an estimation of the efficient performance possibilities for the caseload of a particular jurisdiction. A laboratory's comparison of its own performance to the efficient frontier defines the gap that may be overcome through strategic changes for process improvement.

The end game is a moving target. The change process is dynamic and all too often planners mistakenly prepare for a static environment. By understanding the underlying forces at work, laboratory management may use that knowledge to more accurately forecast the

 DOI: 10.4324/9780429318528-16

outcomes from strategic initiatives. Process improvement also affects the expectations of customers and the resulting demand for laboratory services. Improved forecasts of the customer response will permit the laboratory and its parent funding body to be forewarned and forearmed to best meet the growing demands for its services with the limited resources at its disposal.

There are both direct and indirect predictable effects from process improvement. The indirect consequences need not be surprising; they can be estimated and anticipated in the planning process and the corresponding planning can and should be conducted (Houck 2017). The NFI Lean Six Sigma initiative began with a concentration on the existing level of services and the initiation of process improvement. Too often, however, the initial planning treats potential efficiency gains as a static process. That is, the planning considers the current demands upon the system and self-reflects on how to handle that load better. But the process of "doing it better," has its own effect on the demand for services, because success breeds success as process improvement leads to higher laboratory case throughput and reduced turnaround time (TAT). The planning strategy must include a consideration for the increased requests for testing of current cases and the increase in the number of cases brought to the laboratory as TAT is reduced. The dynamic problem comes in different forms. It includes the higher demand from existing and new cases, but must also deal with the resulting economies or diseconomies of scale.

The complexity of the problem requires planners to first address the inefficiencies associated with the current scale of operations and consider the efficiency gains that might come to that level of output from a productivity improvement process such as a Lean Six Sigma program. The immediate impact from a process improvement program is an increase in output for a given level of expenditures, but that improved performance will have an associated reduction in TAT for casework. Any decrease in TAT triggers a response from the requestors of laboratory services, which can be predicted using a metric, the queuing elasticity of demand. The queuing elasticity of demand suggests the level of demand for services facing the laboratory for any associated TAT. Using forensic industry data, it is possible to estimate the target level of services and associated cost per service that can be expected from achieving greater economies of scale. Using experiential data, the laboratory can begin to anticipate the moving target set in motion by process improvement and then anticipate the outcomes and societal benefit from its internal process improvement activity. Thus, armed with greater forethought of its actions, the laboratory may be better served to meet its mission and to argue effectively for the funding necessary to complete its mandate.

In the sections that follow, we begin with an introduction to a few key economic forces at play for the provision of any good or service, including the services of the forensic laboratory. Following the description of these basic economic forces, a few key performance ratios and their interrelationships are discussed along with data from Project FORESIGHT (Houck. et al. 2009) to illustrate the key behaviors. The econometric model alternatives are introduced and demonstrated for one area of investigation, DNA Casework, and the corresponding tables for efficient performance at various jurisdictional caseloads is presented. A section follows with the tabular representation of the efficient frontier for other areas of investigation follows with corresponding econometric estimates in the Appendix. Concluding comments complete the discussion.

Economic principles and laws in economics

As with the physical sciences, economic behavior is ruled by some immutable laws. Among these laws are two that are relevant to the topic at hand, the Law of Demand and the Law of Diminishing Marginal Returns (LDMR). Understanding the Law of Demand is critical to strategic initiatives that result in changes in the rationing device for scarce resources (e.g., price or wait time). A change in the level of the rationing mechanism will change the quantity of services demanded in response to that change stimulus. The Law of Demand in a for-profit marketplace references price as the rationing mechanism, where the provider continues to raise or lower the price of a service until a level is reached where the quantity of services demanded at that price is equal to the quantity of serves provided. However, the allocation of forensic science services is rarely rationed by price; forensic science services are treated as a "free" service in most situations (Tjin-A-Tsoi 2013). Instead, wait time serves to allocate services and an understanding of a corollary to the Law of Demand with queuing time or TAT replacing price is pertinent to the present discussion.

The Law of Diminishing Marginal Returns may be directly applied from economic theory to the workings of the forensic laboratory. In the provision of any good or service, as more of a particular input is added to a production process, eventually a point is reached where increases in production occur at a diminishing rate. LDMR principles are imbedded in the concepts of economies and diseconomies of scale. That is, there is an ideal scale of operations in the production of any good or service. In the for-profit world, market forces mold businesses into the "right" size, where average total costs of production are minimized and all businesses in a given industry tend towards that ideal size; otherwise they are priced out of business. Without price competition in jurisdictionally based service areas to force forensic laboratories to adjust their size to a cost-minimizing level, there remain untapped economies of scale to experience via expansion of operations (Maguire et al. 2012b). It is critical to the effective planning by the laboratory to understand the magnitude of the reactions of consumers of forensic science services via the Law of Demand and to realize production realities regarding efficiency and cost effectiveness for the laboratory as services are expanded.

Economies of scale and LDMR

A consideration of economic theory in the production of goods or services involves both a sense of efficiency and cost effectiveness. Efficiency refers to the selection of a production process that is best for a given level of activity. The reference to "best" suggests adoption of the cost minimizing process that maintains a given quality standard. The combination of capital and personnel best suited to analyze a small number of cases will be different from the amount and mixture of capital and personnel to analyze a caseload that is ten times larger. Given time, laboratories will gravitate towards the process that permits the greatest amount of case processing (while maintaining quality standards) for the budget at their disposal. With knowledge of LDMR, laboratories are able to expand or contract the use of capital and personnel to find lowest cost alternatives to process a given number of cases.

Process improvement programs, such as Lean Six Sigma, may have a significant impact upon a laboratory's efficiency by identifying the means to make a faster adjustment to the optimal process. However, the degree of success that a laboratory might achieve from

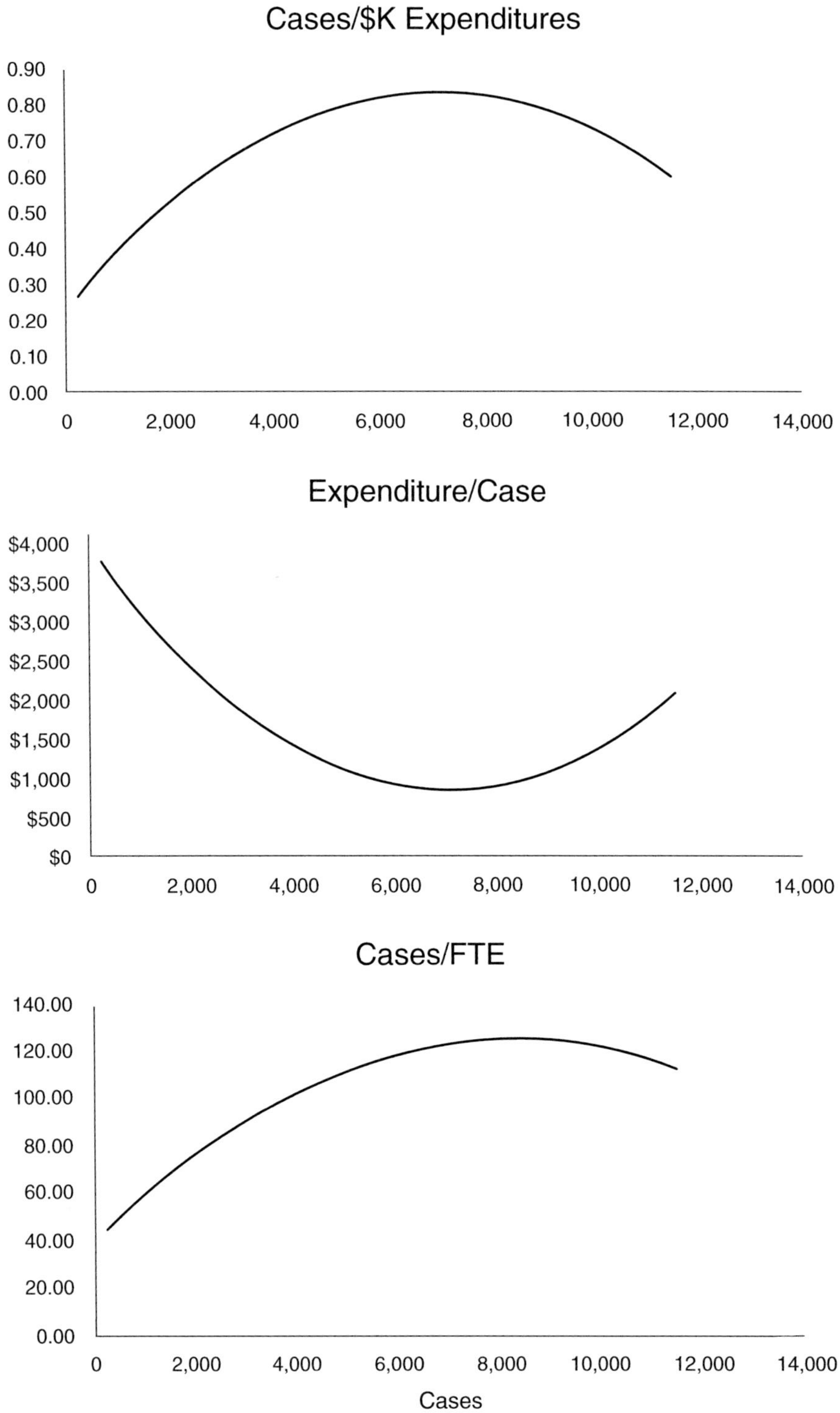

Figure 1 The efficient frontier, average cost, and productivity about here.

a process improvement agenda is limited by the laboratory's relative efficiency prior to consideration of a change process. For a laboratory to achieve success from such a process change program, it must begin from a position of inefficiency if it is to expect gains similar to those achieved by NFI.

Economic theory indicates that as the level of output increases, the average cost to produce that output will fall and the output per dollar spent will increase up to a level of perfect economies of scale; thereafter higher output will result in lower rates of output per expenditure and an associated higher average cost of processing. Figure 1 illustrates this theoretical relationship. The hump-shaped upper graphic demonstrates the rise in output/expenditure as output increases up to the point of perfect economies of scale. When a laboratory is able to produce at a point along the curve, it exhibits efficient production. That is, given its level of casework, a laboratory that is able to reach a point along this "efficient frontier" has provided the best relative output that can be expected. The inverse of that metric, the average cost is illustrated in the second graph with a U-shaped curve. Along the downward portion of the curve we experience greater economies of scale as the caseload increases, but beyond the minimum point, diseconomies of scale are experienced as highlighted in the upward-sloped portion of the curve. The bottom graphic illustrates the relationship between these curves and productivity within the laboratory, where productivity is illustrated as the number of cases processed per full-time equivalent employee (FTE). The rise in productivity that is experienced with reaching the optimal output size is directly related to the changes in average cost of providing that output.

Considerations of efficiency refer to positions along the efficient frontier. Since most laboratory operations are limited to the boundaries of its political jurisdiction, comparisons of a laboratory's performance to the efficient frontier is the most relevant comparison to be made. Cost effectiveness, on the other hand, refers to the point of perfect economies of scale, the maximum point of the upper graph in Figure 1 or the minimum point of the middle graph.

Law of demand

The law of demand recognizes the relationship between the price per unit and the number of units demanded by consumers at each possible price, while holding all other demand influences constant. The price vs. quantity relationship emphasized in the law of demand, indicates that as prices rise, then eventually a price level is reached where the quantity demanded falls. Put simply, there is an inverse relationship between price and the quantity demanded. That relationship between price (P_{own}) and quantity demanded is determined by individuals for given levels of income, wealth, tastes & preferences, and prices of substitute (P_{sub}) or complentary (P_{comp}) goods or services.

$$Q_d = f(P_{own}, Income,\ Wealth,\ Tastes,\ P_{sub},\ P_{comp}) \tag{1}$$

In the for-profit world, producers realize that when they lower price, they expect to sell more units of a good or service; conversely, if they raise price, they expect to sell fewer units. Price serves as a rationing mechanism to allocate scarce resources. While the law of demand indicates the direction of the change in quantity demanded from a given change in price, it does not indicate the magnitude of the change in quantity demanded. The magnitude of the change is more of a reflection of the attitudes of the consumer and the nature of the product of service being offered for sale. Economists use a metric, price elasticity of demand, to predict the impact of a given change in price on the quantity demanded. While the law of demand dictates that a price decrease will result in an increase in quantity demanded, the price elasticity of demand metric permits forecasters to predict whether such a price cut will result in a large enough quantity demanded increase to offset the price decline and lead to an increase in revenue.

Price elasticity of demand:

$$\eta_{price} = \frac{percentage\ change\ in\ quantity\ demanded}{percentage\ change\ in\ price}$$

$$= \frac{\partial Q_d}{\partial P}\frac{P}{Q_d} \tag{2}$$

Because of the inverse relationship between price and quantity demanded in the demand schedule, hprice is a negative number. When $-1 < \eta_{\text{price}} < 0$, then demand is said to be inelastic; that is the percentage change in quantity demanded is small relative to the percentage change in price. A price increase is met with an increase in total revenue (price multiplied by quantity). Conversely, a price decrease would be met by a decline in total revenue. If $\eta_{\text{price}} < -1$, then demand is termed elastic; that is, there is a larger percentage change in quantity demanded than the initiating percentage change in price and an increase in price is met by a decrease in total revenue, while a decrease in price is met by an increase in total revenue. The price elasticity of demand may be estimated by tracking discrete changes in price and observing the corresponding change in quantity demanded.

$$Price\ elasticity\ of\ demand : \eta_{price} = \frac{\%\Delta Q_d}{\%\Delta P} \tag{3}$$

For most forensic science laboratories, there is no explicit price for its services and those services must be rationed by some other mechanism. Often, wait time (as measured by TAT) becomes the rationing mechanism in place of price. If the queue is too long, some cases are never submitted to the laboratory or if submitted, the requests may be limited to a subset of potential items of evidence. However, when a laboratory is able to successfully reduce its TAT, then that serves as a signal to its customers. Laboratory customers respond with an increase in requests to the laboratory. TAT serves much like price as the rationing mechanism. Reactions to changes can be represented by a queuing elasticity of demand.

Queuing elasticity of demand:

$$\eta_{TAT} = \frac{\%\Delta\ quantity\, demand}{\%\Delta TAT} \tag{4}$$

As with the law of demand and price, we know that a reduction in TAT will result in an increased quantity demanded, but hTAT provides an indication of the magnitude of the reaction via quantity demanded to a change in the stimulus.[1]

Economies of scale, demand elasticity, process improvement, and implications for forensic laboratories

For most public sector forensic laboratories, the services of the laboratory are treated as "free" to the customers of the laboratory. Notable exceptions are The Netherlands NFI [(van Asten 2014) and (Tjin-A-Tsoi 2013)] and New Zealand's Institute of Environmental Science and Research (ESR) (Bedford 2011) and (Speaker 2013), where service agreements place the funding in the hands of the corresponding policing organizations who purchase the services of the forensic science provider. For most remaining forensic providers, the users of their services have a seemingly unlimited demand for services which is held in check by the queuing time for the receipt of those services

Will all laboratories have the same experience as the NFI following a process improvement program? What can a laboratory considering the adoption of a process improvement program expect to experience? As the discussion above suggests, the success is highly dependent upon the starting point. Consider laboratories X, Y, and Z highlighted in Figure 2.

Laboratory X begins at a point of inefficiency; that is, for its caseload, it falls below the Cases/$K Expenditures of the efficient frontier in the upper graph and above the efficient average cost in the middle graph. The bottom graph shows that the productivity metric, Cases/FTE, falls below what is possible for that caseload. A process improvement program offers the possibility of higher productivity to be met with a reduction in average cost and a corresponding increase in the inverse of that metric, Cases per thousand dollars of expenditures. But the case output would not be expected to be static for laboratory X. As output productivity increased, given the same level of resources, TAT would decline, prompting an increase in the quantity demanded for casework that could be anticipated from the queuing elasticity of demand from estimation of Equation (4). Thus, laboratory X would not only move closer to the efficient frontier, but would also experience a move along the frontier towards laboratory Y's mapping in each of the graphs

Laboratories Y and Z, however, originate at positions of efficiency. Laboratory Y, although not operating at perfect economies of scale, is experiencing an efficient output for its caseload and no gap in productivity improvement is foreseen. All things equal, a program designed for process improvement would prove to be futile. Any attempt to achieve greater economies of scale would require activity beyond that level dictated by its jurisdiction. Laboratory Z begins from the strongest point where they experience both efficiency and cost effectiveness with perfect economies of scale. A program aimed at process improvement would be fruitless.

Decomposition of returns

Crime laboratories face the challenge of greater public expectations of the capabilities of forensic science (both real and imagined) to assist in the administration of justice. Connecting strategic actions to potential gains has been suggested previously in the literature. This includes considerations of what to measure (Speaker 2009a), how to interpret the measures [(Speaker 2009b) and (Speaker and Fleming 2010)], and implementation of a change process (Newman, Dawley, and Speaker 2011). More recently, a refinement in the decomposition of returns was suggested following an examination of Canadian forensic science laboratories (Maguire et al. 2012a; Houck et al.2015). Refinements from

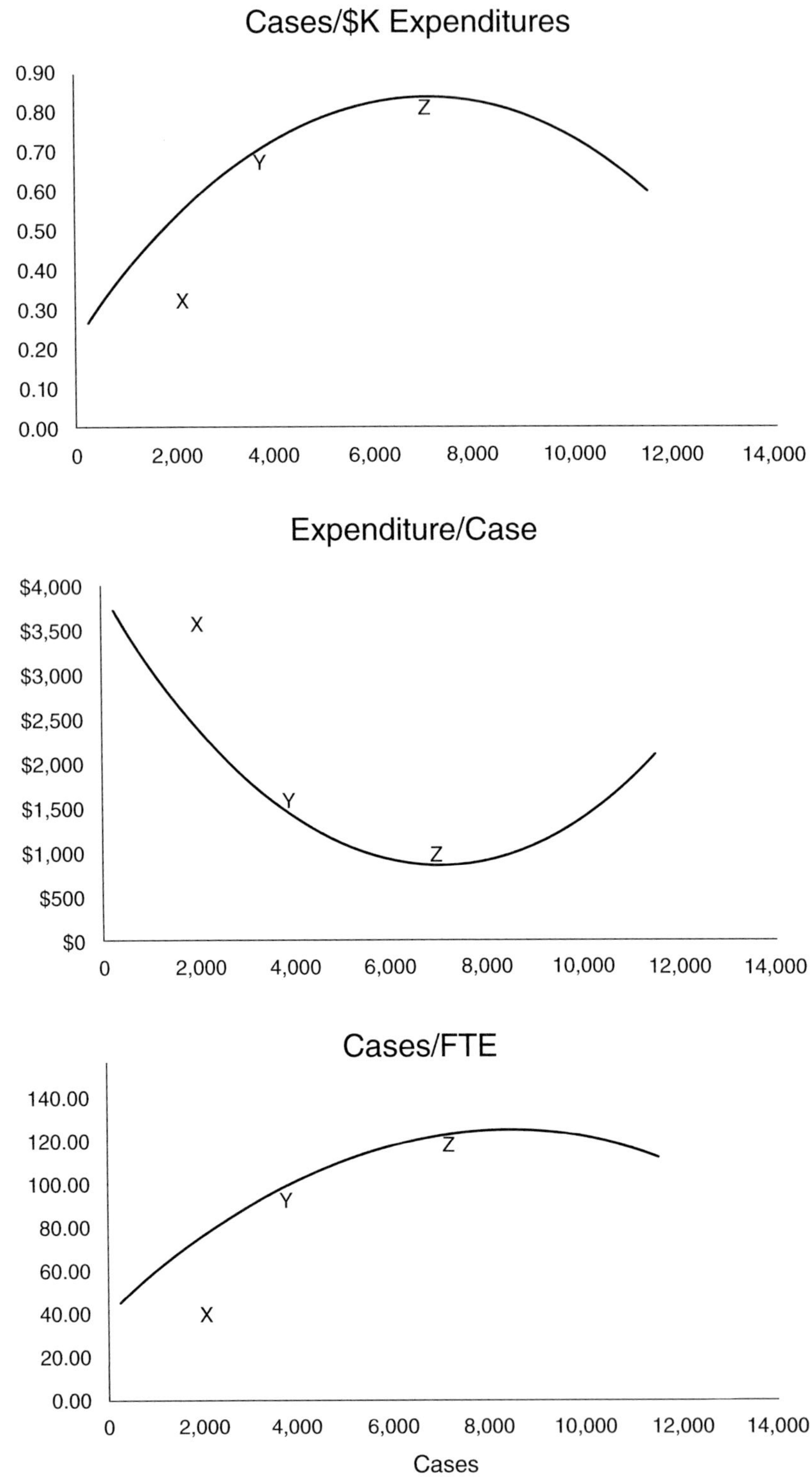

Figure 2 Process improvement, efficiency, and cost effectiveness.

traditional interpretations are necessary because of the differences in economic forces at play between market economies and political jurisdictions (Houck et al. 2015; McAndrew and Roth 2016).

Following the caveats regarding standards for quality (van Asten 2014), forensic laboratories attempt to optimize their processing of casework given their budget constraints. A decomposition process takes the objective function and breaks it down into a series of relevant ratios to explain behavior. These ratios are formed by multiplying the optimization ratio CASES/ TOTEXP (Cases processed divided by total expenditures) by the number one in various formats (Speaker 2009b). Since productivity and market forces have been identified as determinants of laboratory performance, we perform an expansion of the objective function to include consideration for these explanations of performance

$$\frac{CASES}{TOTEXP}=\frac{CASES}{TOTEXP}\times\frac{PEXP}{PEXP}\times\frac{FTE}{FTE}, \qquad (5)$$

where CASES represents cases processed, TOTEXP is the Laboratory's total expenditures, PEXP is the expenditure for personnel, and FTE is the number of full time equivalent employees. Recombining terms yields:

$$\frac{CASES}{TOTEXP}=\frac{CASES}{TOTEXP}\times\frac{PEXP}{PEXP}\times\frac{FTE}{FTE}$$

$$=\frac{\frac{CASES}{FTE}\times\frac{PEXP}{TOTEXP}}{\frac{PEXP}{FTE}} \qquad (6)$$

CASES/FTE is a productivity measure that captures the average number of cases processed per person. PEXP/TOTEXP provides a percentage of expenditures that are spent on personnel and capture spending for the current period rather than investment in capital for future output. PEXP/FTE is a market measure that captures local economic market forces and the average compensation per person in the laboratory. Thus,

$$\frac{CASES}{TOTEXP}=\frac{Productivity\times Percentage\ of\ Expenditures\ on\ Personnel}{Average\ Compensation} \qquad (7)$$

This breakdown suggests that the potential gains to output may be affected by several sources and understanding the potential contribution from each source will enhance the predictability of any strategic change.

The efficient frontier

The potential for gain from process improvement depends upon the starting point of a forensic laboratory, both with respect to efficient operation and with respect to caseload and economies of scale. To illustrate, consider DNA Casework using 2016 fiscal year data from Project FORESIGHT (Houck et al. 2009) in Figure 3.

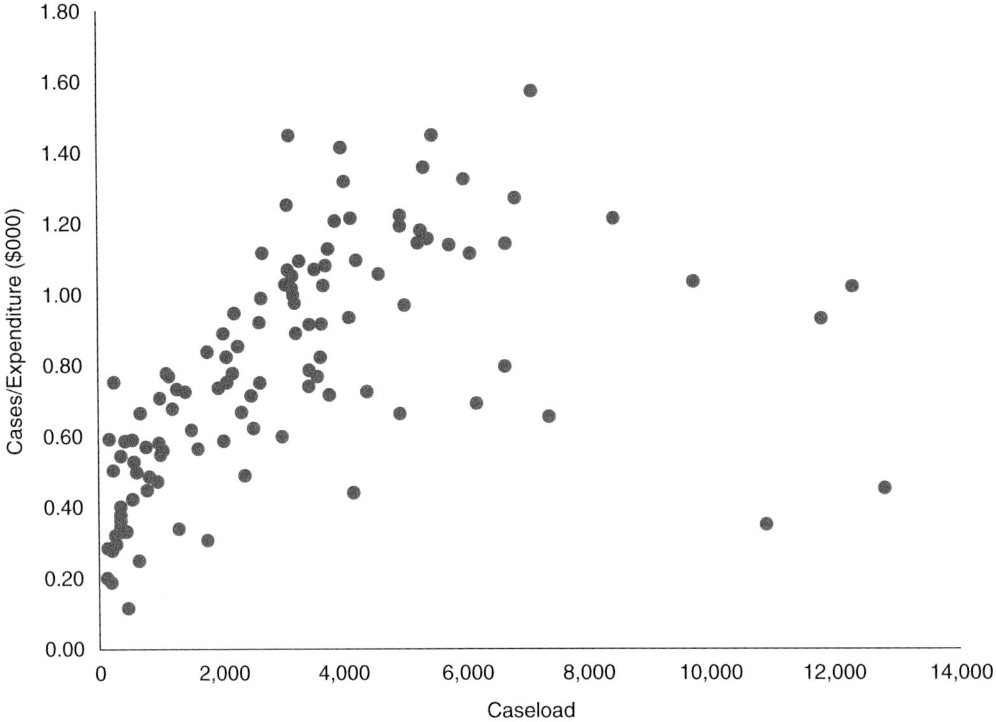

Figure 3 DNA casework—cases/expenditures vs. caseload.

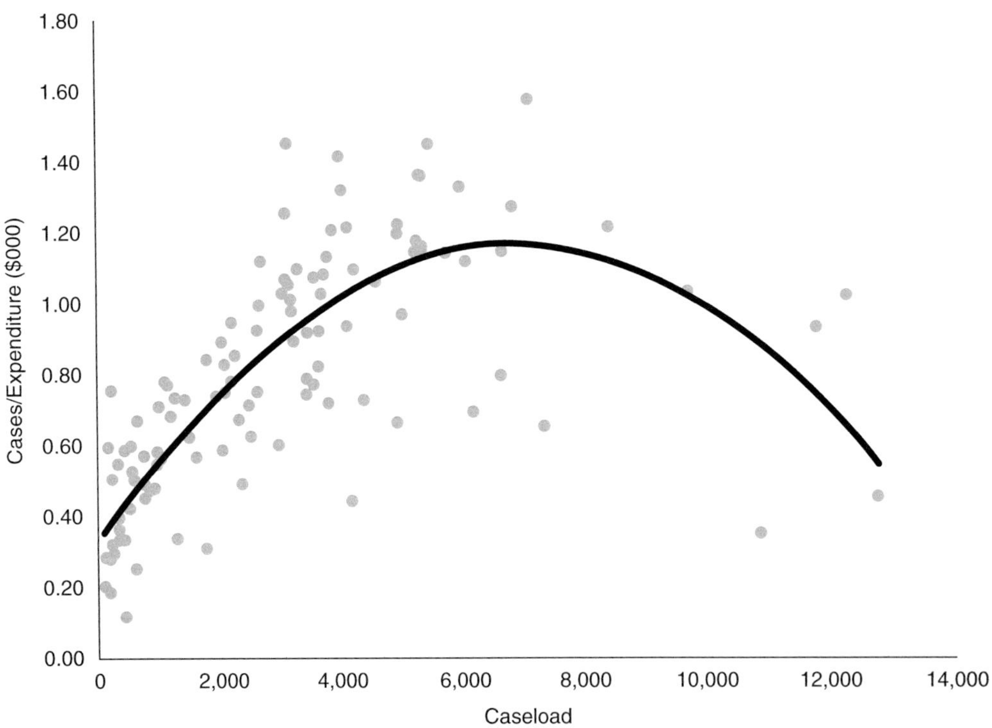

Figure 4 DNA Casework efficient frontier—cases/expenditures vs. caseload.

Economic theory indicates that the relationship between output per expenditure versus size of the operation should be a hump-shaped curve (an inverted U). The inverse of this is a U-shaped economies of scale curve relating average total cost to output level. Perfect economies of scale are associated with the peak of the output per expenditure curve and the trough of the average cost curve (i.e., the most efficient sized laboratory). Figure 4 fits such a Cases per Expenditure curve to the data using a quadratic regression

$$\frac{Cases}{Expenditures} = \beta_0 + \beta_1\, Cases + \beta_2\, Cases^2 + error \tag{8}$$

The significance of this relationship may be found in separation between the efficiency for a given level of activity (caseload) and the fitted curve representing efficiency for that given caseload. The fitted curve offers a sense of the output that can be expected for a given caseload. While a for-profit industry could be expected to see surviving firms produce at an output level associated with the peak of the curve, a jurisdictionally-based enterprise ordinarily is limited to the level of output supported by its jurisdictional boundaries. The efficient frontier maps the associated efficient level of cases per thousand dollars expended for the jurisdiction's caseload. The greater is the distance that a laboratory falls below the efficient frontier, then the greater is the potential gains from a process improvement program. On the other hand, a laboratory that currently performs on or near the efficient frontier is not expected to be able to gain much, if any, from such a program. Thus, before undertaking a program of change, laboratory management should assess what might be gained and whether the cost justifies the benefits from the planned "process improvement" program.

While economic theory proscribes the shape of the efficient frontier, it is possible that the data supporting the estimate of the frontier merely represents a portion of the efficient frontier. That would certainly be the case if jurisdictional restrictions never reach a size consistent with perfect economies of scale. If that were the case, then alternative nonlinear estimations should be considered to capture the downward-sloped portion of the curve. Equation (9) offers on potential estimation using a logarithmic model.

$$\frac{Cases}{Expenditures} = \beta_0 + \beta_1\, LN(Cases) + error \tag{9}$$

Alternatively, a double logarithmic model, as highlighted in equation (10), may also approximate the downward-slopted portion of the curve.

$$LN\left(\frac{Cases}{Expenditures}\right) = \beta_0 + \beta_1 LN(Cases) + error \tag{10}$$

Finally, to offer a point of comparison, a linear model may be considered as highlighted in Equation (11)

$$\frac{Cases}{Expenditures} = \beta_0 + \beta_1 Cases + error \tag{11}$$

Note that the traditional economies of scale interpretation of the performance across levels of output is the inverse of the left-hand-side ratio in Equation (4). Rather than the maximization of Cases per Thousand Dollars Expended, consider the minimization of the inverse Cost/Case

$$\frac{\text{Cost}}{Case} = \alpha_0 + \alpha_1\, Cases + \alpha_2\, Cases^2 + error \tag{8.1}$$

Likewise, the relationships expressed in Equations (9) and (10) may be restated with respect to the inverse ratio, Cost/Case as expressed in Equations (9.1) and (10.1)

$$\frac{\text{Cost}}{Case} = \beta_0 + \beta_1\, LN(Cases) + error \tag{9.1}$$

$$LN\left(\frac{\text{Cost}}{Case}\right) = \beta_0 + \beta_1 LN(Cases) + error \tag{10.1}$$

Since the estimation of (8.1), (9.1), and (10.1) are open to easier interpretation, these equations are estimated in the following section for DNA Casework. The results for additional areas of investigation are presented in the Appendix.

While the quadratic, logarithmic, and double logarithmic relationships in Equations (8), (8.1), (9), (9.1), (10), or (10.1) offer relatively simple measurements of the efficient frontier, more detailed econometric analysis might offer greater insight, particularly at the extreme high and low caseloads (Henrik, Paldam, and Wurtz 2007). The use of these estimations offers individual laboratories a quick response for analysis of their present situation and a sense of potential gains from a process improvement program.

Estimation of the efficient front

To demonstrate how a laboratory might evaluate its present situation and determine whether a process improvement program is worthwhile, we estimate the efficient frontier using 2015–2016 data from Project FORESIGHT. For the 2015–2016 submission year, 139 laboratories worldwide submitted data. The submissions included detailed casework data across nineteen investigative areas and financial data offering a sense of the fully loaded costs to the laboratory including expenditures for personnel, capital, consumables, overhead, and all other direct and indirect costs. Table 1 provides the quartile measures for each of the areas of investigation for the Project FORSIGHT 2015–2016 submissions.

While the data in Table 1 is informative, a more natural metric for interpretaiton is the inverse of this metric, COST/CASE. It is more convenient since the concept of average cost is more easily interpreted and compared by laboratory management within and across activities, by legislatures and other funding bodies, and by stakeholders, in general.

The average cost (COST/CASE) when mapped against the caseload should yield a U-shaped curve with the downward sloped portion of the U highlighing greater economies of scale as caseload increases while the upward-sloped portion shows diseconomies of scale as average cost begins to rise with a rising caseload. Observation of scatter plots of the data for each area of investigation using the 2015–2016 FORESIGHT data suggests this theoretical relationship (e.g., see Figure 3), but not a symmetric curve. The FORESIGHT data reveals that a large portion of laboratories come from jurisdictions with lower caseloads over the declining average cost portion of the efficient frontier. Estimates of the relationship between COST/CASE and caseload using (8.1) provide a good visual fit to the data around the perfect economies of scale (i.e., the central portion of the curve), but show much greater deviation when the observations are compared to the estimated relationship

Table 1 Cases per $1,000 expenditures by investigative area

Area of Investigation	25th percentile	Median	75th percentile
Blood Alcohol	4.34	8.13	11.12
Crime Scene Investigation	0.15	0.25	1.26
Digital evidence - Audio & Video	0.17	0.31	0.64
DNA Casework	0.52	0.75	1.0
DNA Database	6.04	13.50	16.87
Document Examination	0.16	0.29	0.45
Drugs - Controlled Substances	2.36	3.20	4.43
Evidence Screening & Processing	0.56	0.85	2.06
Explosives	0.05	0.08	0.15
Fingerprints	1.03	1.45	2.04
Fire analysis	0.34	0.54	0.95
Firearms and Ballistics	0.33	0.57	1.09
Forensic Pathology	0.33	0.50	0.62
Gun Shot Residue (GSR)	0.28	0.43	0.64
Marks and Impressions	0.11	0.16	0.36
Serology/Biology	0.43	0.68	1.24

at the more extreme caseloads (low and high caseloads). Expressions (9.1) and (10.1), on the other hand, offer much tighter fits to the low and high caseload levels.

Further inspection of the shape of the scatter plot suggests that the lowest caseload portions of the average cost curve are associated with relatively steep curves, but then average cost declines at a declining rate. After bottoming out at perfect economies of scale, the diseconomies enter much more gradually; that is, the upward-sloped portion of the U-shape is much less steep than the downward portion. It appears that greater economies of scale follow a pattern different from the disconomies of scale. This implies that (8.1), (9.1), or (10.1) alone might not define the entire relationship.

Since *a priori*, we do not know the caseload associated with perfect economies of scale, a procedure was followed to let the data inform of the relationship. This is demonstrated for DNA Casework. First, each of the relationships (8.1), (9.1), and (10.1) were estimated over the entire range of submissions for the laboratories reporting DNA Casework for FORESIGHT 2015–2016. The sum of squared deviations were collected for each procedure. Subsequently, the N laboratories were split into two groups, (1, … , x) and (xC1, … , N) for x D 10 through x D N-10 and separate estimations were conducted using (8.1), (9.1), and (10.1) for each subset of the data and the corresponding sum of squared deviations was recorded for each possible paired estimation. The resulting U-shaped "possible efficient frontier" with the lowest sum of squared deviations was selected as the estimated efficient frontier. For DNA Casework, the selected specifications are highlighted in Table 3.

One hundred twenty-three of the FORESIGHT laboratories conducted DNA casework analysis. Table 3 provides the estimation of the efficient frontier for DNA casework. (Note that similar estimations for 15 other areas of investigation are presented in tables located in the Appendix.) The specifications of Equations (8.1), (9.1), and (10.1) that minimized the squared deviations called for the double logarithmic form of Equation (10.1) for the first portion of the efficient frontier and the quadratic for Equation (8.1) the later portion of the frontier.

While the Table 3 estimates are informative, a conversion to the values of the COST/CASE for various caseloads may be more convenient for laboratory use. Table 4 offers the output with some smoothing conducted to connect the two estimate curves from Table 3. [2]

The use of Table 4 (and the related Appendix tables) is relatively straightforward. Suppose, for example, that a laboratory analyzes 700 DNA cases per year at an average cost of $2,000. While Table 2 suggests that this laboratory's average cost approaches the 75th percentile, relative to its caseload, Table 4 indicates that it is performing efficiently; the expected average cost per case for a laboratory with a 700 caseload in DNA Casework is $2,021. Expectations that the laboratory might lower its average cost without a corresponding increase in caseload are unrealistic. Process improvement programs are not the answer. The laboratory is already efficient. It can only expect to be more cost effective if higher volume occurs through another means, such as insourcing or an undesireable higher crime rate that increases the demand for the laboratory's services.

Using either Table 3 or Table 4, a laboratory may substitue its own caseload into the estimated equations from Table 3 or merely referencing their caseload and the efficient average cost from Table 4; the result suggests the efficient level of cost per case for comparison to the laboratory's current position to see potential gains from strategic initiatives.

Table 2 Cost per case by investigative area

Area of Investigation	25th percentile	Median	75th percentile
Blood Alcohol	$90	$123	$230
Crime Scene Investigation	$792	$3,984	$6,765
Digital evidence - Audio & Video	$1,567	$3,188	$5,851
DNA Casework	$943	$1,335	$1,926
DNA Database	$59	$74	$166
Document Examination	$2,213	$3,451	$6,441
Drugs - Controlled Substances	$226	$313	$424
Evidence Screening & Processing	$485	$1,178	$1,777
Explosives	$6,489	$11,940	$20,550
Fingerprints	$490	$692	$975
Fire analysis	$1,054	$1,853	$2,905
Firearms and Ballistics	$920	$1,755	$3,066
Forensic Pathology	$1,602	$2,010	$3,053
Gun Shot Residue (GSR)	$1,560	$2,307	$3,628
Marks and Impressions	$2,751	$6,243	$8,907
Serology/Biology	$810	$1,479	$2,315

Table 3 DNA casework estimated efficient frontier equations

Dependent Variable	Obs.	Constant	Cases	Cases^2	LN (Cases)	F-statistic
Cost/Case	55	1,330.9474	−0.1428	1.41E-05		5.5982
		(355.6750)	(0.1150)	(7.78E-06)		
LN(Cost/Case)	64	9.8396			−0.3402	40.8246
		(0.3611)			(0.0532)	

Table 4 DNA casework efficient frontier values

Caseload	DNA Casework Efficient Cost/ Case	Caseload	DNA Casework Efficient Cost/Case	Caseload	DNA Casework Efficient Cost/ Case
100	$3,917	2,700	$1,277	8,750	$1,164
200	$3,094	2,800	$1,261	9,000	$1,191
300	$2,696	2,900	$1,246	9,250	$1,220
400	$2,444	3,000	$1,232	9,500	$1,251
500	$2,266	3,250	$1,199	9,750	$1,283
600	$2,129	3,500	$1,169	10,000	$1,317
700	$2,021	3,750	$1,142	10,250	$1,353
800	$1,931	4,000	$1,117	10,500	$1,391
900	$1,855	4,250	$1,094	10,750	$1,430
1,000	$1,790	4,500	$1,073	11,000	$1,471
1,100	$1,733	4,750	$1,053	11,250	$1,514
1,200	$1,682	5,000	$1,035	11,500	$1,559
1,300	$1,637	5,250	$1,018	11,750	$1,606
1,400	$1,596	5,500	$1,002	12,000	$1,654
1,500	$1,559	5,750	$987	12,250	$1,704
1,600	$1,525	6,000	$983	12,500	$1,756
1,700	$1,494	6,250	$991	12,750	$1,809
1,800	$1,465	6,500	$1,000	13,000	$1,865
1,900	$1,439	6,750	$1,011	13,250	$1,922
2,000	$1,414	7,000	$1,024	13,500	$1,981
2,100	$1,391	7,250	$1,039	13,750	$2,041
2,200	$1,369	7,500	$1,055	14,000	$2,104
2,300	$1,348	7,750	$1,074	14,250	$2,168
2,400	$1,329	8,000	$1,094	14,500	$2,234
2,500	$1,311	8,250	$1,115	14,750	$2,302
2,600	$1,293	8,500	$1,139	15,000	$2,371

Conclusions

The process improvement program undertaken by the NFI offers a great example of what might be accomplished through the implementation of best practices in management. This article observations and analysis to complement the presentations on the NFI efforts through insights into the economic foundations that provide direction into the forecasting of results from such process improvement (van Asten 2014; Tjin-A-Tsoi 2013). Before adopting similar change processes, forensic laboratory management should first assess the current position of productivity with respect to the efficient frontier, given the laboratory caseload in each area of investigation. Estimates of the efficient frontier are contained in the body of the paper for DNA casework and for other areas of investigation in the Appendix. Comparison to the efficient frontier, either for the output maximization problem or the cost minimization problem, enables management to assess whether an efficiency gap exists and if so, the size of the gap.

Once the size of the gap has been determined with respect to average cost, management can estimate the savings that might be achieved from a process improvement program. That process improvement will be accompanied by a reduction in TAT, which in turn influences the demand for services. Using this information and past experience, the laboratory can anticipate the increase in demand for services as a reaction to its improved productivity. Assessing the future average cost from the efficient frontier, management can then undertake the cost-benefit analysis to determine the worth of the process improvement program.

Notes

1 Office of Justice Programs (2016) suggests that for one area of investigation, DNA Casework, the queuing elasticity of demand falls into the elastic range with a value below -1. That suggests that process improvement efforts that reduce TAT will result in an increase in demand for DNA Casework (i.e., greater submissions) at a higher rate than the reduction in TAT.

2 The smoothing is merely an extension of each of the estimated curves through the overlap.

References

Bedford, K. 2011. Forensic science service provider models – Is there a 'best' option? *Australian Journal of Forensic Sciences* 43 (2–3):147–56. doi:10.1080/00450618.2010.541498.

Castillo-Peinado, L., and L. M. de Castro. 2016. Present and foreseeable future of metabolomics in forensic analysis. *Analytica Chimica Acta* 925:1–15. doi:10.1016/j.aca.2016.04.040.

Henrik, C., M. Paldam, and A. Wurtz. 2007. Public versus private production and economies of scale. *Public Choice* 130 (3–4):311–28.

Houck, M. M. 2017. Strategic leadership through performance management: FORESIGHT as PerformanceStat. *Australian Journal of Forensic Sciences* 1–11. doi:10.1080/00450618.2017.1374457.

Houck, M. M., W. P. McAndrew, M. Porter, and B. Davies. 2015. A Review of Forensic Science Management Literature. *Forensic Science Review* 27 (1):53–68.

Houck, M. M., R. A. Riley, P. J. Speaker, and T. S. Witt. 2009. FORESIGHT: A Business Approach to Improving Forensic Science Services. *Forensic Science Policy & Management: An International Journal* 1 (2):85–95. doi:10.1080/19409040902810723.

Maguire, C. N., M. M. Houck, P. J. Speaker, and R. Williams. 2012a. *A feasible and sustainable model for forensic service delivery in Canada*. Ottawa: Her Majesty the Queen in Right of Canada, Public Safety Canada.

Maguire, C., M. Houck, R. Williams, and P. Speaker. 2012b. Efficiency and the cost effective delivery of forensic science services: In-sourcing, out-sourcing, and privatization. *Forensic Science Policy & Management: An International Journal* 3 (2):62–69. doi:10.1080/19409044.2012.734546.

McAndrew, W. P., and M. G. Roth. 2016. Up from "Arts and Crafts": Division of labor in forensic science laboratories. *Forensic Science Policy & Management: An International Journal* 7(3–4):58–65.

National Research Council of the National Academies of Science (NAS). 2009. *Strengthening forensic science in the United States: A path forward*. Washington, D.C.: The National Academic Press.

Newman, J., D. Dawley, and P. J. Speaker. 2011. Strategic management of forensic laboratory resources: From proect FORESIGHT metrics to the development of action plans. *Forensic Science Policy & Management: An International Journal* 2 (4):164–74. doi:10.1080/19409044.2012.693571.

Office of Justice Programs. 2016, September 9. Looking ahead: The national sexual assault policy symposium. Retrieved from National Institute of Justice: https://nij.gov/topics/law-enforcement/investigations/sexual-assault/Pages/national-sexual-assault-symposium.aspx

Speaker, P. J. 2009a. Key performance indicators and managerial analysis for forensic laboratories. *Forensic Science Policy & Management: An International Journal* 1 (1):32–42. doi:10.1080/19409040802624075.
Speaker, P. J. 2009b. The decomposition of return on investment for forensic laboratories. *Forensic Science Policy & Management: An International Journal* 1 (2):96–102. doi:10.1080/19409040902800260.
Speaker, P. J. 2013. Forensic science service provider models: Data-driven support for better delivery options. *Australian Journal of Forensic Sciences* 45 (4):398–406. doi:10.1080/00450618.2013.773076.
Speaker, P. J., and A. S. Fleming. 2010. Benchmarking and budgeting techniques for improved forensic laboratory management. *Forensic Science Policy & Management: An International Journal* 1 (4):199–208. doi:10.1080/19409044.2010.491894.
Tjin-A-Tsoi, T. 2013. *Trends, challenges and strategy in the forensic science sector*. Gaithersburg, MD: National Institute of Standards and Technology (NIST).
van Asten, A. C. 2014. On the added value of forensic science and grand innovation challenges for the forensic community. *Science and Justice* 54:170–79. doi:10.1016/j.scijus.2013.09.003.

Appendix

The estimation of Equations (8.1), (9.1), and (10.1) was conducted for each area investigation for which sufficient data was available from FORESIGHT 2015–2016. Following the same procedure as described for DNA Casework, Table 5 summarizes the estimations yielding the lowest combined squared deviations for each area of investigation. As with the analysis of DNA casework, a conversion of the equations was conducted to provide the estimated efficient cost per case for various caseloads. Table 6 offers the efficient frontier values for average cost per case across areas of investigation.

Table 5 Efficient frontier equations across areas of investigation

Area of Investigation	Dependent Variable	Obs.	Constant	Cases	Cases^2	LN (Cases)	F-statistic
Blood Alcohol Analysis	LN(Cost/Case)	95	8.0769 (0.3689)			−0.3730 (0.0446)	69.8472
Crime Scene Investigation	LN(Cost/Case)	37	10.6801 (0.5574)			−0.5080 (0.0895)	32.2088
DNA Casework	Cost/Case	55	1,330.9474 (355.6750)	−0.1428 (0.1150)	1.41E-05 (7.78E-06)		5.5982
	LN(Cost/Case)	64	9.8396 (0.3611)			−0.3402 (0.0532)	40.8246
DNA Database	LN(Cost/Case)	51	9.8857 (0.3531)			−0.5833 (0.0393)	220.4712
Document Examination	Cost/Case	67	11,700.4971 (2,219.4849)			−1,467.53 (1,467.5261)	10.6132
Drugs–Controlled Substances	LN(Cost/Case)	95	7.7252 (0.3161)			−0.2268 (0.0364)	38.9251

(*Continued*)

Table 5 (*Continued*) Efficient frontier equations across areas of investigation.

Area of Investigation	Dependent Variable	Obs.	Constant	Cases	Cases^2	LN (Cases)	F-statistic
Explosives Analysis	LN(Cost/ Case)	56	9.8518 (0.2138)	−0.0171 (0.0070)			5.9797
Fingerprint Identification	Cost/Case	50	1,039.8871 (96.9909)	−0.1558 (0.0366)	7.62E-06 (2.16E-06)		13.1753
	LN(Cost/ Case)	37	8.5142 (0.5424)			−0.2575 (0.0826)	9.7032
Fire Analysis	Cost/Case	43	10,939.9131 (1754.2301)	−51.0417 (10.06457)	6.36E-01 (0.0133)		13.7590
	LN(Cost/ Case)	46	9.3391 (0.4726)			−0.3762 (0.1173)	10.2914
Firearms & Ballistics Analysis	LN(Cost/ Case)	82	10.5126 (0.22546)			−0.4604 (0.0331)	193.9070
Gunshot Residue Analysis	Cost/Case	50	3,499.0378 (311.6808)	−7.9045 (1.9200)	6.40E-03 (0.0020)		11.7603
	LN(Cost/ Case)	20	9.5960 (0.2189)			−0.3847 (0.0609)	39.9181
Marks & Impressions Analysis	LN(Cost/ Case)	82	10.3337 (0.2977)			−0.5236 (0.0821)	40.6461
Serology/ Biology Analysis	Cost/Case	14	1,644.8695 (122.7290)	−0.2699 (0.0734)	1.98E-05 (6.76E-06)		7.6499
	LN(Cost/ Case)	80	11.5176 (0.6488)			−0.7079 (0.1261)	31.5317
Toxicology ante mortem Analysis	Cost/Case	50	778.9624 (78.5504)	−0.0841 (0.02666)	5.48E-06 (1.97E-06)		5.3116
	LN(Cost/ Case)	20	11.0139 (1.0015)			−0.5989 (0.1477)	16.4359
Toxicology post mortem Analysis	LN(Cost/ Case)	57	9.2721 (0.2526)			−0.3667 (0.03475)	111.8461
Trace Evidence Analysis	Cost/Case	70	5,305.6021 (338.3597)	−9.1043 (1.6713)	0.0064 (0.0015)		16.5600
	LN(Cost/ Case)	25	10.6627 (0.3439)			−0.4954 (0.10224)	23.5105

Note: Standard errors are shown in parentheses

Table 6 Efficient frontier values for average cost per case across areas of investigation

Caseload	Blood Alcohol Analysis Efficient Cost/Case	Caseload	Blood Alcohol Analysis Efficient Cost/Case	Caseload	Blood Alcohol Analysis Efficient Cost/Case
100	$578	6,500	$122	19,500	$81
200	$446	7,000	$119	20,000	$80
300	$384	7,500	$115	20,500	$79
400	$345	8,000	$113	21,000	$79
500	$317	8,500	$110	21,500	$78
600	$296	9,000	$108	22,000	$77
700	$280	9,500	$106	22,500	$77
800	$266	10,000	$104	23,000	$76
900	$255	10,500	$102	23,500	$75
1,000	$245	11,000	$100	24,000	$75
1,250	$225	11,500	$98	24,500	$74
1,500	$210	12,000	$97	25,000	$74
1,750	$199	12,500	$95	25,500	$73
2,000	$189	13,000	$94	26,000	$73
2,250	$181	13,500	$93	26,500	$72
2,500	$174	14,000	$92	27,000	$72
2,750	$168	14,500	$90	27,500	$71
3,000	$163	15,000	$89	28,000	$71
3,250	$158	15,500	$88	28,500	$70
3,500	$153	16,000	$87	29,000	$70
3,750	$150	16,500	$86	29,500	$69
4,000	$146	17,000	$85	30,000	$69
4,500	$140	17,500	$84	30,500	$68
5,000	$134	18,000	$83	31,000	$68
5,500	$130	18,500	$82	31,500	$68
6,000	$126	19,000	$82	32,000	$67
Caseload	**CSI Efficient Cost/Case**	**Caseload**	**CSI Efficient Cost/Case**	**Caseload**	**CSI Efficient Cost/Case**
5	$19,197	600	$1,687	3,200	$721
10	$13,500	700	$1,560	3,300	$710
15	$10,987	800	$1,457	3,400	$699
20	$9,493	900	$1,373	3,500	$689
25	$8,476	1,000	$1,301	3,600	$679
30	$7,726	1,100	$1,240	3,700	$670
40	$6,676	1,200	$1,186	3,800	$660
50	$5,960	1,300	$1,139	3,900	$652
60	$5,433	1,400	$1,097	4,000	$644
70	$5,024	1,500	$1,059	4,100	$635
80	$4,694	1,600	$1,025	4,200	$628
90	$4,422	1,700	$994	4,300	$620
100	$4,191	1,800	$965	4,400	$613

(*Continued*)

Table 6 (*Continued*) Efficient frontier equations across areas of investigation

125	$3,742	1,900	$939	4,500	$606
150	$3,411	2,000	$915	4,600	$599
175	$3,154	2,100	$893	4,700	$593
200	$2,947	2,200	$872	4,800	$587
225	$2,776	2,300	$852	4,900	$580
250	$2,632	2,400	$834	5,000	$575
275	$2,507	2,500	$817	5,250	$560
300	$2,399	2,600	$801	5,500	$547
325	$2,303	2,700	$786	5,750	$535
350	$2,218	2,800	$771	6,000	$524
375	$2,142	2,900	$758	6,250	$513
400	$2,073	3,000	$745	6,500	$503
500	$1,851	3,100	$732	6,750	$493

Caseload	DNA Casework Efficient Cost/ Case	Caseload	DNA Casework Efficient Cost/ Case	Caseload	DNA Casework Efficient Cost/ Case
100	$3,917	2,700	$1,277	8,750	$1,164
200	$3,094	2,800	$1,261	9,000	$1,191
300	$2,696	2,900	$1,246	9,250	$1,220
400	$2,444	3,000	$1,232	9,500	$1,251
500	$2,266	3,250	$1,199	9,750	$1,283
600	$2,129	3,500	$1,169	10,000	$1,317
700	$2,021	3,750	$1,142	10,250	$1,353
800	$1,931	4,000	$1,117	10,500	$1,391
900	$1,855	4,250	$1,094	10,750	$1,430
1,000	$1,790	4,500	$1,073	11,000	$1,471
Caseload	DNA Casework Efficient Cost/ Case	Caseload	DNA Casework Efficient Cost/ Case	Caseload	DNA Casework Efficient Cost/ Case
1,200	$1,682	5,000	$1,035	11,500	$1,559
1,300	$1,637	5,250	$1,018	11,750	$1,606
1,400	$1,596	5,500	$1,002	12,000	$1,654
1,500	$1,559	5,750	$987	12,250	$1,704
1,600	$1,525	6,000	$983	12,500	$1,756
1,700	$1,494	6,250	$991	12,750	$1,809
1,800	$1,465	6,500	$1,000	13,000	$1,865
1,900	$1,439	6,750	$1,011	13,250	$1,922
2,000	$1,414	7,000	$1,024	13,500	$1,981
2,100	$1,391	7,250	$1,039	13,750	$2,041
2,200	$1,369	7,500	$1,055	14,000	$2,104
2,300	$1,348	7,750	$1,074	14,250	$2,168
2,400	$1,329	8,000	$1,094	14,500	$2,234
2,500	$1,311	8,250	$1,115	14,750	$2,302
2,600	$1,293	8,500	$1,139	15,000	$2,371

(*Continued*)

Table 6 (*Continued*) Efficient frontier equations across areas of investigation

Caseload	DNA Database Efficient Cost/ Case	Caseload	DNA Database Efficient Cost/ Case	Caseload	DNA Database Efficient Cost/ Case
100	$1,339	18,000	$65	44,000	$38
200	$894	19,000	$63	45,000	$38
300	$705	20,000	$61	46,000	$37
400	$596	21,000	$59	47,000	$37
500	$524	22,000	$58	48,000	$37
600	$471	23,000	$56	49,000	$36
700	$430	24,000	$55	50,000	$36
800	$398	25,000	$53	51,000	$35
900	$372	26,000	$52	52,000	$35
1,000	$350	27,000	$51	53,000	$34
2,000	$233	28,000	$50	54,000	$34
3,000	$184	29,000	$49	55,000	$34
4,000	$156	30,000	$48	56,000	$33
5,000	$137	31,000	$47	57,000	$33
6,000	$123	32,000	$46	58,000	$33
7,000	$112	33,000	$45	59,000	$32
8,000	$104	34,000	$45	60,000	$32
9,000	$97	35,000	$44	61,000	$32
10,000	$91	36,000	$43	62,000	$31
11,000	$86	37,000	$43	63,000	$31
12,000	$82	38,000	$42	64,000	$31
13,000	$78	39,000	$41	65,000	$31
14,000	$75	40,000	$41	66,000	$30
15,000	$72	41,000	$40	67,000	$30
16,000	$69	42,000	$40	68,000	$30
17,000	$67	43,000	$39	69,000	$30

Caseload	Document Examination Efficient Cost/Case	Caseload	Document Examination Efficient Cost/ Case	Caseload	Document Examination Efficient Cost/ Case
5	$9,339	170	$4,164	430	$2,802
10	$8,321	180	$4,080	440	$2,768
15	$7,726	190	$4,000	450	$2,735
20	$7,304	200	$3,925	460	$2,703
25	$6,977	210	$3,853	470	$2,671
30	$6,709	220	$3,785	480	$2,640
35	$6,483	230	$3,720	490	$2,610
40	$6,287	240	$3,658	500	$2,580
45	$6,114	250	$3,598	510	$2,551
50	$5,960	260	$3,540	520	$2,523
55	$5,820	270	$3,485	530	$2,495
60	$5,692	280	$3,431	540	$2,467
65	$5,574	290	$3,380	550	$2,441

(*Continued*)

Table 6 (*Continued*) Efficient frontier equations across areas of investigation

70	$5,466	300	$3,330	560	$2,414
75	$5,364	310	$3,282	570	$2,388
80	$5,270	320	$3,235	580	$2,363
85	$5,181	330	$3,190	590	$2,338
90	$5,097	340	$3,146	600	$2,313
95	$5,018	350	$3,104	610	$2,289
100	$4,942	360	$3,062	620	$2,265
110	$4,802	370	$3,022	630	$2,241
Caseload	Document Examination Efficient Cost/ Case	Caseload	Document Examination Efficient Cost/ Case	Caseload	Document Examination Efficient Cost/ Case
120	$4,675	380	$2,983	640	$2,218
130	$4,557	390	$2,945	650	$2,195
140	$4,449	400	$2,908	660	$2,173
150	$4,347	410	$2,872	670	$2,151
160	$4,253	420	$2,836	680	$2,129
Caseload	Drugs-Controlled Substances Analysis Efficient Cost/Case	Caseload	Drugs-Controlled Substances Analysis Efficient Cost/Case	Caseload	Drugs-Controlled Substances Analysis Efficient Cost/Case
100	$797	11,000	$274	37,000	$208
200	$681	12,000	$269	38,000	$207
300	$621	13,000	$264	39,000	$206
400	$582	14,000	$260	40,000	$205
500	$553	15,000	$256	41,000	$204
600	$531	16,000	$252	42,000	$203
700	$513	17,000	$249	43,000	$201
800	$497	18,000	$245	44,000	$200
900	$484	19,000	$242	45,000	$199
1,000	$473	20,000	$240	46,000	$198
1,500	$431	21,000	$237	47,000	$197
2,000	$404	22,000	$235	48,000	$197
2,500	$384	23,000	$232	49,000	$196
3,000	$369	24,000	$230	50,000	$195
3,500	$356	25,000	$228	51,000	$194
4,000	$345	26,000	$226	52,000	$193
4,500	$336	27,000	$224	53,000	$192
5,000	$328	28,000	$222	54,000	$191
5,500	$321	29,000	$220	55,000	$191
6,000	$315	30,000	$219	56,000	$190
6,500	$309	31,000	$217	57,000	$189
7,000	$304	32,000	$215	58,000	$188
7,500	$299	33,000	$214	59,000	$188
8,000	$295	34,000	$212	60,000	$187
9,000	$287	35,000	$211	61,000	$186
10,000	$280	36,000	$210	62,000	$185

(*Continued*)

Table 6 (*Continued*) Efficient frontier equations across areas of investigation

Caseload	Explosives Analysis Efficient Cost/Case	Caseload	Explosives Analysis Efficient Cost/Case	Caseload	Explosives Analysis Efficient Cost/Case
2	$18,352	28	$11,752	54	$7,525
3	$18,040	29	$11,552	55	$7,397
4	$17,733	30	$11,355	56	$7,271
5	$17,432	31	$11,162	57	$7,148
6	$17,136	32	$10,973	58	$7,026
7	$16,844	33	$10,786	59	$6,907
8	$16,558	34	$10,603	60	$6,789
9	$16,277	35	$10,423	61	$6,674
10	$16,000	36	$10,245	62	$6,561
11	$15,728	37	$10,071	63	$6,449
12	$15,461	38	$9,900	64	$6,339
13	$15,198	39	$9,732	65	$6,232
14	$14,939	40	$9,566	66	$6,126
15	$14,685	41	$9,404	67	$6,022
16	$14,436	42	$9,244	68	$5,919
17	$14,190	43	$9,087	69	$5,819
18	$13,949	44	$8,932	70	$5,720
19	$13,712	45	$8,781	72	$5,527
20	$13,479	46	$8,631	74	$5,341
21	$13,250	47	$8,485	76	$5,161
22	$13,025	48	$8,340	78	$4,987
23	$12,803	49	$8,199	80	$4,819
24	$12,586	50	$8,059	82	$4,656
25	$12,372	51	$7,922	84	$4,499
26	$12,161	52	$7,788	86	$4,348
27	$11,955	53	$7,655	88	$4,201

Caseload	Fingerprint Identification Efficient Cost/Case	Caseload	Fingerprint Identification Efficient Cost/Case	Caseload	Fingerprint Identification Efficient Cost/Case
100	$1,523	1,700	$797	7,750	$290
125	$1,438	1,800	$784	8,000	$281

Caseload	Fingerprint Identification Efficient Cost/Case	Caseload	Fingerprint Identification Efficient Cost/Case	Caseload	Fingerprint Identification Efficient Cost/Case
150	$1,372	1,900	$771	8,250	$273
175	$1,319	2,000	$759	8,500	$266
200	$1,274	2,250	$728	8,750	$260
225	$1,236	2,500	$698	9,000	$255
250	$1,203	2,750	$669	9,250	$251
275	$1,174	3,000	$641	9,500	$248
300	$1,148	3,250	$614	9,750	$245

(Continued)

Table 6 (*Continued*) Efficient frontier equations across areas of investigation

350	$1,103	3,500	$588	10,000	$244
400	$1,066	3,750	$563	10,250	$244
450	$1,034	4,000	$539	10,500	$244
500	$1,006	4,250	$515	10,750	$246
550	$982	4,500	$493	11,000	$248
600	$960	4,750	$472	11,500	$256
650	$941	5,000	$451	12,000	$268
700	$923	5,250	$432	12,500	$283
800	$920	5,500	$414	13,000	$303
900	$906	5,750	$396	13,500	$326
1,000	$892	6,000	$379	14,000	$353
1,100	$878	6,250	$364	14,500	$383
1,200	$864	6,500	$349	15,000	$418
1,300	$850	6,750	$336	15,500	$456
1,400	$837	7,000	$323	16,000	$498
1,500	$823	7,250	$311	16,500	$544
1,600	$810	7,500	$300	17,000	$594

Caseload	Fire Analysis Efficient Cost/ Case	Caseload	Fire Analysis Efficient Cost/ Case	Caseload	Fire Analysis Efficient Cost/ Case
10	$4,784	140	$1,773	390	$702
15	$4,107	145	$1,749	400	$694
20	$3,686	150	$1,727	410	$699
25	$3,389	160	$1,686	420	$716
30	$3,164	170	$1,648	430	$746
35	$2,986	180	$1,613	440	$788
40	$2,840	190	$1,580	450	$844
45	$2,717	200	$1,550	460	$912
50	$2,611	210	$1,522	470	$993
55	$2,519	220	$1,495	480	$1,086
60	$2,438	230	$1,471	490	$1,192
65	$2,366	240	$1,447	500	$1,311
70	$2,301	250	$1,425	510	$1,443
75	$2,242	260	$1,404	520	$1,587
80	$2,188	270	$1,385	530	$1,744
85	$2,139	280	$1,366	540	$1,914
90	$2,093	290	$1,348	550	$2,096
95	$2,051	300	$1,331	560	$2,292
100	$2,012	310	$1,226	570	$2,500
105	$1,975	320	$1,116	580	$2,720
110	$1,941	330	$1,019	590	$2,953
115	$1,909	340	$934	600	$3,200
120	$1,878	350	$862	610	$3,458
125	$1,850	360	$803	620	$3,730
130	$1,823	370	$757	630	$4,014
135	$1,797	380	$723	640	$4,311

(*Continued*)

Table 6 (*Continued*) Efficient frontier equations across areas of investigation

Caseload	Firearms & Ballistics Analysis Efficient Cost/ Case	Caseload	Firearms & Ballistics Analysis Efficient Cost/ Case	Caseload	Firearms & Ballistics Analysis Efficient Cost/ Case
10	$12,740	550	$2,013	1,850	$1,152
20	$9,259	600	$1,934	1,900	$1,138
30	$7,682	650	$1,864	1,950	$1,124
40	$6,729	700	$1,802	2,000	$1,111
50	$6,072	750	$1,745	2,100	$1,086
60	$5,583	800	$1,694	2,200	$1,063
70	$5,201	850	$1,648	2,300	$1,042
80	$4,891	900	$1,605	2,400	$1,022
90	$4,633	950	$1,565	2,500	$1,003
100	$4,413	1,000	$1,529	2,600	$985
125	$3,982	1,050	$1,495	2,700	$968
150	$3,662	1,100	$1,463	2,800	$952
175	$3,411	1,150	$1,434	2,900	$936
200	$3,207	1,200	$1,406	3,000	$922
225	$3,038	1,250	$1,380	3,250	$889
250	$2,894	1,300	$1,355	3,500	$859
275	$2,770	1,350	$1,332	3,750	$832
300	$2,661	1,400	$1,309	4,000	$808
325	$2,565	1,450	$1,288	4,250	$785
350	$2,479	1,500	$1,268	4,500	$765
375	$2,401	1,550	$1,249	4,750	$746
400	$2,331	1,600	$1,231	5,000	$729
425	$2,267	1,650	$1,214	5,250	$713
450	$2,208	1,700	$1,197	5,500	$697
475	$2,154	1,750	$1,182	5,750	$683
500	$2,104	1,800	$1,166	6,000	$670
Caseload	**Gunshot Residue Efficient Cost/ Case**	**Caseload**	**Gunshot Residue Efficient Cost/ Case**	**Caseload**	**Gunshot Residue Efficient Cost/ Case**
2	$11,264	110	$2,707	475	$1,198
4	$8,628	120	$2,643	500	$1,157
6	$7,382	130	$2,580	525	$1,125
8	$6,609	140	$2,519	550	$1,100
10	$6,065	150	$2,458	575	$1,084
12	$5,654	160	$2,399	600	$1,075
14	$5,329	170	$2,341	625	$1,075
16	$5,062	180	$2,285	650	$1,083
18	$4,838	190	$2,230	675	$1,099
20	$4,646	200	$2,176	700	$1,122
25	$4,264	210	$2,123	725	$1,154
30	$3,975	220	$2,072	750	$1,194
35	$3,746	230	$2,022	775	$1,242
40	$3,558	240	$1,973	800	$1,298

(Continued)

Table 6 (*Continued*) Efficient frontier equations across areas of investigation

45	$3,401	250	$1,926	825	$1,362
50	$3,266	260	$1,879	850	$1,434
55	$3,148	270	$1,834	875	$1,515
60	$3,048	280	$1,791	900	$1,603
65	$3,012	290	$1,748	925	$1,699
70	$2,977	300	$1,707	950	$1,804
75	$2,942	325	$1,610	975	$1,916
80	$2,908	350	$1,522	1,000	$2,036
85	$2,874	375	$1,441	1,025	$2,165
90	$2,840	400	$1,368	1,050	$2,301
95	$2,806	425	$1,303	1,075	$2,446
100	$2,773	450	$1,246	1,100	$2,599
Caseload	Marks & Impressions Efficient Cost/ Case	Caseload	Marks & Impressions Efficient Cost/ Case	Caseload	Marks & Impressions Efficient Cost/ Case
3	$17,300	81	$3,080	260	$1,672
6	$12,035	84	$3,022	270	$1,640
9	$9,733	87	$2,967	280	$1,609
12	$8,372	90	$2,915	290	$1,580
15	$7,448	95	$2,833	300	$1,552
18	$6,770	100	$2,758	310	$1,525
21	$6,245	105	$2,689	320	$1,500
24	$5,823	110	$2,624	330	$1,476
27	$5,475	115	$2,564	340	$1,453
30	$5,181	120	$2,507	350	$1,431
33	$4,929	125	$2,454	360	$1,410
36	$4,710	130	$2,404	370	$1,390
39	$4,516	135	$2,357	380	$1,371
42	$4,344	140	$2,313	390	$1,353
45	$4,190	145	$2,271	400	$1,335
48	$4,051	150	$2,231	410	$1,318
51	$3,924	160	$2,157	420	$1,301
54	$3,809	170	$2,089	430	$1,285
57	$3,702	180	$2,028	440	$1,270
60	$3,604	190	$1,971	450	$1,255
63	$3,513	200	$1,919	475	$1,220
66	$3,429	210	$1,870	500	$1,188
Caseload	Marks & Impressions Efficient Cost/ Case	Caseload	Marks & Impressions Efficient Cost/ Case	Caseload	Marks & Impressions Efficient Cost/ Case
69	$3,350	220	$1,825	525	$1,158
72	$3,276	230	$1,783	550	$1,130

(*Continued*)

Table 6 (*Continued*) Efficient frontier equations across areas of investigation

75	$3,207	240	$1,744	575	$1,104
78	$3,142	250	$1,707	600	$1,079

Caseload	Serology/ Biology Efficient Cost/ Case	Caseload	Serology/ Biology Efficient Cost/ Case	Caseload	Serology/ Biology Efficient Cost/ Case
10	$19,683	375	$1,546	3,300	$970
20	$12,050	400	$1,540	3,550	$937
30	$9,043	425	$1,534	3,800	$905
40	$7,377	450	$1,527	4,050	$877
50	$6,299	475	$1,521	4,300	$851
60	$5,536	500	$1,515	4,550	$827
70	$4,964	550	$1,502	4,800	$806
80	$4,516	600	$1,490	5,050	$787
90	$4,155	650	$1,478	5,300	$771
100	$3,856	700	$1,466	5,550	$757
110	$3,605	800	$1,442	5,800	$746
120	$3,389	900	$1,418	6,050	$737
130	$3,203	1,000	$1,395	6,550	$727
140	$3,039	1,100	$1,372	7,050	$727
150	$2,894	1,200	$1,350	7,550	$737
160	$2,765	1,300	$1,328	8,050	$757
170	$2,649	1,400	$1,306	8,550	$786
180	$2,544	1,500	$1,285	9,050	$826
190	$2,448	1,600	$1,264	9,550	$875
200	$2,361	1,700	$1,243	10,050	$934
225	$2,172	1,800	$1,223	10,550	$1,004
250	$2,016	2,050	$1,175	11,050	$1,083
275	$1,884	2,300	$1,129	11,550	$1,172
300	$1,772	2,550	$1,086	12,050	$1,271
325	$1,674	2,800	$1,045	13,050	$1,498
350	$1,589	3,050	$1,006	14,050	$1,765

Caseload	Toxicology ante mortem Efficient Cost/Case	Caseload	Toxicology ante mortem Efficient Cost/Case	Caseload	Toxicology ante mortem Efficient Cost/Case
300	$1,994	2,900	$581	6,250	$468
400	$1,678	3,000	$576	6,500	$464
500	$1,468	3,100	$571	6,750	$461
600	$1,316	3,200	$566	7,000	$459
700	$1,200	3,300	$561	7,250	$458
800	$1,108	3,400	$556	7,500	$457
900	$1,032	3,500	$552	7,750	$457
1,000	$969	3,600	$547	8,000	$457
1,100	$916	3,700	$543	8,250	$458

(*Continued*)

Table 6 (*Continued*) Efficient frontier equations across areas of investigation

1,200	$869	3,800	$539	8,500	$460
1,300	$828	3,900	$534	8,750	$463
1,400	$792	4,000	$530	9,000	$466
1,500	$760	4,100	$526	9,250	$470
1,600	$732	4,200	$523	9,500	$475
1,700	$705	4,300	$519	9,750	$480
1,800	$682	4,400	$515	10,000	$486
1,900	$660	4,500	$512	10,250	$493
2,000	$640	4,600	$508	10,500	$500
2,100	$627	4,700	$505	10,750	$509
2,200	$621	4,800	$502	11,000	$517
2,300	$615	4,900	$499	11,500	$537
2,400	$609	5,000	$496	12,000	$559
2,500	$603	5,250	$489	12,500	$584
2,600	$597	5,500	$482	13,000	$612
2,700	$592	5,750	$477	13,500	$643
2,800	$587	6,000	$472	14,000	$676

Caseload	Toxicology ante mortem Efficient Cost/Case	Caseload	Toxicology ante mortem Efficient Cost/Case	Caseload	Toxicology ante mortem Efficient Cost/Case
10	$4,573	950	$861	3,750	$520
20	$3,546	1,000	$845	4,000	$508
30	$3,057	1,050	$830	4,250	$497
40	$2,751	1,100	$816	4,500	$487
50	$2,534	1,150	$803	4,750	$477
60	$2,371	1,200	$790	5,000	$468
70	$2,240	1,250	$779	5,250	$460
80	$2,133	1,300	$767	5,500	$452
90	$2,043	1,350	$757	5,750	$445
100	$1,966	1,400	$747	6,000	$438
150	$1,694	1,450	$737	6,250	$432
200	$1,525	1,500	$728	6,500	$425
250	$1,405	1,600	$711	6,750	$420
300	$1,314	1,700	$696	7,000	$414
350	$1,242	1,800	$681	7,250	$409
400	$1,182	1,900	$668	7,500	$404
450	$1,132	2,000	$655	8,000	$394
500	$1,089	2,100	$644	8,500	$386
550	$1,052	2,200	$633	9,000	$378
600	$1,019	2,300	$623	9,500	$370
650	$990	2,400	$613	10,000	$363
700	$963	2,500	$604	10,500	$357
750	$939	2,750	$583	11,000	$351
800	$917	3,000	$565	12,000	$340
850	$897	3,250	$548	13,000	$330
900	$878	3,500	$534	14,000	$321

(Continued)

Table 6 (*Continued*) Efficient frontier equations across areas of investigation

Caseload	Trace Evidence Analysis Efficient Cost/Case	Caseload	Trace Evidence Analysis Efficient Cost/Case	Caseload	Trace Evidence Analysis Efficient Cost/Case
5	$19,253	135	$4,193	380	$2,766
10	$13,658	140	$4,156	390	$2,724
15	$11,173	145	$4,119	400	$2,683
20	$9,689	150	$4,083	410	$2,643
25	$8,675	160	$4,012	420	$2,605
30	$7,926	170	$3,942	430	$2,568
35	$7,343	180	$3,873	440	$2,533
40	$6,873	190	$3,806	450	$2,498
45	$6,484	200	$3,739	475	$2,418
50	$6,154	210	$3,675	500	$2,346
55	$5,870	220	$3,611	525	$2,281
60	$5,623	230	$3,549	550	$2,225
65	$5,404	240	$3,487	575	$2,176
70	$5,209	250	$3,428	600	$2,136
75	$5,034	260	$3,369	625	$2,103
80	$4,876	270	$3,312	650	$2,079
85	$4,732	280	$3,256	675	$2,062
90	$4,599	290	$3,201	725	$2,053
95	$4,498	300	$3,148	775	$2,075
100	$4,459	310	$3,095	825	$2,130
105	$4,420	320	$3,044	875	$2,216
110	$4,381	330	$2,995	925	$2,334
115	$4,343	340	$2,946	1,025	$2,665
120	$4,305	350	$2,899	1,125	$3,124
125	$4,267	360	$2,853	1,225	$3,710
130	$4,230	370	$2,809	1,325	$4,424

Strategy
Choosing What Not to Do

6

FORESIGHT is based on data, which provides you with information that describes your FSP. In your particular context, such as political, economic, social, technical (PEST), that information gives you knowledge about your FSP and how to manage the strengths, weaknesses, opportunities, and threats (SWOT) that you face. This chapter discusses how to use strategic thinking and planning to better prepare your FSP for success and resilience.

Thinking or Planning?

Strategic thinking versus strategic planning: What's the difference? A good example is Edwin Land. Land was a scientist and inventor who worked for the Polaroid Corporation. When Land worked there (starting in 1937), photographs were taken on film and it took several days to get it developed and any photographs produced. While on vacation with his family in 1943, Land's three-year-old daughter asked why she could not see the picture he had just taken of her right then. Land went for a walk along the beach for about an hour, considering her questions. Within the hour, Land conceived a camera and film process that would forever change Polaroid: The SX-70, the world's first instant camera. The camera was compact, foldable, and used a proprietary film packet that contained all the chemicals necessary to develop a picture in one minute.

Did Land have meeting after meeting, structured sessions with posted ideas and focused discussion, that is, a strategic planning process? No, although he did immediately call his attorney to get the patent paperwork started (smart). Land's vision combined the insightful naivete of his daughter's question with his technical knowledge of engineering and optics. Land said that on his walk, "Within an hour, the camera, the film, and the physical chemistry became so clear" that he could see the whole process. *That's strategic thinking.*

Henry Mintzberg, the *enfant terrible* of management, argues that *strategic planning is analysis*, disassembling a target goal into steps or phases, working out how each step happens, and forecasting the expected outcomes from each phase and the overall goal (and goals are important: "A goal without a plan is just a wish," said Antoine de Saint-Exupéry). On the other hand, *strategic thinking is about synthesis.* Synthesis uses creativity and intuition to generate an integrated perspective of where the organization needs to go. Mintberg says that strategic planners assume that analysis (breaking apart) encompasses synthesis (creative reconstitution). To strategic planners, planning, thinking, and making are all bundled together. Why do they think this? For Mintzberg, it's because they think that prediction is possible and that strategy-making can be formalized. Not true, says Mintzberg, because "[s]trategy is not the consequence of planning, but the opposite: its starting point."

DOI: 10.4324/9780429318528-17

Figure 6.1 Polaroid SX-70 (Wikimedia Commons).

A Journey, Not an Event

Strategic planning is a calculated, structured kind of management. As odd as it may sound, managers who commit, who engage their people on a journey to an envisioned goal, create more enthusiasm and help build excitement than those who only step through a formal strategic planning session every year. By collaborating, all personnel are involved in the process (echoes of Deming's Point #14: "Put everybody in the company to work to accomplish the transformation. The transformation is everybody's job") and help to make it work out. Planners anchor to a goal and calculate what needs to be done to get there, not taking into account the preferences or insights of those involved (those people are *knowledge* workers, after all, so you need their knowledge). Calculated strategies have no value in and of themselves. Just because a goal is mandated does not mean the people will buy into it. They may dig their heels in and resist the change, saying, "That's not how we do it around here." Unless you have a clear focus on strategy and performance, your FSP can develop a pathological culture preventing it from ever succeeding. Public sector employees buy into an organization's mission; they believe. However, this makes nonprofits susceptible to being insular and start to see the agency as an end in itself. Employees stop asking if something services the mission and instead ask if it fits their rules. This inhibits performance by eroding vision and dedication. As Peter Drucker said, "Culture eats strategy for breakfast." Getting your personnel on board as collaborators gets them involved, gives them a sense of ownership, and helps you get at what really needs to change.

It's not the plans that need to change, but *the mental models the employees have about the organization*. Where are the mental models? At the bottom. Think of an iceberg. The majority of an iceberg's mass is below the water. Likewise, the majority of what constitutes an organization's culture is also below what you can see. In system's thinking, the

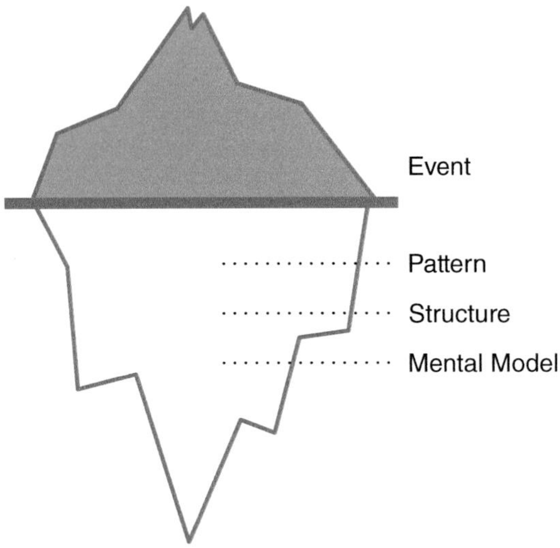

Figure 6.2 The iceberg model for events, patterns, structure, and mental models in organizational behaviors.

iceberg model is used to help communicate the relationship between what is seen and what is “really going on,” the underlying causes of behavior.

The **event** level is what people see happening on a daily basis, their experiences about how things happen. Harry is always late for work, for example, and everyone knows this. The **pattern** level is where the relationship between events is seen as patterns or trends. Everyone notices that Harry is specifically late for work on Mondays, for example. At the **structure** level, the causal connections and interactions between the above patterns are seen. Harry’s lateness on Mondays? Why late and only on Mondays? Maybe his childrens’ school opens an hour later on Mondays or has another Monday errand to run before work. Regardless, the structure level demonstrates how events, organizations, and structures are put together and interconnected.

Finally, at the **mental model** level, the deepest level, people’s most closely held assumptions, values, and beliefs about the organization reside. These models are very difficult to change because they incorporate personal worldviews and beliefs, so it’s not just mental models about work but mental models about everything. Harry knows that he has to drop off his children at school and meet their teacher briefly; that makes him an hour late every Monday. Because Harry thinks that his children come first, showing up a “little late” on Mondays is acceptable to him. How does that correlate with the beliefs and values of others at the organization? Good question. What seems acceptable to Harry may not be to his coworkers. This is culture change, not strategic change. To make sure strategy is not on culture’s menu, the mental models of the people need to change. The planner de Geus said, “The real purpose of effective planning is not to make plans but to change the…mental models that…decision makers carry in their heads.”

Strategic planning operates at the event level; strategic thinking takes in *all of the levels*. Think of strategic planning as building blocks that can be taken apart and reassembled differently. Take this block from here and put it over here—click! Fixed, structured, and limited. Strategic thinking, however, is more like painting with oils: Fluid, changeable, mixable, smearing one color into another.

Figure 6.3 Strategic planning is like modular building blocks, while strategic thinking is more like painting with oils (Duplo Image by Bruno /Germany, Pixabay; Oil paint Image by lee seonghak, Pixabay).

Roll Up Your Sleeves and Get to Work

Real strategic thinkers roll up their sleeves and dig into processes, relationships, interactions, and values. Strategic thinkers don't sit back and "think big thoughts"; instead, they become involved in the daily activities and understand what happens at the other three levels. These details are gathered, considered, abstracted, and then synthesized into a bigger picture. Some things may need to be done away with, edited out of the organization, as it were. Just because it is something that your FSP has always done doesn't mean you must keep doing it. As Drucker notes, "There is nothing so useless as doing efficiently that which should not be done at all"; this is echoed by Mintzberg: "what is not worth doing is not worth doing well."

A Fictitious Example

Assume your FSP has average costs and serves a population of 100,000 people. Below are listed your cost per case and the number of submissions you receive each year:

This means your budget is just under $488,000 for casework. Plot this out, however, and a few things become clear:

There is a small amount of low-level casework that is costing you quite a bit of money:

- Crime scene investigation
- Digital evidence
- Document examinations
- Gunshot residue
- Trace evidence

If you could remove those services, you'd save over $83,000 per year, or 17% of your total budget. This is a simplified example, of course, and other complexities, like what do you do with the staff conducting those services and will your jurisdiction allow you to remove the

	Cost/Case	Cases/100 K
Blood Alcohol	$236	68
Crime Scene Investigation	$3,867	6
Digital evidence	$3,695	6
DNA Casework	$1,488	88
DNA Database	$78	138
Document Examination	$5,783	1
Drugs	$404	235
Fingerprints	$987	32
Fingerprints Database	$532	40
Firearms	$2,272	24
Gun Shot Residue (GSR)	$3,314	4
Marks and Impressions	$8,852	1
Serology	$1,114	40
Trace Evidence	$4,936	2

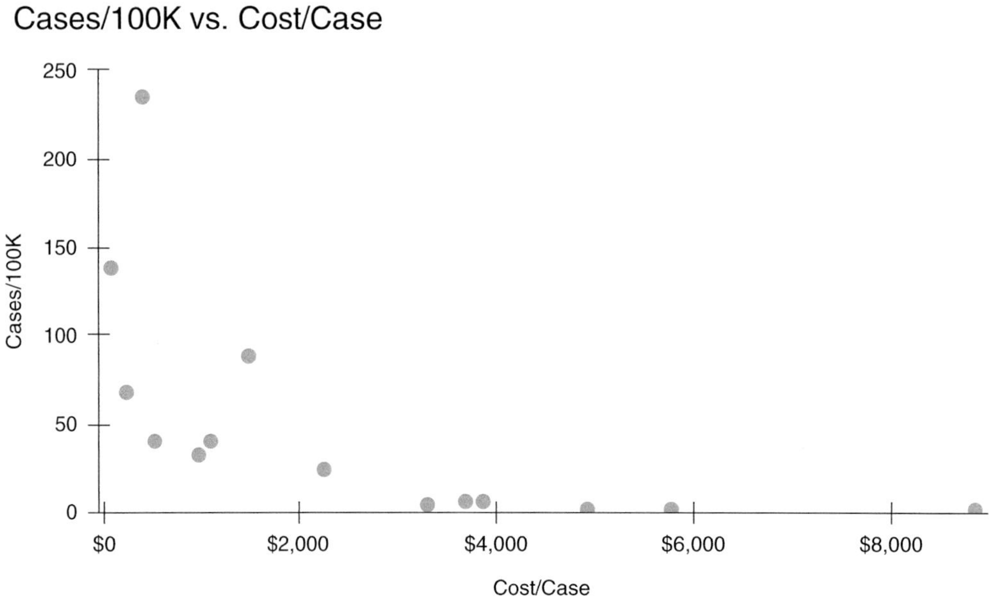

Figure 6.4 A plot of the Cost/Case and Cases/100,000 population served is instructive about expensive low-level casework.

services. But this example demonstrates a key strategic question: *How much good are you doing?* Is it worth it for your FSP to continue offering those services? Would that money be better spent somewhere else, that is, reallocating those resources to do more good? How could you argue for that reallocation? How would you know if you had done more good? This is where strategic thinking meets FORESIGHT. As Churchill said, "However beautiful the strategy, you should occasionally look at the results." Your strategic thinking now has data to inform the strategic plan; the information that comes back to you is feedback, letting you see if the plan worked or not.

A Real Example

The Centre for Forensic Sciences (CFS) in Toronto, Canada, took the challenge of strategic change and benchmarking head-on. They first looked internally at the laboratory's value chain, the sequentially linked set of activities that define the laboratory's purpose: Evidence collection, case triage, sample analysis, and sample reporting. Value chain analysis examines each activity and asks the question, Is this activity adding value. If the answer is no, the manager must find a way to either add value or, as in the fictitious example above, edit it. The CFS effort, then led by Jonathan Newman, asked the following questions about the organization's strengths and weaknesses (the S and W of SWOT):

- Which links in that value chain can be called weaknesses? Can they be fixed?
- What are the closest upstream and downstream linkages to the activity under question?
- Are the linkages identified above partially responsible for deficiencies in the focal link? Is there a potential for synergies between the focal link and links with the other value chains? For example, can any links between the value chains of serology be combined into that of DNA analysis?

After the internal analysis, they looked outwards to the environment external to the laboratory, the political, economic, social, and technological (PEST) factors mentioned earlier. Newman and his colleagues viewed the PEST factors as opportunities and threats (the OT of SWOT) and, once identified, then asked:

- What is the probability of occurrence?
- What is the probability of impact on the laboratory?

The CFS produced a matrix detailing the results of their SWOT analysis:

Working off this, the CFS implemented a two-year strategic management cycle. Significant results came back by the end of the first year and were systematized by the end of the second year. The improvements were seen first in toxicology, both ante-mortem and post-mortem.

	STRENGTHS – S • Quality Assurance • Compensation and Benefits • Scientific staff is young and highly educated • Low employee turnover • Upward trend in productivity • Data collection and monitoring	**WEAKNESSES – W** • Limited Cross training • Some aging equipment • Capacity utilization • Staff Absences • Several analytical processes with low productivity • Unfilled positions
OPPORTUNITIES – O • Federal Government funding initiatives • Technological advances • Funding for outsourcing Toxicology postmortem testing. • Growth in forensic management literature	**SO STRATEGIES** **(examples)** • Use Federal government funding for new equipment to meet increased demands on DNA testing due to changes to the CCC • Improve data quality to ensure accurate accounting of expenditures and casework output.	**WO STRATEGIES** **(examples)** • Use temporary funding for outsourcing testing of a historical backlog of Toxicology postmortem cases. • Improve cost per case and efficiency in Toxicology Ante Mortem through batching of same case type for processing and shifting case related duties.
THREATS – T • Government budget constraints • Privatization of services • Changes to the Criminal Code of Canada (CCC) • Redistribution of the demand for services across investigative areas • Declining demand for some services • Generation Y work expectations	**ST STRATEGIES** **(examples)** • Develop and implement new methods eg. new combined drug assays in Toxicology. • Restructuring of operations in Toxicology Ante Mortem and Post Mortem and streamlining of testing processes.	**WT STRATEGIES** **(examples)** • Refurbish older SEM/EDX instruments to extend life without need for full replacement • Reduce Trace cost per case by using technologists rather than scientists for evidence collection

Figure 6.5 The CFS SWOT analysis (Newman et al., 2011).

Measure	Metric	Toxicology ante mortem	Toxicology post mortem
ROI	Δ Cost/Case	2.02%	−11.98%
Demand	Δ Cases	−4.6%	4.6%
Demand	Δ Items	5.5%	−33.5%
Productivity	Δ Cases/FTE	6.6%	16%
Productivity	Δ Tests/FTE	87%	−21%
Testing Intensity	Δ Tests/Case	75%	−30.42%
Labor	Δ Staff FTE/area	−9%	−9%
Labor expense ratio	Δ Capital Expense/Total Expense	49.54%	40.13%
Turnaround time	Δ Median TAT	−51%	−52%

Figure 6.6 Improvements at the CFS through benchmarking and FORESIGHT.

Newman and co-authors (2011) describe the changes in more detail:

> In the toxicology ante-mortem investigative area, the average cost per case increased (5%). However, the return on this modest increase was a 51% decrease in the median turnaround time required to complete the analysis required and to produce a report. Productivity improved by a 6.6% increase in the number of cases per person and an 87% increase in the number of tests per person. The improvement in the tests per person measure was important for the impaired driving by drugs cases where testing for a wide range of drugs is needed. The increase in cost per case was also due in part to increases in the investment in capital as a contribution to the higher cost. In the toxicology post-mortem investigative area, the average cost per case decreased by almost 12%. Changing processes had a significant impact by reducing testing intensity (tests/case) by over 30% and productivity (cases/FTE) increased by 16%. Most importantly for the customer was the 52% decrease in the median turnaround time required to complete the analysis required and to produce a report.

Productivity, as Cases per FTE, *increased* by 6.6% for ante-mortem and 16% for post-mortem. Turnaround times dropped by half–*half!* That's some pretty exciting news to tell your supervisors and stakeholders.

Another Strategic Example

Lease or buy? That may sound like a pitch heard at a car dealership but it pertains to many products, including laboratory equipment. As demands for FSP services increase and budgets stall out or are reduced, you will inevitably look for ways to cut costs. Improving productivity, cutting services, and putting off buying equipment or hiring people are all ways to reduce costs, if unpleasant. But what if you could get the equipment or instrumentation you need and stay within budget? Will McAndrew and Paul Speaker have shown that there may be as much as 25% of the cost of purchasing equipment (capital investment) that may be shared between a lessee laboratory and an equipment lessor. The lease could include maintenance costs, which are bundled into the regular payment; maintenance contracts are often 10% or more of the instrument's purchase price and must be paid annually. Thus, instead of searching for money to make one big purchase and then paying out maintenance annually, an FSP could merely make the annual budget align with the lease payments. To get into the nitty-gritty, read Will and Paul's paper but suffice it to say, leasing is a solution most FSP managers probably haven't considered. There are always more solutions than problems, as the saying goes.

Go Deeper

De Geus, A.P., 1988. Planning as learning. *Harvard Business Review*, March/April, pp.70–74.

Drucker, P.F., 2010. *Managing the non-profit organization: Principles and practices.* New York, NY: HarperCollins.

Houck, M.M. and Speaker, P.J., 2014. Developing new business models for forensic laboratories. In Strom, K.J. and Hickman, M.J. eds., *Forensic science and the administration of justice: Critical issues and directions* (pp. 221–234). Sage Publications.

Houck, M.M., 2019. Strategic leadership through performance management: FORESIGHT as PerformanceStat. *Australian Journal of Forensic Sciences*, 51(3), pp.348–358.

Kaplan, R.S., 2001. Strategic performance measurement and management in nonprofit organizations. *Nonprofit Manage Leadership*, 11(3), pp.353–370.

Marschall, L.A., 1999. Insisting on the impossible: The life of Edwin Land. *The Sciences*, 39(1), pp.46–47.

McAndrew, W.P., 2012. Are forensic science services club goods? An analysis of the optimal forensic science service delivery model. *Forensic Science Policy & Management: An International Journal*, 3(4), pp.151–158.

McAndrew, W.P., 2012. Is privatization inevitable for forensic science laboratories? *Forensic Science Policy & Management: An International Journal*, 3(1), pp.42–52.

McAndrew, W. and Speaker, P.J., 2012. Expanding budgets via strategic use of leasing. *Forensic Science Policy & Management: An International Journal*, 3(4), pp.169–179.

Mintzberg, H., 1994. The fall and rise of strategic planning. *Harvard Business Review*, 72(1), pp.107–114.

Newman, J., Dawley, D. and Speaker, P.J., 2011. Strategic management of forensic laboratory resources: From project foresight metrics to the development of action plans. *Forensic Science Policy & Management: An International Journal*, 2(4), pp.164–174.

Porter, M., 1985. *Competitive advantage: creating and sustaining superior performance.* New York: The Free Press.

Porter, M., 1996. What is strategy? *Harvard Business Review*, November–December, pp.61–78.

Article: Are Forensic Science Services Club Goods? An Analysis of the Optimal Forensic Science Service Delivery Model

6.1

WILLIAM P. McANDREW
Department of Economics, West Virginia University, Morgantown, West Virginia, USA

Introduction

The institutional frameworks under which forensic science services are provided vary widely among countries and within countries in terms of form and function, with each system having its own unique characteristics. The United States, for example, has multiple frameworks for forensic science service delivery and the provision of those services is primarily offered by government-run forensic laboratory systems (and most of these public laboratories are subsidiaries of policing agencies) at the federal, state, and local level levels of government, while a smaller number of laboratories are privately run (Houck and Siegel 2010). The United Kingdom in contrast has recently become the only leading industrialized nation to be served entirely by private forensic laboratories when in December 2010 the U.K. Home Office closed the FSS, a government-owned forensic laboratory that for a time operated as a virtual monopoly with a 'preferred status' designation until the market was opened to private competition (Adetunji 2011). As a third example, consider the Institute of Environmental Sciences and Research (ESR), which was established under an act of Parliament in New Zealand as a Crown Research Institute (CRI) in 1992 and behaves as a 'pseudo-commercial' entity. Although ESR is incorporated, the controlling owners are other government bodies. However, ESR is required under the CRI Act to provide an adequate rate of return to its owners by charging its only customer, the New Zealand Police, in a fee-for-service business model (Bedford 2011). Fee-forservice is a business model seen in many other industries, but currently most forensic laboratories do not practice charging customers directly. Rather, most forensic laboratories operate on a budget for a given period of time and are expected to provide as much service as budgets permit. To a great extent this diversity among service delivery models in the forensic science industry comes from varied historical legacies, respective government policies, and the respective legal systems within which forensic laboratories operate (Bedford 2011). It is difficult then to determine whether or not there is an overall 'best' universal service delivery model for forensic science since these models are customized to the specific needs of the jurisdictions they serve. However, it is also unlikely that any service delivery model used today

DOI: 10.4324/9780429318528-18

is perfect. Indeed, many service delivery models used today are dynamic with marginal changes continuing to be made in order to improve the delivery of forensic science services (Bedford 2011).

How then might a given laboratory system be improved if change can and is being made? There are many topics that could be analyzed related to possible advancements in the delivery of forensic science services. However, to move forward with broad changes to service delivery models one must first consider an important and often misunderstood debate, which is whether or not forensic science services are public or private goods,[1] the answer to which will ultimately determine the manner in which these services ought to be provided if efficiency is a chosen goal. If the economic nature of forensic science service output is misunderstood, it is unlikely that the institutions under which laboratories operate would form in such a way to allow for the optimal achievement of the specific objectives of forensic laboratories. Thus, by understanding forensic science services under the framework of economics, the goals of fiscal responsibility, optimal output levels, and efficiency can be better pursued.

In the remaining sections of this paper a formal explanation of what constitutes private and public goods in the economic sense will be detailed, followed by an analysis of the degree to which forensic science services do or do not behave as public or private goods. How public or private forensic science services are will determine the optimal level of output and the consumption-sharing framework under which these services ought to be provided to achieve an acceptable level of efficiency.

Are Forensic Science Services Public or Private Goods?

The evolution of the forensic science industry is neither unexpected nor necessarily unwelcomed, as change is a constant factor in many if not all industries. Indeed, how people produce and consume many products today could not be imagined even a few decades ago. Movies can be purchased at home without receiving a physical copy, groceries can be delivered to your door via Internet options, and the Internet allows video chat through Skype and similar products on phones and computers. How industry level change is managed by forensic laboratories is important since failures in this industry do not simply result in product lines shutting down or lower profits, but may cause the criminal justice system to deteriorate. Academicians and practitioners who recognize the importance of managing changes in forensic science delivery models have debated how to better organize the forensic science industry (e.g., McAndrew 2012; Jobling 2011; Roberts 1996). An important question asked concerns about whether private or government efforts best provide forensic science services. However, with different nomenclature between different fields of expertise, communications can struggle between forensic scientists and outside observers.

Forensic science has been described as a 'public good' by several practitioners, legal professionals, and scholars (e.g., Jobling 2011; Daeid 2010; Roberts 1996), the majority of which suggest that forensic science is simply something good for the public. Indeed it would be hard to argue otherwise. In an economic sense the concept of a public good is defined differently from this colloquial meaning, leading to confusion in discussions between forensic scientists and business consultants concerning how to evaluate laboratory performance and ultimately consider strategic change from an economic or efficiency perspective. If forensic science were in reality a public good in the economic sense, there would be a strong argument for its provision by government. McAndrew (2012) however argues that forensic science is not a public good in the economic sense, but remains silent on whether it then is a private good in the economic sense. An important question to ask then is what type of economic good is forensic science? The answer to this will help determine the appropriate institutional framework that would optimally provide forensic science services, where optimal provision implies a maximization of social welfare through the criminal justice system, subject to budget constraints.

In order to answer this question, a basic understanding of how economists categorize private and public goods is needed. Within the “pure theory of public goods” developed by Samuelson (1954) a private good is a good that can be divided out among individuals.[2] An example of a private good could be a glass of wine. Although people can easily share a glass of wine with another individual if they desire to do so, those people cannot share without reducing the total amount of wine that they personally consume. Once parceled out to the ultimate size intended for consumption, only one person could consume the wine. This example demonstrates two characteristics of a private good, which are: it is rival in consumption, meaning one person’s consumption of a private good affects the opportunity for another person to consume the same good, and it is also excludable, meaning that the owner of the good can decide to share or not to share the private good they own.

Because only one person ultimately consumes private goods, the appropriate institutional framework for determining total production size for society is a market where buyers and sellers voluntarily come together and trade goods and services at mutually agree upon prices and quantities that make both parties better off. As markets reveal information about social value—the value of a good or service when its alternative uses are considered—by allowing a multitude of consumers and producers to determine an equilibrium price and quantity for a specific private good, social value or happiness can be maximized. This is achieved when markets permit consumers to individually demand the specific quantity of the private goods and services they desire at each possible price. Individuals make decisions to maximize satisfaction by comparing the relative price of one good versus all other goods. Prices are information and a way to deal with the economic problem of scarcity, or that there are unlimited human wants but limited resources with which to fulfill these wants. In this way prices act as a rationing mechanism, which in efficient markets will move goods to consumers who value them most. As an explicit example, consider goods A, B, and C with prices of P_A, P_B, and P_C, respectively. Assume that the additional happiness, which economists term marginal utility (MU), from consuming one more unit of each of these goods for a given person is MU_A for A, MU_B for B, and MU_C for C. Market exchange continues to occur until the market comes to rest at a point where producers have realized all potential gains in the form of profits and consumers have realized all potential gains in satisfaction from trade. At that point we have market equilibrium. At that market equilibrium, $MU_A/P_A = MU_B/P_B = MU_C/P_C$, which implies that the marginal utility per dollar spent is the same across all products, giving no reason for the consumer to purchase more or less of any good. If there were a situation where $MU_A/P_A > MU_B/P_B = MU_C/P_C$, it would be in the interest of the consumer to purchase more of good A as good A is providing a higher marginal utility per dollar spent. If the consumer then acts on the information of market-determined prices, they increase their individual welfare, which contributes a portion toward total societal welfare. It is important to note that this behavior is only possible when individuals can decide the quantity of their desired consumption of a good after reacting to its market prices and the expected marginal utility of its consumption relative to all other goods.

Governments can and often do provide many private goods (e.g., health care and housing; Besley and Coate 1991); however, market solutions are frequently ignored by governments when determining the distribution of scarce resources. When market prices are not used to ration scarce resources and thereby to determine the quantity consumed, the provision of private goods will result in quantities that are either too high or too low for optimal social welfare to be achieved. This has been the case for DNA analysis in the United States. Although three quarters of the 2011 DNA funding received by the National Institute of Justice from Congress has gone toward backlog reduction (Sternstein 2013), it is unlikely that laboratories will operate at a desirable level of output if budgets rather than prices are continued to be used in determining production. Without market prices that serve as information on the value of a good relative to other goods, the marginal comparison of satisfaction per dollar spent is not possible. Furthermore, even if market prices were somehow known in the absence of the actual markets, governments would still be unlikely to know or provide for the individualized preferences that suit all consumers' desires.

A public good according to Samuelson (1954) differs from a private good in that a public good can be shared in consumption without causing a reduction in the amount consumed by any individual.[3] An often used example of a public good is national defense. National defense is not parceled out to individuals but rather a set amount of the good is provided and consumed jointly by all regardless of individual preferences. The consumption by one person of national defense is not lessened when additional consumers begin to share and consume the same good, and furthermore it is impractical to prevent the consumption of national defense by anyone even if they do not pay. As specific example, consider a missile defense shield covering the 48 continental states. This shield can defend 300,000,000 people just as easily as 350,000,000 without a change in defense spending or a change in the quantity or quality of defense each individual consumes. This example demonstrates two characteristics of a public good that contrasts with the characteristics of a private good. That is, first, public goods are *non-rival in consumption*, meaning one person's consumption of a public good cannot affect the opportunity for another person to consume that same good, and second, a public good is *non-excludable*, meaning that no one can be denied access to consumption of a public good. Theoretically this implies that the optimal sharing arrangement for the consumption of purely public goods would be infinitely large, or at least the entire planet's population.

In general, markets will underprovide public goods when the individual consumption of these goods is not dependent upon the individual's contribution toward the payment for the public good. When people stop paying for their consumption, preferences are not revealed and the information on the value of these goods, normally revealed through market prices, is lost to society. This is known as the free rider problem, where individuals pay too little or not at all in a market setting for the public good with the expectation that others will pay for them, resulting in a socially suboptimal provision of the public good. The market failure from free riding may create the need for government involvement if it can be anticipated that a government can reach a point closer to the socially optimal provision of the public good at an acceptable level of expenditure. There are historical examples of the successful private provision of public goods, such as the lighthouses privately provided in Britain during the 19th century (Coase 1974), but these historical examples are the exceptions rather than the norm. The exceptions do however illustrate an important point. Public goods and government-provided goods are not synonymous as governments and private companies can provide both public and private goods. The institutional efficiencies of providing these goods by public and private entities may be different for reasons already discussed. In Table 1, a comparative description of public and private goods is illustrated, where the distinctions between these two types of goods lies in how rivalrous and excludable they are.

So what type of economic good or service is forensic science, public or private? Is forensic science a shared sense of safety as with defense spending or is it something else? These questions are not easily answered. First, forensic science is not simply one good or service, but a multitude of goods and services including reports from analysis of test kits, autopsies, expert testimony in courtrooms, etc., and more broadly forensic science provides the service of bringing safety and justice to society. The degree to which forensic science is public or private then will depend upon the specific product to be considered. Second, determining whether a good is public or private largely depends upon its consumption arrangement, specifically whether the good is non-rival or non-excludable in consumption.[4] Determining how forensic science is consumed will require defining the consumer, which is awkward for this industry. The consumer can be considered the general public, a police force, the prosecution, or some other option. How these goods are shared and thus whether or not the good is public or private will then depend upon which definition of consumer is accepted.

In some cases it is clear that forensic science output behaves as a private good. The production of autopsy reports, for example, require resources that at the time of autopsy cannot be easily shared with other jurisdictions. The technicians, pathologist, and equipment are able to produce only so many autopsy reports in a given time period for a given jurisdiction with quality standards remaining the same. In other words, one autopsy technician cannot work in two jurisdictions on two separate cases at the same moment in time. Any increase in the number of autopsy reports consumed by whoever is the defined consumer would therefore require additional resources and spending as autopsy reports are indeed rival and excludable in consumption. Autopsy reports, however, contribute to the broader service of safety and criminal justice, which likely do not behave as a private good. A small marginal increase in consumption of safety and criminal justice by the addition of one person will not measurably affect the level of consumption by the original members of a jurisdiction, and furthermore consumption of safety and justice would be difficult to deny to those living in the jurisdiction served. Nonetheless, very large changes in population would require additional resources to retain the level and quality of forensic science services, suggesting the broadly defined good of safety and justice may not be a purely public good either.

Table 1 Private versus public goods

Private Goods	Public Goods
Rival: One's consumption of goods and services affects the availability to others; individual preferences determine output and consumption.	Non-Rival: Shared consumption that does not appreciably change with the addition of consumers; individual preferences are not considered.
Excludable: The owner of the good or service can decide not to share.	Non-excludable: Access to service is universal.
Examples: Glass of wine or finite DNA analysis capacity.	Example: National defense.

Unlike a missile defense shield, which is not affected by population growth in terms of consumption quantity or quality, larger populations do lead to a greater number of crimes and will require increased spending on forensic science. Some benefits from forensic science such as solving one crime by gathering evidence from a second, removing perpetrators from the streets, and reducing the rate of repeat offenders can be perceived as having characteristics of public goods in that it would be difficult to exclude someone from these benefits and the general level of increased safety can be shared in consumption. However, unlike national defense, which is global in nature, crime is often concentrated to regions and thus the spending and benefits of forensic science services will also be concentrated similarly. Deutsch et al. (1984) explains that, although there are certainly spatial spillovers in crime, such as a gang member from Detroit being arrested in Pittsburgh and thus leading to solved crimes in both cities, the average criminal tends to commit crimes near their residence. The degree to which criminals operate regionally rather than nationally or even globally will determine how relatively private forensic science is as an economic good and how concentrated the benefits of a laboratory are on the jurisdiction it serves.

So what then can be said about forensic science, how it behaves as an economic good, and thus what the optimal framework under which it should be provided? It is likely that forensic science services are not purely public or private goods but some combination of the two. Buchanan (1965) in his theory of clubs states there is a whole spectrum of possibilities between those goods that are purely public or private, suggesting that there are consumption ownership options other than one person or an infinitely large sharing group consuming a good provided by either the private market or government respectively, as seen in Figure 1. Goods located toward the left of the spectrum in Figure 1 are private, with consumption decisions solely resting on individual consumers and should thus be provided by markets that best match consumer preferences. As one moves from left to right on the spectrum (from pure private goods toward pure public goods), the goods represented are consumed through larger sharing arrangements such as community pools, homeowners association services, etc., where consumption decisions are now made by groups or clubs. Goods scaled along a point further toward the right will be consumed by even larger sharing arrangements, eventually approaching and becoming governments. The point on the spectrum where a club good should be provided by some level of government is not explicitly clear, but theoretically exists (Holcombe 1994). With large fixed costs to begin operation, forensic science service output must have some "publicness," where the optimal sharing groups or clubs are larger than one person, but exactly how large these

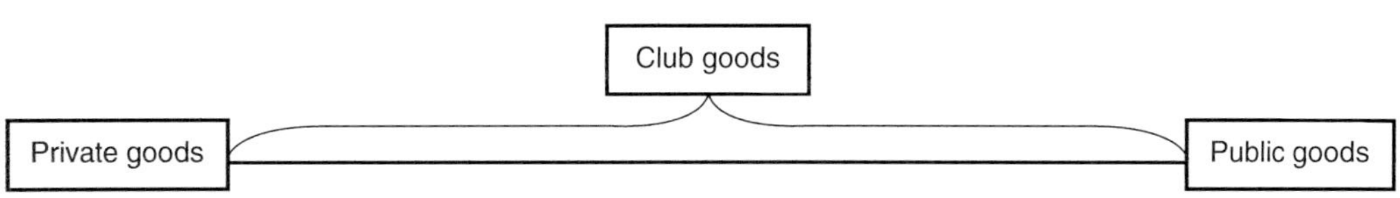

Optimal sharing: unity	Optimal sharing: x where $1<x<\infty$	Optimal sharing: infinity
Optimal framework: markets	Optimal framework: unknown	Optimal framework: government

Figure 1 The spectrum of possibilities between those goods that are purely public or private is continuous. Points that lie toward the left of the spectrum have more qualities of private goods than public goods. Points that lie toward the right of the spectrum have more qualities of public goods than private goods. Goods toward the center of the spectrum, however, are not clearly public or private and are termed *club goods*.

clubs should grow and where they are on the spectrum between private and public goods is not known *a priori*. For example, an expensive mass spectrometer, which is used to test for drugs in drug-related crimes, cannot be realistically employed to benefit only one person, but also cannot be realistically employed to benefit the entire world. Somewhere in between one person consumption and global consumption rests an optimal framework for the provision of this service. The degree of publicness of forensic science services, once determined, will help determine the optimal institutional framework under which these services ought to be provided and thus the rationing mechanism that will likely be put in place to deal with scarcity. The two available options for rationing mechanisms used in the industry today are prices or a queuing system, although other options are possible.

To answer the question of how public or private forensic science services are empirically, data must be analyzed to see how output responds to the size of the sharing arrangements we see in reality. Holcombe and Sobel (1995) and King (2007) each analyze empirically the degree to which the government provided goods of state legislation and K-12 education, respectively, behave as private or public goods to determine a coefficient number representing where the good lies on the spectrum seen in Figure 1. A similar study on the forensic science industry can be done with the proper data in order to determine the optimal framework to provide these services. Such a study will enhance the interpretation of where and when governments or private companies ought to be involved; it will also help determine the optimal sharing arrangement or jurisdiction size to provide these goods. Indeed the most important question to be answered in the theory of club is, *what is the optimal size of the consumption-sharing arrangement* (Buchanan 1965)?

Fiscal Federalism and Forensic Science

The theory of fiscal federalism developed by Oates (1972), Musgrave (1959), Buchanan (1965), and Tiebout (1965), among others, explores two main questions. First, what is the optimal club or jurisdiction size to provide a given level of goods or services, and second, what is the optimal level of output to be provided. In answering these questions a tradeoff must be considered. Ceteris paribus, consumers prefer the government of a jurisdiction to be smaller as smaller governments tend to be more responsive to the needs of voters and are better able to match expenditures to the preferences of voters. However, larger jurisdictional sizes and thus governments experience economies of scale, or lower average total costs for the output provided. The optimal jurisdiction size for providing any good, including forensic science service output, occurs, according to Buchanan (1965), at the point where the difference between the benefit and the cost of the sharing arrangement is maximized and individuals are therefore happiest. In Figure 3 the cost function represents the per person cost of providing forensic science services at different levels of sharing, and the benefit function represents the individual level benefit of these services at different levels of sharing as well. The net cost and benefit is an individual's level of utility or happiness. When there is only one individual, the cost of providing forensic science services would be exorbitantly high due to large fixed costs in providing forensic services. The total expenses are so high that the cost to the individual is more than the benefit received and they will choose not to provide forensic science services as depicted at point X. As more individuals are added to the sharing arrangement the cost per person from providing these forensic services will decline. The benefit function will also increase as access to previously unaffordable resources becomes available to all individuals in the club under a larger sharing arrangement. Setting up a forensic laboratory is costly, but as jurisdictions grow or collaborate with other jurisdictions to create cross-jurisdictional laboratories, more and perhaps better technology can be afforded that will increase the benefit to all individuals. Eventually if this sharing arrangement continues to grow the benefit function will peak and then decrease as congestion in consumption sets in, where further sharing no longer adds additional benefits to individuals. Congestion can happen when a growing sharing arrangement becomes too large causing bureaucratic inefficiencies. Alternatively, congestion occurs if individuals (now with a smaller influence on the group) no longer consume close to their preferences when the group as a whole has decided upon some level, type, or quality of consumption that is undesirable to the individual. This situation is depicted at point Y where benefits are decreasing and the costs increasing.

Consider the simple example given in Maguire et al. (2012) where a small county laboratory has a high level of average total cost due to large fixed costs and low levels of submissions. The county may consider outsourcing their DNA casework to a larger state-run laboratory in order to take advantage of the state's economies of scale. The cost per individual in the county for DNA casework will certainly fall as the sharing arrangement increases with the state laboratory. However the county's influence on the laboratory performing this work will decrease. If state-level DNA casework submissions are proffered a priority over county-level submissions, the benefits of average cost reduction for the county may be smaller than the increased cost of unfair queuing and the corresponding increased wait time that results. In this case the sharing arrangement for individuals in the small county should be smaller than the state-level laboratory system.

The point N* in Figure 2 represents the optimal club or sharing arrangement size where the difference between the benefit function and cost function is maximized. The determination of N* for a laboratory jurisdiction would be complex, but would be determined

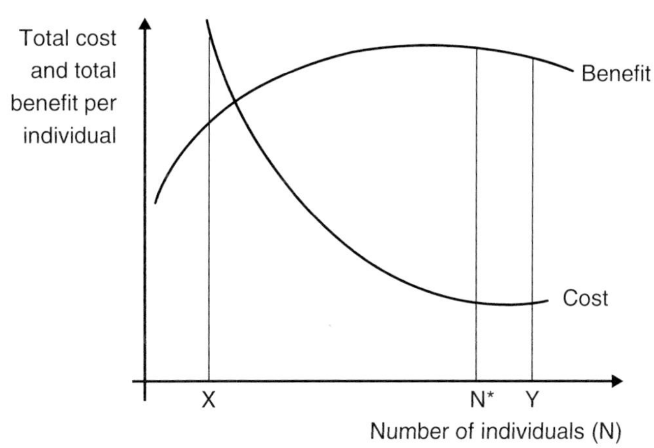

Figure 2 The optimal jurisdiction size as measured by population, N, served by a forensic laboratory is determined by the tradeoffs of costs and benefits. When there are relatively few people served by a laboratory, such as at point X, the minimum fixed cost of equipment and personnel will be very large relative to the benefit per person. As more individuals are added to the jurisdiction, the cost per person drops and benefits rises up to the optimal point N*, where the difference between benefit and cost is maximized. Further increases in population served past N* will lead to congestion in consumption with a decrease in benefits and an increase in costs per person.

in part by expected wait times, control of resources, burden of cost, shifts in technology, exogenous changes in capital, etc. All of these factors change dynamically, and so N* is not a static goal to be achieved but rather something that will need readjusted as the world changes. Exactly where N* is located today in reality will determine the optimal jurisdiction size to provide forensic services; that is, whether these services ought to be provided at the local, state, or federal level, or provided by some other options including outsourcing services, insourcing, or even privatizing if these options are better able to operate near the optimum N*.

The second point from the theory of fiscal federalism seeks to determine the optimal level of output of a good, and is parallel to determining the optimal club or jurisdiction size. Within the framework of Buchanan's theory of clubs, finding the optimal level of forensic output is more practical than determining the optimal jurisdiction size due to the type of data that is available, such as that found in project FORESIGHT (Houck et al. 2009). The benefit and cost to individuals from larger or smaller sharing arrangements is inherently difficult to measure when preferences are subjective and not easily comparable. It is simply impossible to measure how happy or satisfied people are in an objective and comparable way (Arrow 1951). Output levels and their costs conversely are objective measures that forensic laboratories can record explicitly as numbers.

Speaker (2013) and Witt and Speaker (2012) demonstrate that when the level of forensic science service output, such as DNA casework analysis, is mapped against the average total cost from providing the output, a U-shaped function results as depicted in Figure 3. This average total cost function can help laboratory directors determine whether bigger is better, or whether more or less output would lower average total cost. If a forensic laboratory that initially operated at a low level of output began to produce more, perhaps by insourcing from bordering jurisdictions, it would begin to experience economies of scale, or a decrease in average total cost. Reasons for this could come from increased use of idle laboratory resources, better management, or specialization as forensic scientists perform more

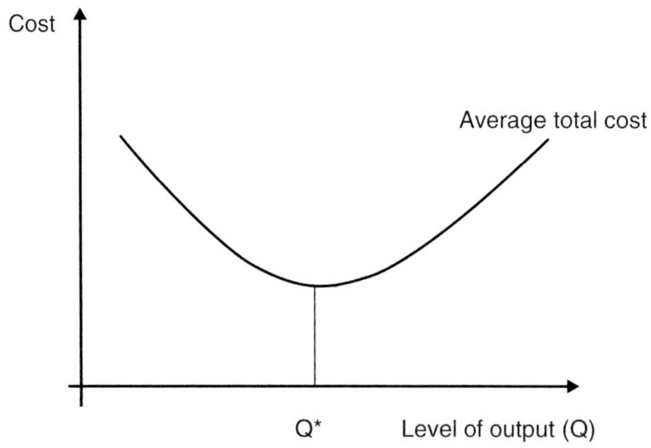

Figure 3 Q* represents the point of perfect economies of scale where average cost per sample processed is minimized, a competitive advantage point for a private laboratory. Laboratories operating to the left of Q* can reduce their average costs by increasing capacity and output, thus lower average total costs through economies of scale by moving rightward on the curve. However, rightward movements past Q* will result in diseconomies of scale as congestion in production eventually sets in.

specific tasks to improve efficiency given a larger output level. Bigger will not always be better, and eventually as output continues to increase diseconomies of scale, or an increase in average total cost, will result. This can occur as bureaucratic inefficiencies result from the growing complexity of managing larger output levels, or from the congestion of limited laboratory resources.

Operating at the minimum point on the average total cost function, thus achieving perfect economies of scale, is desirable if cost-efficiency is a goal. Although government-operated laboratories can operate at this point, it would likely be coincidental if achieved as a result of the normally suboptimal sharing arrangements that government-run laboratories operate under. If sharing arrangements are largely determined by political borders rather than economic efficiency and competition, one would expect an output level that is either too large or too small (Poole 1983; Benson 1998; McAndrew 2012). Unlike government laboratories, private laboratories do not operate under any political jurisdiction and therefore output is not related to a politically created sharing arrangement. Private laboratories compete for business, and one important way they compete is through cost reduction. This is why private companies almost always operate at the minimum point on the average total cost function, as any deviation to the left or right would decrease their competitiveness. If government laboratories can copy the behavior of private firms through insourcing, outsourcing, or contracting out of services to private firms in order to achieve an optimal sharing arrangement, they will achieve cost savings. This cost savings could then be used to help in other areas, such as to deal with backlogs.

Conclusion

There is scarcity in this world, meaning that there are a limited amount of resources with which to provide all goods, including forensic science services. With forensic laboratory budgets being cut in an era of austerity for many countries, understanding where to find efficiency gains is of growing importance for forensic laboratories. The theory of clubs developed by Buchanan (1965) suggests that creating the "best" forensic laboratory system is not simply about having the best equipment, smartest scientists, and largest budgets, but should also examine the optimal jurisdiction size and the optimal level of output of forensic services to be provided to these jurisdictions. Laboratories that are "best" will efficiently provide the quantity of forensic science output at the appropriate quality standards society desires. This will be determined by how public or private forensic science goods are in an economic sense. As governments continue to 'right-size' the jurisdictions that forensic laboratories serve and the output they provide, considerations for contracting with other governments, or even privatization, may improve efficiencies from economies of scale.

Endnotes

1 For simplicity the word "good" is used generally throughout this paper to mean anything that is produced by a firm. This could include physical goods such as a drug test kit or nonphysical services such as the analysis of evidence by a forensic scientist. The economic theory used to model goods and services is identical, so no information is lost and any conclusions will be the same.

2 More explicitly, consider the purely private goods (X_1,..., X_n), where the private good X_j can be divided out to persons i = 1, 2,..., s where every individual i will receive X^i and also $X_j = L^s X^i$. For good X_j, the optimal arrangement for consuming the good is one person, where any increased consumption from person i subtracts from the X_j available to the rest of society.
3 More explicitly, consider the purely public goods (X_{n+1},..., X_{n+m}). If i consumes X^i, the total original amount of X_{n+j} is still available to consume by other individuals, so that $X^i = X_{n+j}$ for all individuals. Theoretically, the optimal sharing arrangement for these goods would be an infinitely large number of sharing members.
4 Where non-rival implies that the consumption of the good does not affect any other persons consumption of that same good and non-excludable implies that no one can be denied access to consuming the good.

References

Adetunji, J. 2011. Forensic science regulator could gain statutory powers. *The Guardian*. Accessed April 28, 2013, http://www. guardian.co.uk/public-leaders-network/2011/apr/28/forensicscience-regulator-statutory-powers

Arrow, K.J. 1951. *Social choice and individual values*. New York: Wiley.

Bedford, K. 2011. Forensic science service provider models: Is there a 'best' option? *Australian Journal of Forensic Sciences* 43:147–156.

Benson, B. 1998. *To serve and protect: Privatization and community in criminal justice*. New York: New York University Press.

Besley, T., and S. Coate. 1991. Public provision of private goods and the redistribution of income. *The American Economic Review* 81(4): 979–984.

Buchanan, J.M. 1965. An economic theory of clubs. *Economica* 32(125): 1–14.

Coase, R.H. 1974. The lighthouse in economics. *The Journal of Law and Economics* 17(2): 357–376.

Daeid, N.N. 2010. *Fifty years of forensic science: A commentary*. West Sussex, UK: John Wiley and Sons Ltd.

Deutsch, J., S. Hakim, and J. Weinblatt. 1984. Interjurisdictional criminal mobility: A theoretical perspective. *Urban Studies* 21(4): 451–458.

Holcombe, R.G., and R.S. Sobel. 1995. Empirical evidence on the publicness of state legislative activities. *Public Choice* 83: 47–58.

Holcombe, R. 1994. Chapter 5: The distinction between clubs and governments. In *The Economic Foundations of Government*, 73–91. New York: New York University Press.

Houck, M.M., and J.A. Siegel. 2010. *Fundamentals of forensic science*. Waltham, MA: Academic Press.

Houck, M.M., R.A. Riley, P.J. Speaker, and T.S. Witt. 2009. FORESIGHT: A business approach to improving forensic. *Forensic Science Policy & Management* 1:85–95.

Jobling, M.A. 2011. Elementary, my dear Cameron. *Investigative Genetics* 2:1–2.

King, K.A. 2007. Do spillover benefits create a market inefficiency in K-12 public education? *Cato Journal* 27(3): 447–458.

Maguire, C., M.M. Houck, R. Williams, and P.J. Speaker. 2012. Efficiency and the cost-effective delivery of forensic science services: Insourcing, outsourcing, and privatization. *Forensic Science Policy & Management* 3:62–69.

McAndrew, W.P. 2012. Is privatization inevitable for forensic science laboratories? *Forensic Science Policy & Management* 3:42–52.

Musgrave, R.A. 1959. *The theory of public finance*. New York: McGraw-Hill.

Oates, W.E. 1972. *Fiscal federalism*. New York: Harcourt Brace Jovanovich.

Poole, R. 1983. Why not contract policing? *Reason* 14:10.

Roberts, P. 1996. What price a free market in forensic science services? The organization and regulation of science in the criminal process. *The British Journal of Criminology* 36(1): 37–60.

Samuelson, P.A. 1954. The pure theory of public expenditure. *The Review of Economics and Statistics* 36(4): 387–389.
Speaker, P.J. 2013. Forensic science service provider models: Data-driven support for better delivery options. *Australian Journal of Forensic Sciences* 45(2): 1–9.
Sternstein, A. 2013. DNA backlog stymies development of 'CSI'-like inventions. *Nextgov*. Accessed February 8, 2013, https://www.nextgov.com/emerging-tech/2013/02/dna-backlog-stymies-development-csi-inventions/61187/.
Tiebout, C.M. 1965. A pure theory of local expenditures. *The Journal of Political Economy* 64:416–424.
Witt, T.S., and P.J. Speaker. 2012. The power of information. *Forensic Magazine*, April 10, 1–5.

Article: Is Privatization Inevitable for Forensic Science Laboratories?

6.2

WILLIAM P. McANDREW
West Virginia University, Morgantown, West Virigina, USA

Introduction

The organizational structure of the forensic science industry and the structure of individual forensic science laboratories vary widely between different countries, regions, and jurisdictions (Houck & Siegel 2010). A historical commonality of the administration of forensic science services among developed countries has been that of public ownership at some level of government (Koppl 2005). The public provision of many government services, including forensic science services, has come under scrutiny as a reaction to the changing economic and political climates worldwide, resulting in various austerity programs in many countries. Indeed, in the United Kingdom an austerity response included policy that contributed to the closure of the government-owned company, Forensic Science Services (FSS) (Boudowle, Kayser, & Sanjantila 2011).

It has often been the case in the past that when forensic scientists or other public officials considered changing some aspect of the forensic science industry, they analyzed the alternatives in such a way as to discount or to completely overlook a very large possibility: privatization. Although the benefits of privatization are clear in many industries (Mueller 2003), the question often asked is: How can *government* bring about change to improve the effectiveness and efficiency of the forensic science industry? This attention to a government-led solution may seem like a natural response to the question, given that most forensic scientists have or do work for publicly run institutions, but it too often dismisses the possibility that there are alternatives to a government-initiated solution. Furthermore, as countries implement their austerity programs, government-led solutions may not be financially viable or politically feasible moving forward.

Although degrees of privatization in the forensic science industry have existed for years (Christian 2011), the U.K. Home Office has uniquely explored the privatization of their entire forensic science industry as an option with the closure of the FSS in December 2010, after it suffered continual monthly losses of £1 to £2 million. After the FSS's closure, the UK became the only leading industrialized nation without a government-operated forensic service agency. The forensic analysis that had previously been conducted by the FSS was redistributed among police laboratories and companies such as LGC, Orchid Cellmark, and Key Forensic Services, among others, all of which are privately registered corporations. Several scholars have expressed serious concerns about privatization and, using a business model for forensic science, claim that using markets in this industry, rather than

DOI: 10.4324/9780429318528-19

a government agency, may lead to catastrophe (e.g., Roberts 1996; Boudowle, Kayser, & Sanjantila 2011; Jobling 2011). Many of their arguments are powerful and include passionate calls for retained government control in order to "stick in the gizzard of true free marketers" (Roberts 1996, p. 52).

There is no special status for the forensic sciences such that this industry does not follow the laws of economics. Although privatization and the use of markets will never lead to perfect allocation of scarce resources, very often privatization has benefits that government solutions simply lack. Economists have argued for centuries that given the right conditions, markets offer the best quality goods and services at the lowest possible prices to consumers as a response to society's demands. An important question to ask then is: what are these conditions, and how would they work in the forensic science industry? Having a goal to understand how privatization works in order to harness the power of markets and improve the forensic science industry could be an important step for those working in the field. Embracing nontraditional approaches to improve the forensic science industry may be the only choice in the future given the state of many national economies and the necessary tradeoffs governments face in determining policy, along with the benefits that come with privatization (Maguire, Houck, & Speaker forthcoming; Christian 2011). Indeed, the widening of the abyss between political factions may force privatization solutions as any consideration for compromise continues to fall out of favor for politicians.

The paper begins with a review of the economic arguments outlining the general benefits of privatization. This includes foundational comments on the nature of markets in an economy and interpretations of market-determined equilibriums. That is followed with an outline of the common criticisms of privatizing the forensic science industry. Next is an exploration of the options for privatizing the forensic science industry. This is followed by some concluding remarks.

The General Benefits of Privatization

Market Prices Reflect Relative Social Value

The fundamental economic problem is how to deal with scarcity. There are simply not enough goods, services, and resources freely available in nature to satisfy all human desire; therefore, some sort of rationing mechanism will eventually need to be put in place. There are various mechanisms available for rationing scarce goods and services, and every society or economy addresses this issue of scarcity through a choice of whether to organize their economy through market mechanisms, a political process, or a combination of the two. Regardless of choice of organizational system for economic interaction, a form of planning occurs. The difference between choices is that, in market systems, individuals plan for themselves, while in political processes government experts plan for the collective (Hayek 1945). For rational economic choices to be made in either case, the proper accumulation of knowledge relevant to the circumstance is essential. However, the degree of quality in the information will influence the outcome of the economic choice.

The manner in which prices are established will determine the quality of information in these prices and the quality of decisions that economic agents (producers and consumers) make. When there is sufficient knowledge for proper social choice to occur, it becomes possible to perform what is termed *economic calculation*, or the judgment of the relative value of two or more goods based on a common unit, normally money (Mises 1932). This economic calculation, which in practice occurs through trade or exchange, allows for prices to serve as information and a rationing mechanism for goods and resources. With this price information, social choice can be made.

If trade occurs through private markets, prices will be determined by how much society values a particular good or service relative to all other goods and services. If the price of a good or service is too high relative to the value of that good or service when compared to the prices and values of all other goods and services, some consumers would simply stop purchasing that good or service until the price declines and reflects the concept of relative value. Consequently, markets provide an efficient rationing mechanism that allows for rational tradeoff decisions to be made. Goods and services that are valued more highly by society, relative to their degree of scarcity, will command a relatively higher price, and goods and services that are less valued by society, relative to their degree of scarcity, will see a relatively lower price. With this information economic agents will tend to produce more toward the optimal amounts, and thus avoid shortages and surpluses that would reduce the welfare and efficiency of society.

Under political planning, prices are not market determined and do not reflect value, but rather some political mandate or regulation. Without market-determined prices, there is a lack of definitive information on preferences, which can lead to a misunderstanding of the relative values of various goods, services, and resources. It would be impossible to compare the value of one good versus another if prices do not reflect value, and therefore production will be suboptimal and shortages and surpluses are expected to result. As a result, political planning does not lead to a rational economic order, since the knowledge to understand the tradeoffs between scarce goods, services, and resources is minimal without market prices. In the case of forensic science services that rarely are priced by markets, the result has been large casework backlogs for many forensics laboratories (e.g., Alabama Live 2011; Michigan State Police Forensic Science Division 2011; Office of the Auditor General of Ontario 2009; National Institute of Justice 2003, 2011).

Markets Provide Different Incentives Than Government Bureaus

There are no secret methods of running an operation that private companies have of which the government is not aware (Poole 1983). The technical knowledge and desire to do good in the minds of managers is likely to be the same in both public and private institutions, but the organizational framework with which they operate can make it more or less difficult to provide goods and services. The resulting measures of efficiency for public and private firms will likely be different as a result. The incentives found within an organization are a powerful force that can affect the quality of outcomes, and these incentives are expected to be different in public versus private institutions (Benson, Rasmussen, & Sollars 1995). Within a public bureau, these organizational incentives will likely be governed by politics, while within private firms the incentives are largely determined by market forces.

Although there is a growing trend toward outsourcing, many government outputs are still provided by government-controlled bureaucracies. These bureaus are often not compensated on a unit-by-unit basis, but rather are given a large budget and expected to produce some level of output in the form of goods or services (Niskanen 1971). The managers of these bureaus normally are not compensated in income for the level, quality, or efficiency of this output, and therefore may develop other goals not seen in market organizations (Niskanen 1971; Weber 1947). For example, a bureau manager's goals may be to increase his or her public reputation and political clout, often accomplished by increasing the allotted budget through the political process and thus increasing the output of the bureau. This can result in overproduction of the good or service in question. The problem is further exasperated by the tendency for bureaus to be granted monopoly power for the good or service provided. The reason for giving bureaus monopoly power within their market is often to avoid wasteful duplication, but the unavoidable consequence of this is that bureaus are no longer under the competitive pressure of other firms or competing bureaus (Mueller 2003). The lack of competition then makes it difficult for the government to effectively gauge the performance of a particular bureau if there is no comparison.

Unlike government bureaus, private firms normally do not have a monopoly over a market and will therefore have to compete by offering the best quality-price ratio. Those entities that are best able to compete in their given market will be the firms that add the most value to the resources use in the minds of their clients, and therefore create profit or the excess of revenue after accounting for all costs. Therefore, to increase profits, private firms will attempt to either increase revenues or lower costs, or a combination of the two.

Entrepreneurs are the individuals who drive the competitive process to add value to resources and create economic profit (Shumpeter 1942; Kirzner 1997). Economic profit not only accounts for accounting costs, but also indirect or "opportunity costs." These opportunity costs refer to the next best alternatives to the resources that the entrepreneurs employ. In other words, in a competitive market a successful investment needs to create an accounting profit, but must also create a risk-adjusted rate of return that is comparable with other investments the entrepreneur could have chosen. This contrasts sharply with a government bureau that normally has no comparable alternative with which to compare rates of return.

Entrepreneurs will innovate to attract customers with new and improved products, as well as find new ways to produce the same or better goods at lower costs (Shumpeter 1942; Kirzner 1997). Although it is true that entrepreneurs can increase profits by cutting the quality and thus lower the cost of producing their products, an entrepreneur will only be able to do so if cutting quality does not result in an offsetting decrease in demand and revenue for a firm (Benson 1998). The more competitive a market is, the less an entrepreneur will be able to cut quality to increase profits since consumers will simply take their business elsewhere. This competitive process therefore will lead to higher quality products at lower prices to consumers.

In cases where entrepreneurs and firms do not compete effectively, and therefore do not create an economic profit, eventually they will be forced to leave the market and will either go bankrupt or be sold to more efficient companies. Poorly performing firms eventually do go out of business. Thus there is a strong incentive in private markets for good business behavior. Poorly performing government bureaus in contrast, rather than going bankrupt, are often granted larger budgets to fix problems.

Economies of Scale and Scope

Perfect economies of scale occur when a producer finds a scale of operation that allows it to produce at a level that minimizes the average cost per unit. Producers with levels of operation that are too small or too large will find that per unit costs are not minimized (Mas-Colell, Whinston, & Green 1995). Perfect economies of scope occur when a producer finds the optimal combination of goods or services to provide in their portfolio of products in order to make the most effective use of their resources and also minimize average cost per unit. Although it is theoretically possible for a government to find the level and scope of operations to minimize average cost, it is unlikely (Poole 1983; Benson 1998). Geographically defined political entities such as a city, county, state, or country rarely correspond with the most efficient size for the provision of various government services. In particular for forensic science laboratories, a recent investigation suggests significant cost reductions could be obtained industrywide if some consolidation of services were to take place (Maguire, Houck, & Speaker forthcoming). Without a decisive move from within the industry to address its inefficiencies, political processes and economic forces may well arise to force the decision toward consolidation, elimination of services, or privatization. Neither is there reason to believe that these political entities would find the optimal mix of services to provide in their portfolio of forensic science services to achieve the best mix of services for the clientele served (i.e., perfect economies of scope). If some level of government did provide a given service at the minimum average cost, it would largely be coincidental, and would likely be true for only a small percentage of the total government services provided.

To illustrate the issue of the volume of activity in a political jurisdiction, consider the Utah Bureau of Forensic Services, an accredited state-run laboratory system that provides services to the police agencies of Utah. The laboratory facilities are state of the art and offer high-quality scientific investigation for the state of Utah. Among the many investigative areas covered by the laboratory is trace evidence.[1] An analysis of the cost-effectiveness of the trace evidence area for this laboratory provides a first glance suggestion that this laboratory is much more costly than other trace evidence analysis. With an in-depth decomposition of the average cost (Speaker 2009), the data reveals that this laboratory compares well with other laboratories with respect to market conditions and processes. However, the other decomposition variables depicting productivity and process are out of sync with other laboratories. Further inspection reveals that the higher costs are reflective of a low case volume rather than problems with process or productive ability. The only "problem" is that there is just not enough crime in Utah or enough submissions of evidence to support the infrastructure, both capital investment and minimum personnel, resulting in very large average costs per case processed. However, once adjusted for case submissions, the laboratory performs at a high rate of efficiency. That leaves the laboratory with a variety of choices if it wishes to overcome the "problem" of a low Utah crime rate. Other than incite more crime (and hence laboratory submissions), realistic alternatives call for the consideration of insourcing, outsourcing, or privatization of this service (Maguire, Houck, & Speaker forthcoming).

Benson (1998) explains how private firms do not have these political boundaries to worry about when deciding what their scale of operations ought to be, and are thus more flexible. In other words, private firms can decide to operate across many different political boundaries if doing so creates economies of scale and thus lowers average cost and increases profit. Entrepreneurs will compete to lower costs through innovation, but they will also lower cost through the discovery of the most efficient scale and scope of production. Their motive to increase profit by lowering cost is a powerful incentive, resulting in most private firms operating at their minimum average cost level. If governments wish to have some sort of cost efficiency, contracting with private companies may be an effective response. This may occur by having one large firm produce for several different political jurisdictions, or by having several small firms produce for one political jurisdiction. In either case the private firms are more likely to be operating at their efficient scale and scope compared to government bureaucracies (Benson 1998). An alternative may be for the public laboratory to mimic the market behavior of the private sector and take the initiative to achieve economies of scale through insourcing or outsourcing alternatives (Maguire, Houck, & Speaker forthcoming).

Criticisms of Privatizing the Forensic Science Industry

Private Markets Cannot Provide the Same Quantity and Quality of Service

Several authors have expressed serious concerns about using market models in managing forensic science laboratories (e.g., Roberts 1996; Gill 2011; Boudowle, Kayser, & Sanjantila 2011; Jobling 2011). They express the concern that privatization will lead to a system where profits are the main goal, which could take priority away from the goals of quality service, innovation, and social welfare. "A market based on the cheapest possible provision of standard forensic services will fail [in] the... criminal justice system" (Jobling 2011, p. 5). Professor Peter Gill, a world leader in DNA profiling, agrees and recently warned that the trend toward privatization of forensic services in the UK, as well as the recent closure of the government-owned FSS, is leading to catastrophe (Peachy 2012). Indeed the private forensic services contractor LGC, which took over 65% of the work the FSS formerly conducted, has come under recent scrutiny after there was discovery of DNA sample contamination in a sexual assault case (BBC News 2012).

Markets are surely not perfect, and free market economists do not claim that they are. The issue here is not how to achieve perfection, but how to achieve the best possible form of forensic science services in a given jurisdiction. Private companies make mistakes, and if they make too many of them profits may fall, staff may be laid off, and in extreme circumstances bankruptcy could result. In the case of LGC, their contract with the UK government might not be renewed if mistakes continue. Governments, however, also make mistakes. In 1997 the U.S. Department of Justice investigated some of the operations of the Federal Bureau of Investigation (FBI), including forensic science service practices, and discovered severe malpractice that bordered on the fabrication of evidence (Bromwich 1997; Hsu, Jenkins, & Mellnik 2012). Unlike LGC, which is a private company, it seems unlikely that the FBI will ever go bankrupt, as profit is simply not a bureaucracy's motive. In fact, the FBI's budget has grown even with the large amount of evidence that the Department of Justice collected that shows they are not performing as well as expected (Bromwich 1997).

Although both private and public systems can make mistakes, one reason some proponents of government-provided forensic service might still argue against the private provision of forensic services is that it is simply too complex. Gill (2011), for example, claims, "The complexity of forensic science is underestimated. It is unrealistic to suppose that the necessary infrastructure can be supplied solely by private companies" (p. 1). It is true that the provision of forensic science services is complex, but it is not clear why this would be an argument against its provision being done privately. Indeed, markets successfully provide many goods and services that could easily be described as very complex. Furthermore, many economists would argue that the opposite is true, or that forensic science is too complex to leave to governments to provide.

The power of expertise is a common argument made by bureaucrats who claim superior knowledge or ability to provide a good or service in order to gain power or control (Mueller 2003; Weber 1947). Indeed, if a bureaucracy has a monopoly or near-monopoly power over an industry, it would be unlikely that any private company would have the ability to develop the skills or infrastructure needed to compete on the same level as a government bureau. It would be expected that the FSS, for example, which enjoyed many years of government protection and the absence of any sort of significant competition, would have dominated the market before privatization. Therefore, the claim by Gill (2011) seems legitimate in the short run. But this does not mean that, in the absence of government-provided forensic services, private companies would not effectively provide the service. Markets have shown countless examples of their ability to provide complex outputs if there is a level playing field, competition, and the time to develop (Scherer & Ross 1990). However, if a market is not allowed to be competitive due to government favoritism, restrictive contracts, and high legal barriers to entry for new firms entering the market, the outcome may not be as desired. The National Forensic Framework Agreement (NFFA), for example, which was completed in the summer of 2008 by the National Policing Improvement Agency of the UK, significantly reduced the level of competition in the market for forensic service providers by making it nearly impossible for new firms to legally enter the market for police business (National Policing Improvement Agency 2008). Simply having private companies provide forensic science services will not necessarily cause all the benefits of privatization to occur if there is a low degree of competition.

Another reason that proponents of government-provided forensic services argue against the private provision of forensic science services is that privatization will lead to low quality and low innovation when companies seek profit as their goal (Roberts 1996; Boudowle, Kayser, & Sanjantila 2011). If there were a reasonable degree of competition in the market, however, most economists would reject this claim. Entrepreneurs have a strong incentive to create profit by innovating, lowering prices to consumers, and increasing the quality of their product. They do these things not to be benevolent, but rather as a reaction to market forces. If they can attract more consumers by having better business practices, they will expand their market share and profit. LGC, for example, spends between 5%–10% of its forensic revenues on research and development to improve its business. Professor Bernard Silverman in a recent report states, "If you ask the question, 'is there R&D in forensic science other than in the Forensic Science Service,' the answer is definitely there is, yes" (House of Commons Science and Technology Committee 2011, p. 1). Benson (1998) explains that entrepreneurs, including those in the forensic science industry, are exceedingly aware that companies that provide a low quality- to-price ratio and that do not innovate will eventually be removed from the market. It thus becomes their goal to innovate and make customers happy by providing high- quality, low-priced goods or services in order to survive. Furthermore, it may be the case that private companies using company money to perform research and development will perform more effectively than government bureaus because they will have to bear the financial burden of unsuccessful research and development.

Forensic Science is a Public Good, Thus Ought to be Publicly Provided

It is often argued that forensic science is a public good, and therefore ought to be provided publicly by the government, as seen in Jobling (2011) and Roberts (1996). This assertion that forensic science services and forensic research is a public good is a very strong claim for economists, especially given the extreme rarity of occurrences of pure public goods in an economic sense. The theory of public goods developed by Samuelson (1954) is one example of a market failure, which can be an argument for the public provision of a given good or service if the cost of public provision is less than the cost of market failure. In his theory Samuelson describes public goods, or collective goods, as goods "which all enjoy in common in the sense that each individual's consumption of such a good leads to no subtraction from any other individual's consumption of that good" (Samuelson 1954, p. 387). In other words, a change in the size of a population does not affect the quality or quantity of each individual's consumption of a set level of output of the public good. This contrasts directly with private goods, which Samuelson states can be "parceled out among different individuals" (Samuelson 1954, p. 387). With private goods, as population rises the output of the private good must also rise if individuals are to receive the same quality and quantity of consumption. The day-to-day operations of processing case loads in forensic laboratories undoubtedly behave as private goods, seeing that as population increases so too should output. The physical materials and labor hours used to analyze one case submission for one police jurisdiction cannot be used on another case submission for another police jurisdiction, and therefore cannot be enjoyed in common. The degree to which a publicly provided good or services qualifies as a public good can be measured by economists, which will be a first step toward analyzing this issue more formally.[2]

But what of the critical R&D performed by public laboratories such as the FSS? The claim by Roberts (1996) that free riding in research and development for forensic science services would cause a socially suboptimal level of research being done is another important concern. However free riding in research and development is foreign to free markets if the proper institution of property rights is in place. Private firms who can register patents are in most contexts more than willing to do research for new products and processes, which often can increase profits (Levin et al. 1987). If the LGC and other forensic laboratories in the UK were able to retain property over new innovations in forensic services, they could use these new ideas to create profit. Under NFFA, however, police in the UK define the type of services required, quality standards, turnaround times, and product volume, and thus forensic laboratories within the country can only compete on price that limits the industry competition greatly.

The argument made that forensic science is a public good and therefore should be publicly provided is often not claimed in the sense that Samuelson (1954) described, but rather stated to imply that forensic science services benefit a large portion of the public. It is true that forensic science services benefit a large number of people, but it is unclear why then it should be provided publicly. Many privately provided goods and services such as groceries, the internet, cars, college education, etc., also benefit a large number of people, but are provided successfully by private companies. Supreme Court Justice Antonin Scalia, in a 2012 Department of Health and Human Services Supreme Court hearing, argued this very point and declared that individuals rather than government ought to make decisions about how much of a given good or service ought to be purchased and thus produced (Department of Health and Human Services v. Florida 2012). Therefore, in either the Samuelsonian sense or in the sense of being beneficial for a large number of people, there is not a strong argument that forensic science must necessarily be a public good and therefore publicly provided.

Privatization of Forensic Services will Lead to a Monopoly

Another objection to the privatization of forensic services is that privatization may create a market with a low level of competition, with one dominant company acting as a near monopoly with many smaller less competitive companies in the market (Roberts 1996).

In the 1990s this was the case with the UK's FSS being the dominant player and enjoying government support. Today, Environmental Science and Research Limited (ESR) also has monopoly control over the forensic services market in New Zealand. However, it would be inaccurate to describe this as a market outcome, since the FSS and ESR do not operate under normal market forces, seeing that they were created and were or are owned by their respective governments. It is certainly true that competition may be lacking, but this is a design from political forces, not market forces.

Most monopolies are the result of government favoritism and protection, which contrasts sharply with how economists describe private markets. Under normal market conditions without government support, monopolies are rare and unstable precisely because of direct or latent competition (Sobel 2004). Even the largest of companies such as General Electric or Wal-Mart would not be considered monopolies, and many large companies of the past such as Westinghouse or K-mart, which formerly held a position of dominance, have subsided in power. Mizutani and Uranishi (2003) give an example of competition of small private companies reducing the Japanese government's monopoly power over postal services, and show that the monopoly control of the delivery of packages in Japan by the post office was eroded significantly by the introduction of a smaller private company, Yamato, in 1976, leading to cost reduction due to competition in Yamato but not in the government post office.

Markets are known for creating not destroying competition, thus it is difficult to make the case that market organization of forensic science services would cause any less competition in the historically government-controlled near monopoly of the UK's FSS, New Zealand's ESR, or similar government-controlled forensic science laboratories. Indeed, an increased level of competition is precisely what happened within the United Kingdom in the 1990s and early 2000s, affecting the FSS's market share. When the FSS was deregulated, it dominated the market share for a short time; however, this was a temporal state as smaller laboratories have innovated, competed, and grew, creating more price competition and choice in the market.

Privatization of Forensic Science Services Will Cause Availability and Equity Concerns

A fourth, and what Roberts (1996) believes to be the most serious objection toward privatizing forensic services, is that markets as they exist are imperfect and will therefore not lead to perfect outcomes. As a result of imperfect markets, the defined consumers of forensic services, being police or defendants, may not make the best choices when deciding what forensic services to purchase. In the case of both police and defendants, Roberts (1996), Boudowle, Kayser, and Sanjantila (2011), and Jobling (2011) all argue that substandard levels of forensic services will be purchased due to market forces that focus on costs and benefits. In the case of police, if direct charging is in place, they may decide to lower the quantity of forensic services purchased to save money given a limited budget, although the quality of service should be standard under NFFA in the United Kingdom. Defendants, many of whom rely on public funds, may also decide to purchase a lower quantity of forensic services as well due to inadequate knowledge or ability to pay. Therefore, in this view, markets will reduce the availability of forensic science compared to a government bureau which, as Roberts (1996) seems to suggest, could somehow be able to purchase the highest possible quality and quantity of forensic services required.

There are flaws in this argument. First, history shows that competitive markets have a way of expanding access to products and services that society desires and were initially only available to the very rich. Cannon and Tanner (2007), for instance, explain that healthcare costs for Americans in areas with a high degree of market competition and consumer choice (e.g., Lasik eye surgery) have experienced a reduction in prices. This has resulted in healthcare services that rely more on market mechanisms to be more available, not less. Alternatively, healthcare services that rely less on markets and more on government intervention and highly regulated third-party payment methods generally saw increasing costs and prices (Cannon & Tanner 2007). If the forensic science market is allowed to be competitive, with little government control, favoritism, or contracts that increase barriers to entry, economists would expect availability to both police and defendants to rise and prices to fall in the long run.

Second, Robert's assumption (1996) that, if private provision of forensic science services does not meet a perfectly idealized state, government control would be preferable, is a straw-man argument. He is correct in stating that markets will not provide the highest quality goods and services at whatever quantity desired to any and all consumers regardless of their ability to pay. However, he ignores the fact that it is equally unlikely that a government could perform this action effectively. In terms of quality of forensic services, there is no reason to imagine that governments can consistently provide high-quality forensic science services for those unable to pay (Scheck et al. 2003; Garrett 2011; Giannelli 2011). Additionally, "in forensic science, cost-effectiveness and budgetary constraints are constant concerns" (National Institute of Justice 1999, p. 13), which shows an understanding that even governments cannot reach an idealized perfect state of production and distribution. Boudowle, Kayser, and Sanjantila (2011) recognize this, but still reject a cost-benefit model. This rejection, however, is a rejection of an understanding of relative social value in a good or service, which would make socially optimal choices very difficult. If the expected marginal cost of catching a criminal is less than the expected marginal benefit, then money should be spent on catching the criminal. However, if the expected marginal cost of catching a criminal is more than the expected marginal benefit, economists would suggest that resources are better placed elsewhere (Gwartney et al. 2013).[3] Unfortunately, the socially optimal percent of convictions from crimes is not 100%, as there are simply not enough resources to properly investigate all crimes. Forensic science cannot escape the fundamental problem of scarcity, and therefore tradeoffs must be made. The use of trace evidence in the United States' courtrooms, for example, has been declining (Koch 2007), which may have resulted due to trace evidence not having a high enough benefit-to-cost ratio compared to other forms of forensic evidence. The best mechanism for understanding these tradeoffs so far is the market price system which, given limited budgets, would give understanding to police or defendants about the relative value of a forensic test.

How Might Privatization Work in the Forensic Science Industry?

Privatization of the Supply and Demand of Forensic Services

When discussing privatization of forensic science services, one must consider both privatization of the supply of forensic science services, privatization of the demand of forensic science services, or some combination of the two. The privatization of supply is manifested as a contracting out of the provision of forensic services to private companies. The privatization of demand of forensic services is manifested when the consumers of forensic services decide on the quality and quantity of these services they wish to consume given the prices of various services—in other words, a payment-for-service model. The consumers of forensic services are often going to be government police agencies working alongside of prosecutors, but there could be a situation where defendants are given forensic service vouchers to pay for the analysis of their evidence as well (Koppl 2005).

Privatization of the supply of forensic services has been occurring for decades in many countries. In the UK, the forensic science industry has seen a series of steps decided upon by government to move toward privatization, culminating ultimately in the dissolution of the FSS to be replaced entirely by private contractors such as LGC. In New Zealand, the ESR was set up as a Crown Research Institute that practices a payment-for-service model with the New Zealand Police and is expected to produce an adequate rate of return, currently around 8%, similar to the expectations owners may have of a private company (ESR Annual Report 2011). This trend toward privatization of the provision of forensic services is not unique to governments. The American retailing company Target, for example, has an in-house forensics laboratory accredited by the American Society of Crime Laboratory Directors/Laboratory Accreditation Board (ASCLD/LAB) to investigate crimes committed in Target retail outlets and to protect the company's interests.[4]

When traditional government services are contracted out to private companies, cost savings by the government and profit for the contractor are normally the result (Poole 1983). These cost savings are very often underestimated since private contractors normally pay taxes to their government, and therefore the attractiveness of privatization is often more than it seems on the surface of a given study (Benson 1998). Because of the financial appeal of privatizing the supply of forensic services, an increase in the contracting out of forensic services by many countries may very well continue as a response to austerity measures to deal with the current economic climate.

The privatization of demand for forensic science services is somewhat different. In this case, the privatization of demand implies that the consumers of forensic services are not politicians, and furthermore these consumers make decisions on how much forensic science services to consume based upon a per unit benefit to cost evaluation using prices. This contrasts sharply with some systems of forensic science in which demand is still determined by politicians as some set budget for an expected level of output in a given time period. In cases where demand is privatized, normally consumers will take the form of police, as in the case of the New Zealand Police acting as customers to ESR. Koppl (2005) suggests that defendants could act as consumers of forensic services as well if they use their own private funds or are given forensic science vouchers as a way to challenge the forensic science experts on the side of the prosecution. This could add cost to the overall system of justice, but it will also increase the competitiveness of the forensic science industry if forensic laboratory experts challenge each other publicly. Reducing the prosecution's monopoly on expert opinion may reduce the error rate in convictions and provide improvements to the system of justice.

Although quality standards are in place for many countries' forensic science industries, improvements in the quality and not just quantity of service would be desirable, suggesting these standards ought to change dynamically as appropriate. It is important to note then that simply privatizing the demand of forensic science services will not necessarily lead to increased quality or quantity of output on the supply side. As Benson (1998) explains, quality of output depends more on the demand side than on the supply side. If consumers do not demand high quality or are unable to pay for it, suppliers generally will correctly respond by providing a lower quality or quantity of goods. It is therefore important that the budgets of police agencies, which act as consumers of forensic services, are at the level that can afford the quality defined in the standard requirements and quantity of forensic services desired by society. If a police agency's budget for forensic services is cut for the government to save money, its ability to pay for the amount of service needed at the standardized level of quality may be reduced. Furthermore, police agencies working with politicians must rightly gauge what society desires out of forensic science services to determine budgets and the portfolio of forensic services they wish to purchase from a forensic service provider. One cannot expect the suppliers of forensic services, whether public or private, to be able to provide the high quality and quantity of service needed if they are not compensated or directed appropriately by the consumers of forensic service. In short, you get what you pay for.

Challenges of Privatizing the Forensic Science Industry

There are many benefits to privatizing the forensic science industry of a country. However, privatization, if not done in the right conditions, may not necessarily result in desirable outcomes. There are many challenges that must be overcome for the benefits of markets to be experienced in the forensic science industry. A few of the more central challenges are explained below.

First, the benefits of privatization come from creating a free market system with very limited government involvement. Simply creating a legally private forensic company, without actually removing most of the influence of political forces, will not allow markets to work naturally. The beauty of markets is that they are natural systems that no single human or small group of humans has designed, but which still form what is called a spontaneous order (Hayek 1988). This spontaneous order develops from many individuals using dispersed knowledge to make decisions to better themselves—not out of the design of a small number of individuals or bureaucrats. Through this process society can utilize higher quality information through market prices, which will often lead to superior outcomes over a system designed by one or a few individuals. Therefore, if the forensic science industry is to capture the benefits of privatization, political forces must largely be removed.

A second challenge to overcome will occur if privatization only occurs on the supply side, but not on the demand side. If government continues to be the sole purchaser of forensic services, a monopsony will persist. Monopsony, or the situation where there is only one purchaser of a service, will result in the monopsonist having more power than any individual seller and the monopsonist will therefore be able to dictate the terms of a contract, which is the case in the UK as a result of NFFA. It is therefore important that governments who are monopsonists in the forensic industry behave in such a way as to make it attractive for providers of forensic services to operate effectively. If contracts are unattractive, provide little assurances, or are not available for a long enough period of time to encourage investment, providers of forensic services will struggle, choose not to enter the market, or choose to exit the market.

A third challenge with privatization is that, as firms are allowed to compete, new and varying products will enter the market that may be difficult for judges, attorneys, or juries to understand and compare the advantages and disadvantages of these products from different forensic firms (Gill 2011). Understanding what a particular forensic test can and cannot do is certainly important, but if privatization truly occurred, certain forensic tests from various forensic laboratories may likely become the standard precisely because they were the best, similar to the way that Microsoft and Apple have become the standard in computers or other companies have become the standard in their industry (Stango 2004). Nonetheless, there will likely be a lag between when privatization occurs and when standardization occurs in a significant sense, creating a possible short-run challenge. However, it may be the case that standardization is not desirable across broad geographical areas with diverse demographics and crime statistics. In the United States, no standardization between individual state forensic laboratories has been attempted to meet the need of each particular state (Houck & Siegel 2010).

These three challenges seem to be the most glaring, but there are no doubt many other challenges that will need to be overcome to give the forensic science industry the ability to operate efficiently in a market system. Further details are outside of the scope of this paper, but Lawless (2010) examines additional details on the challenges to the forensic science industry, specifically in liberalizing their organization by using market approaches. Vickers and Yarrow (1991), Fama (1980), and Coase (1937) analyze challenges to privatization more generally.

Conclusion

The forensic science industry is surely at crossroads in many countries in determining whether to move toward privatization or public control. Is privatization of forensic science inevitable? This issue will likely need to be one of comparison of the cost and benefits of the alternatives and not a rejection of a system when it fails to reach a utopian dream. Generally speaking, privatization has benefits that public responses simply lack. As forensic science laboratories continue their trend toward greater reliance on markets as austerity measures are implemented, data will be collected for analysis of this issue more formally (Houck et al. 2009). With data, an *a posteriori* analysis of the benefits and costs of privatization will be possible, which may be used to make changes to the forensic science industry. These changes will likely include nontraditional approaches, including privatization, to achieve the best possible level of forensic services for society. Many challenges will surely present themselves in implementing market approaches in the forensic science industry, but if a spirit of entrepreneurship is embraced along with a desire to improve the criminal justice system, society will surely reap the benefits.

Endnotes

1 The investigative area "trace evidence" follows the definitions provided by Project FORESIGHT's LabRAT tool, in which the area includes hairs and fibers, paint, and glass; other traditional trace evidence descriptions are included.

2 Good models exist to test the "publicness" of a good or service as has been shown for other economic markets. Using an empirical median voter demand curve model such as the ones implemented in Holcombe and Sobel (1995), King (2007), or Mathers (forthcoming), it would be straightforward to estimate the degree to which forensic science services behave as private goods. If forensic services do indeed behave as private goods, the argument for public provision of these services would be weak.

3 It is important to note that cost and benefit includes not only the financial cost and benefit, but also the emotional, social, and opportunity costs and benefits of an act. Furthermore, the benefits of catching a criminal not only include the initial benefits associated with bringing justice to the original crime committed, but also with the crimes prevented that may have occurred had the criminal not been apprehended.
4 http://abullseyeview.com/an-unexpected-career-target-forensic-services-laboratory/

References

Alabama Live. 2011. Budget cuts cause delays in Alabama forensic tests. Accessed May 24, 2012, https://www.al.com/wire/2011/12/budget_cuts_cause_delays_in_al.html.
BBC News. 2012. Rape case reopened after DNA forensics blunder. *BBC News Manchester*, March 9. Accessed March 10, 2012, http://www.bbc.co.uk/news/uk-england-manchester-17311101.
Benson, B.L., D.W. Rasmussen, and D.L. Sollars. 1995. Police bureaucracies, their incentives, and the war on drugs. *Public Choice* 83:21–45.
Benson, B. 1998. *To serve and protect: Privatization and community in criminal justice*. New York: New York University Press.
Boudowle, B., M. Kayser, and A. Sanjantila. 2011. The demise of the United Kingdom's forensic science service (FSS): Loss of the world's leading engine of innovation and development in the forensic sciences. *Investigative Genetics* 2:4.
Bromwich, M. 1997. *The FBI laboratory: An investigation into laboratory practices and alleged misconduct in explosives-related and other cases*. Washington, DC: Department of Justice, Office of the Inspector General. Accessed April 18, 2012, http://www.fas.org/irp/agency/doj/oig/fbilab1/00exesum.htm.
Cannon, M.F., and M.D. Tanner. 2007. *Healthy competition: What's holding back health care and how to free it*. Washington, DC: Cato Institute.
Christian, D. 2011. How the clinical laboratory improvement amendments (CLIA) can improve forensic laboratory quality. *Forensic Science Policy and Management* 2:18–27.
Coase, R. 1937. The nature of the firm. *Economica* 4:386–405.
Department of Health and Human Services v. Florida 613 U.S. 80 (2012).
The Institute of Environmental Science and Research (ESR). 2011. *ESR annual report*. Auckland, Australia: ESR. http://www. esr.cri.nz/SiteCollectionDocuments/ESR/Corporate/PDF/ESR AnnualReport2011webversion.pdf.
Fama, E. 1980. Agency problems and the theory of the firm. *Journal of Political Economy* 88:288–307.
Garrett, B. 2011. *Convicting the innocent: Where criminal prosecutions go wrong*. Cambridge MA: Harvard University Press.
Giannelli, P.C. 2011. Daubert and forensic science: The pitfalls of law enforcement control of scientific research. *University of Illinois Law Review* 1:53–90.
Gill, P. 2011. Forensic science service. Accessed May 2, 2012, http://www.publications.parliament.uk/pa/cm201011/cmselect/cmsctech/writev/forensic/m19.htm.
Gwarthey, J.D., R.L. Stroup, R.S. Sobel, and D.A. Macpherson. 2013. Price takers and the competitive process. In *Economics: Private and public choice* 442–444. Mason, OH: South-Western.
Hayek, F.A. 1988. *The fatal conceit: The errors of socialism*. Chicago: Unversity of Chicago Press.
Hayek, F. 1945. The use of knowledge in society. *The American Economic Review* 35(4): 519–530.
Houck, M.M., R.A. Riley, P.J. Speaker, and T.S. Witt. 2009. FORESIGHT: A business approach to improving forensic science services. *Forensic Science Policy and Management* 1: 85–95.
Houck, M.M., and J.A. Siegel. 2010. *Fundamentals of forensic science*. Waltham, MA: Academic Press.
House of Commons Science and Technology Committee. 2011. Research and development (R&D): Home office review of R&D. Accessed August 6, 2012, http://www.publications.parliament. uk/pa/cm201012/cmselect/cmsctech/855/85507.htm.

Hsu, S., J. Jenkins, and T. Mellnik. 2012. DOJ review of flawed FBI forensics processes lacked transparency. *The Washington Post*, April 17. Accessed April 2012, http://www.washingtonpost. com/local/crime/doj-review-of-flawed-fbi-forensics-processes-lacked-transparency/2012/04/17/gIQAFegIPT story.html.

National Institute of Justice. 1999. Status and needs of forensic science service providers: A report to Congress. Accessed August 6, 2012, https://www.ncjrs.gov/pdffiles1/nij/213420.pdf.

Jobling, M. 2011. Elementary, my dear cameron. *Investigative Genetics* 2:5.

Kirzner, I.M. 1997. Entrepreneurial discovery and the competitive market process: An austrian approach. *Journal of Economic Literature* 35:60–85.

Koch, S. 2007. Trace evidence symposium. *Forensic Science Communications* 9(1). Accessed August 6, 2012, http://www.fbi.gov/about-us/lab/forensic-science-communications/fsc/jan2007/communications/2007 01 communications02.htm.

Koppl, R. 2005. How to improve forensic science. *European Journal of Law and Economics* 20:255–286.

Lawless, C. 2010. *A curious reconstruction? The shaping of 'marketized' forensic science*. London: Centre for Analysis of Risk and Regulation

Levin, R., A. Klevorick, R. Nelson, S. Winter, R. Gilbert, and Z. Griliches. 1987. Appropriating the returns from industrial research and development. *Brooking Papers on Economic Activity* 3(Special Issue on Microeconomics): 783–831.

Maguire, C., M. Houck, and P. Speaker. Forthcoming. Cost effectiveness in the forensic sciences: Insourcing, outsourcing, and privatization. *Forensic Science Policy & Management.*

Mas-Colell, A., M.D. Whinston, and J.R. Green. 1995. *Microeconomic theory*. New York: Oxford University Press.

Michigan State Police Forensic Science Division. 2011. Forensic science for Michigan: Backlog prevention. Accessed May 24, 2012, http://michiganforensics.org/Backlog Prevention.php.

Mises, L. 1932. *Socialism: An economic and sociological analysis. Liberty Fundamentals* 97:105.

Mizutani, F., and S. Uranishi. 2003. The post office vs. parcel delivery companies: Competition effects on costs and productivity. *Journal of Regulatory Economics* 23(3): 299–319.

Mueller, D. 2003. *Public Choice III*. Cambridge, MA: Cambridge University Press.

National Institute of Justice. 2003. *Report to the attorney general on delays in forensic DNA analysis.* Washington, DC: Author.

National Institute of Justice. 2011. *Making sense of DNA backlogs, 2010—Myths vs. reality*. Washington, DC: Author.

National Policing Improvement Agency. 2008. National forensic framework agreement. Accessed April 18, 2012, http://www. npia.police.uk/en/9398.htm.

Niskanen, W. 1971. *Bureaucracy and represnetative government*. Chicago: Aldine-Atherton.

Office of the Auditor General of Ontario. 2009. Ministry of Community Safety and Correctional Services: Centre of Forensic Sciences, follow-up on VFM section 3.02, 2007 annual report. Accessed May 24, 2012, https://www.crto.on.ca/pdf/Reports/2009AuditorsReport.pdf.

Peachy, P. 2012. Privatisation is a catastrophe, warns godfather of forensics. *The Independent*, April 2. Accessed April 3, 2012, http://www.independent.co.uk/news/uk/crime/privatisation-is-a-catastrophe-warns-godfather-of-forensics-7606789.html.

Poole, R. 1983. Why not contract policing? *Reason* 14:10.

Roberts, P. 1996. What price a free market in forensic science services? The organization and regulation of science in the criminal process. *British Journal of Criminology* 36: 37–60.

Samuelson, P.A. 1954. The pure theory of public expenditure. *The Review of Economics and Statistics* 36:387–389.

Scheck, B., P. Neufeld, and J. Dwyer. 2003. *Actual innocence: When justice goes wrong and how to make it right*. New York: New American Library.

Scherer, F.M., and D. Ross. 1990. *Industrial market structure and economic performance*. Boston: Houghton Mifflin Company.

Schumpeter, J.A. 1942. *Capitalism, socialism, and democracy*. New York: Harper & Brothers.

Sobel, R. 2004. Welfare economics and pubic finance. In *Handbook of public finance*, edited by J.G. Backhaus and R.E. Wagner, 19–51. New York: Springer.
Speaker, P.J. 2009. The decomposition of return on investment for forensic laboratories. *Forensic Science Policy and Management* 1(2): 96–102.
Stango, V. 2004. The economics of standards wars. *Review of Network Economics* 3:1.
Vickers, J., and G. Yarrow. 1991. Economic perspectives on privatization. *The Journal of Economic Perspectives* 5:111–132.
Weber, M. 1947. *The theory of social and economic organization*. New York: Free Press.

Article: Expanding Budgets via Strategic Use of Leasing

6.3

WILLIAM McANDREW
AND PAUL J. SPEAKER
College of Business and Economics, West Virginia University, Morgantown, West Virginia, USA

Introduction

If necessity is the mother of invention, perhaps sequestration fits the role of the wicked stepmother. The Great Recession of 2008 resulted in severe tax revenue reductions for governments around the world. As a result of the loss in revenues, governments have reacted with corresponding cutbacks in the allocations to agencies in the public sector. These austerity measures have been highlighted by the infectious damage centered in Portugal, Ireland, Greece, and Spain (the so-called PIGS[1]). Seemingly more stable countries have faced their own austerity measures, including the plight of the forensic sciences in the United Kingdom with the closure of Forensic Science Services (FSS) in 2010. In the United States the inability to find a political solution to declining revenues gave way to the distasteful agreement of sequestration[2] and the corresponding automatic cuts in most public sector areas, including funding for forensic science laboratories. Such cuts have exacerbated the funding strains felt across the forensic science industry.

The automatic across-the-board reductions in expenditures that came from sequestration left little room for laboratories to make choices. With personnel expenditures comprising roughly 70% of the budget for an average laboratory (Newman, Dawley, & Speaker, 2012), many laboratories have reluctantly looked to unpleasant choices, such as reductions in staffing as the logical response to a call for budget cuts. As demand for forensic laboratory services has grown, this simultaneous strain on the ability to supply services has stressed numerous laboratory systems.

But what if laboratory budget cuts could be avoided or lessened, regardless of national policy? An examination of the budgets of forensic laboratories in the FORESIGHT study (Houck, et al 2009) reveals that most laboratories have an unused or underused tool at their disposal, equipment leasing, which offers an opportunity for a unilateral increase in the purchasing power of existing budgets and an immediate response to austerity measures. Rather than react to sequestration with reductions in force, shared furloughs, or the forfeiture of unfilled positions, a laboratory director can forestall such measures and even see an effective increase in disposable income through a planned use of operating leases. Consider the net expenditure differences between a public laboratory and a private laboratory involving the purchase of an identical piece of laboratory equipment. If the public laboratory makes an equipment purchase, the cost will be the full list price from the

DOI: 10.4324/9780429318528-20

equipment supplier. When the private laboratory makes the same equipment purchase, it pays the supplier the full list price, but is able to deduct the expense from its income when it calculates its corporate income tax and ends up with a final expense, net of taxes, that is considerably less than the cost to the public laboratory.

Although the public laboratory cannot receive the wealth transfer from the tax deduction, it can reap the same benefits of the lower after-tax net cost by finding a partner in the private sector, the lessor, who purchases and then leases the desired equipment to the public laboratory. Leasing offers an opportunity for a tax-paying private company to transfer tax benefits to a public sector laboratory through their lower net cost for equipment, while simultaneously keeping some of the benefits from the increased profit from a reduction in its tax base. These gains can be substantial. An example below shows potential capital budget savings of nearly 25%.

In addition, other benefits beyond the cost of capital equipment may be realized from the adoption of a leasing strategy. The operating lease can reduce equipment maintenance costs and improve quality as leases permit the laboratory to speed up replacement of dated equipment in favor of immediate adoption of technological improvements. Thus the laboratory may be able to employ more technologically advanced procedures and indirectly expand its budgets simultaneously.

In the sections to follow, the realities of corporate tax laws are presented and it is shown how laboratories in the private sector are able to acquire their equipment at a significantly lower net cost than public laboratories. Tax laws provide certain benefits to the individual business that result in additional benefits to the general economy through the enhancement of economic activity and the creation of additional employment. While the same incentives are not directly applicable to the public sector agency, leasing offers an opportunity for the public sector laboratory to effectively trade tax positions with a private sector lessor, thereby extending those benefits to, and stretching the budget of, the public laboratory. In the sections to follow, a brief introduction to the tax laws, including the specific laws that outline public sector agencies use of leasing and the qualifications of a lease agreement are presented. This is followed by an example for a lease versus purchase decision for a piece of laboratory equipment. While the example is based in the U.S. tax code, the principles may be easily extended to other countries with a consideration of the nuances of each country with respect to treatment of the lessee and the lessor. When the example is extended to the entire capital budget of the laboratory, the full magnitude of the lease option is revealed, and the extent of the indirect expansion of the budget may be realized.

Some Leasing Basics

The potential benefits from the tax deductibility of lease payments only accrue when the lease terms have met certain provisions. Leases generally fall into two broad categories, financial leases and operating leases.[3] Financial leases, alternatively termed capital leases, represent a method of financing equipment that provides a substitute for direct borrowing as a means to acquire equipment.

Operating leases, on the other hand, offer the use of capital assets for a time period that is less than the economic life of the asset, and such leases also include maintenance as the lessor is the owner of the equipment and bears the responsibility of upkeep. The lease contract may also include cancellation clauses before the end of the contract period. It is this latter category, the operating lease, which presents the laboratory with the potential budget gains as long as the terms of the lease meet Internal Revenue Service (IRS) guidelines.[4] These IRS guidelines determine which party, lessee or lessor,[5] is able to stake claim to the benefits of ownership, depreciation and/or investment tax credits. Guidelines specify:

- The term of the lease must be less than or equal to 80% of the engineering life of the equipment;
- At the end of the lease, the expected salvage value of the equipment must equal or exceed 20% of the starting value;
- No predetermined price may be established for the sale of the equipment from the lessor to the lessee at the end of the lease;
- The lessee's regular payments to the lessor may only include the lease payments; and
- The equipment cannot be so specialized that the lessee is the only potential customer for sale of the equipment when the lease ends.

If the lease contract meets the IRS guidelines, then there are several ways in which net benefits could accrue to the laboratory (lessee) as described by Lewellen et al. (1976). For each of the benefits to the lessor, the lessee gains through a contract in which the lessor passes on a portion of the gains to the lessee. These different potential transferable gains include:

- Effectively exchanging tax positions where the lessor (with the higher tax rate) receives a greater tax reduction from the expense;
- Aggressive use of depreciation and amortization (D&A) to experience tax benefits at a faster pace for the lessor;
- Market advantages to the lessor as a more frequent participant in the marketplace, where volume discounts may be negotiated for purchase price or the bundling of maintenance contracts;
- After market (resale) advantages to the lessor via releasing or sale of used laboratory equipment; and
- Differences in after-tax costs of borrowing to purchase laboratory equipment.

Each of these potential contributions to the lease versus purchase discussion is isolated and examined in turn in the succeeding sections. The discussion is supported by a numerical example to highlight the potential size of any net advantage to leasing over purchase. The magnitude of the leasing gains is provided in the text that follows with the detailed calculations relegated to the appendix.

Tax Laws, Incentive Systems, and Time Value of Money

Investments in capital assets provide services that are spread over several years. Accounting conventions provide the rules by which the initial investment expenditure may be assigned to each accounting year of potential use of the capital asset through the assignment of depreciation.[6] Among the alternative methods of depreciation is straight-line depreciation, where an equal portion of the cost is assigned over a term of years. Other depreciation methods accelerate the assignment of costs so that the tax advantages from depreciation deductions may be realized sooner rather than later. For the public laboratory, the method of depreciation chosen has little direct impact on the operation of the laboratory since tax deductibility of expenses is irrelevant to a tax-exempt organization. Depreciation offers a means to determine a book value of assets, but does not have a meaningful impact on the public sector laboratory's income statement since the public laboratory does not pay taxes. For a private laboratory, however, the choice of depreciation method has a direct impact on the value of the laboratory, where the value generally increases as more accelerated methods of depreciation are used (if permitted). This benefit occurs since there is a time value to money and, *ceteris paribus*, the earlier deductions are taken, the higher will be the value of those deductions at the time of the investment decision.

The magnitude of the benefit from depreciation may be realized through a consideration of the time value to money, where benefits received sooner may be reinvested to provide greater returns in the future. Consider first the public laboratory that purchases some equipment at a particular upfront price and plans to use that equipment over the following N years. Additionally, suppose that the laboratory purchases a maintenance contract for which it pays a contracted amount at the beginning of the year for each of those N years. Further, at the end of the N periods, the laboratory may choose to sell the equipment for the market value of that equipment, which represents a positive cash flow in the Nth period. Then the total cash flows for the public laboratory across all periods may be described as the cash outflow for the equipment expense at time 0, the maintenance contract at the beginning of each period t = 1,..., N, and any cash inflow at the end of period N from the sale of the used equipment. Depreciation, whether straight-line, accelerated, or otherwise, will not affect the cash flow of the public laboratory since the public entity does not pay taxes.[7] Because the cash flows occur in different periods, a present value comparison is calculated to discount the cash flows to the present using a rate that reflects the financial market's interpretation of the risk-adjusted return for the political entity in which the laboratory operates. Such a discount rate is easily inferred from the trading of general obligation bonds for that political jurisdiction.

Alternatively, examine a private laboratory that considers the purchase of identical equipment and associated maintenance contracts, but which is assessed a positive tax rate on ordinary income and a positive tax rate on capital gains income. The cash flows for the private laboratory will be the same as the public laboratory for equipment and maintenance with a few exceptions. The private laboratory will have the upfront equipment expense and the beginning of period maintenance expenditure, but unlike the public sector laboratory, it will deduct depreciation at the end of each period and affect its after-tax net income.[8] That depreciation deduction provides a tax shield for each period t, which is effectively a subsidy via a reduction in tax liability.

The present value cost to the public laboratory represents the amount that the jurisdiction requires in today's currency to pay for the investment including the initial investment and annual maintenance contract, net of any resale of the equipment at the end of the contract period. The appendix contains the detailed formula for the representation of the present value cost. Additionally, the appendix includes a series of examples with detailed calculations of the potential gains from leasing over purchase.

For the private entity, expenses are tax deductible. As a result of the tax effects, the present value cost for the private entity is offset by reductions in their tax bill. The private entity, whether it is a laboratory, bank, or leasing company, ends up paying less for equipment and maintenance than is experienced by the public laboratory. The tax benefits to the maintenance charge are taken each year when maintenance is paid, while the tax benefits from the equipment expense is spread over time through deductions for depreciation following applicable tax code. These gains are partially offset through potential capital gains on the resale of equipment after the lease term ends.

In a comparison of the present value cost for the public laboratory versus the private sector purchaser of equipment, the potential gains from leasing emerge. As noted in the previous section, if the present value cost is less for the private entity, then by making the equipment purchase in the private sector, the reduced cost from the tax deductibility of depreciation and maintenance can potentially be shared with the public sector laboratory through annual lease payments, which are lower than the annualized costs for a public laboratory.

There are legal requirements that keep the potential gains in check. In 1984 the U. S. federal government passed the Deficit Reduction Act (DEFRA), which limited this benefit of a leasing arbitrage opportunity. The limitation presents itself when a private lessor leases an asset to a tax-exempt entity such as a state or county in the form of a financial lease. Before this legislation, tax exempt entities would often pursue long-term projects using leases that at a local or agency level may have benefits greater than their costs, but at a societal level would be more expensive than traditional debt financing as a result of federal tax deductions being used to subsidize the asset acquisition of tax exempt entities.[9] If one assumes the lessor and the tax exempt lessee differ only in their income tax rate, then the value gained by leasing rather than debt financing for the lessor and lessee before 1984 is equal to the subsidy or loss that the government directly pays to the lessor and indirectly pays to the lessee through income tax deductions.

With the passing of DEFRA and subsequent laws that add more detail to the permissible lease activities involving tax-exempt entities, it is clear that the federal government attempted to stop some indirect subsidies to government and other tax exempt entities' investment through financial leases. Indeed before DEFRA often undesirable projects were undertaken through leasing, which cost taxpayers more money. However, there are still legally permissible ways for public forensic laboratories to take advantage of lease situations through disparate taxes with private lessors if certain IRS guidelines on leasing are followed advantageously. These situations are highlighted above as guideline leases that have met the criteria of performing as a lease with the ownership burden on the lessor, rather than a disguised loan that circumvents the intention of the law.

The Size of the Gains—Example

Consider the net expenditure differences between a public laboratory and a private laboratory involving the purchase of an identical piece of laboratory equipment and the associated maintenance contract for that equipment. To isolate each of the potential gains from leasing as detailed above, consider a series of scenarios, which isolate the contributions that make leasing a viable consideration for the public laboratory. (The detailed calculations, corresponding to each of these scenarios, appear in the Appendix.)

Suppose that a laboratory is considering the purchase of scientific instrumentation with a list price of $100,000. That list price is the same for either a public sector laboratory or a private sector laboratory. Additionally, the laboratory would purchase an annual maintenance contract on the capital equipment at a cost of $5,000 per year and with the maintenance contract payments due at the beginning of each year. The instrument falls into the Internal Revenue Service (IRS) 5-year depreciation class even though it has an engineering life of 10 years. Suppose further that the bonds of the public sector are issued with a rate of return that is the same as the rate for the private entity. Finally, suppose that the public laboratory or private entity is able to sell the equipment at the end of four years[10] for $20,000. Given this information consider the present value cost to the public laboratory versus the present value cost in the private sector for the purchase and maintenance of the scientific instrumentation for a four-year period.

Scenario 1

First consider the present value cost for a direct purchase of equipment and maintenance versus a leasing contract that includes use and maintenance under the following five assumptions. This baseline scenario suggests an environment without any of the conditions under which the aforementioned leasing advantages exist. This will help to isolate the gains evidenced in the subsequent scenarios.

Assumption 1 (tax rates): There are no taxes.
Assumption 2 (volume discounts): The price of equipment is the same for either the public or private entity.
Assumption 3 (maintenance contract): Maintenance contracts cost the same for either a public sector or private sector entity.
Assumption 4 (resale): Expected equipment resale values at time N are the same for either the public or private entity.
Assumption 5 (discount rate): The cost of debt is the same for either the public sector or private sector entity.

In this scenario, since the equipment price and maintenance contracts are the same, the cash flows to the public entity and the private entity are identical and there would be no advantage for leasing over purchase. The lack of a tax structure results in identical treatment for the public sector and the private sector. That is, the minimum price that the private entity could charge in a lease would just cover all costs and would have an identical present value to the purchase costs for the public laboratory. (See the Appendix for the details of the present value cost comparison.) In this baseline scenario, the present value cost of the equipment and associated maintenance at a discount rate of 5.5% is $102,345 for either the public or private sector entity.

With this baseline established, the contributions from the tax structure, depreciation method, deductibility rules, and discount rates may be highlighted. The following scenarios attempt to isolate these contributions to the potential differences in equipment purchase versus leasing.

Scenario 2

Suppose that the tax rate assumption of this baseline scenario is relaxed and income taxes may exist at the national, state, and local levels. Potential gains emerge for leasing versus purchase of laboratory equipment. To demonstrate the potential gains, consider next a scenario where assumption 1 is relaxed.

> Assumption 1 (tax rates): There are no taxes for the public entity, but an ordinary income tax rate $T_I > 0$ and a capital gains tax rate $T_C \geq 0$ are assessed to the private entity.
> Assumption 2 (volume discounts): The price of equipment is the same for either the public or private entity.
> Assumption 3 (maintenance contract): Maintenance contracts cost the same for either a public sector or private sector entity.
> Assumption 4 (resale): Expected equipment resale values at time N are the same for either the public or private entity.
> Assumption 5 (discount rate): The after tax cost of debt is the same for either the public sector or private sector entity.

The difference in the present value costs between leasing and ownership using the assumed tax rates in the appendix is $38,196, which represents the opportunity for gains from leasing and trading tax positions between the private and public sectors. Note that the final amount of the gains to the public laboratory is offset by the tax liability to the lessor as lease payments are received.

Scenario 3

While the trading of tax positions offers the greatest contribution to cost reduction and potential gains, other market advantages may provide room for additional gains. Consider the potential bargaining power for higher volume activity for a private firm that specializes in the leasing of laboratory equipment. The third scenario takes this volume activity into account.

> Assumption 1 (tax rates): There are no taxes for the public entity, but an ordinary income tax rate $T_I > 0$ and a capital gains tax rate $T_C \geq 0$ are assessed to the private entity.
> Assumption 2 (volume discounts): Equipment prices are x% lower for the private entity versus the public entity via volume discounts.
> Assumption 3 (maintenance contract): Maintenance contracts costs are y% lower for the private entity versus the public entity via volume discounts.
> Assumption 4 (resale): Expected equipment resale values at time N are the same for either the public or private entity.
> Assumption 5 (discount rate): The after tax cost of debt is the same for either the public sector or private sector entity.

To demonstrate the impact, begin with the conditions of scenario 2, where the laboratory is considering the purchase of some scientific instrument, but add the conditions where the private entity has negotiated an x = 10% discount in the equipment price for volume purchases and y = 20% discount for the maintenance contract by diversifying the risk across the high-volume purchases. The previous difference in the present value costs of $38,196 from scenario 2 grows to a difference of $47,066 and represents the reduction in present value costs from leasing, trading tax positions, and bargaining positions between the private and public sectors.

Scenario 4

Relax the assumptions further with the following scenario. Suppose that an infrequent seller of used equipment has less expertise to identify potential purchasers of used equipment while a regular seller in the resale (or re-lease market) may possess an informational advantage to secure a higher resale price. This is analogous to the differences in the used car market where dealer sales of used cars command a higher price than that obtained by individual sellers. Assume the private entity is able to command a higher salvage value along with the other assumptions from Scenario 3.

> Assumption 1 (tax rates): There are no taxes for the public entity, but an ordinary income tax rate $T_I > 0$ and a capital gains tax rate $T_C \geq 0$ are assessed to the private entity.
> Assumption 2 (volume discounts): Equipment prices are x% lower for the private entity versus the public entity via volume discounts.
> Assumption 3 (maintenance contract): Maintenance contracts costs are y% lower for the private entity versus the public entity via volume discounts.
> Assumption 4 (resale): Salvage values at time N are z% higher for the private entity versus the public laboratory.
> Assumption 5 (discount rate): The after tax cost of debt is the same for either the public sector or private sector entity.

Such an advantage in the resale market could emerge for a frequent participant to the resale market. Frequent participants could include specialized leasing companies or even the equipment manufacturer. The present value cost for the private entity falls and the difference in the present value costs from leasing jumps to $53,637 if a 50% salvage value advantage for the private sector is assumed.

Scenario 5

In the fifth scenario all of the assumptions are relaxed. Consider the potential advantages for the private sector lessor versus the public sector entity with respect to the cost of borrowing. Suppose that the discount rate assumption is relaxed and the after-tax cost of borrowing for the private entity was lower than the cost of borrowing to the public jurisdiction.[11]

> Assumption 1 (tax rates): There are no taxes for the public entity, but an ordinary income tax rate $T_I > 0$ and a capital gains tax rate $T_C \geq 0$ are assessed to the private entity.
> Assumption 2 (volume discounts): Equipment prices are x% lower for the private entity versus the public entity via volume discounts.
> Assumption 3 (maintenance contract): Maintenance contracts costs are y% lower for the private entity versus the public entity via volume discounts.
> Assumption 4 (resale): Salvage values at time N are z% higher for the private entity versus the public laboratory.
> Assumption 5 (discount rate): The cost of debt is lower for the private sector entity than it is for the public sector entity.

Suppose that the pre-tax interest rate for the private entity is 4.5% compared to the 5.5% rate for the public sector entity. In this case, since the interest payments on debt are tax deductible, the after-tax cost of debt, given the tax rates used previously for local, state, and federal levels, is lower for the private entity. The present value cost for the private entity falls and the difference in the present value costs from leasing jumps to $54,501.

Negotiations and Realized Gains

Granof (1984) notes that when a lease contract is executed, there is an explicit agreement that the costs, risks, and benefits of the asset in question are to be split up in a specific fashion between the lessor and lessee. A leased asset will always represent a basket of these costs, risks, and benefits that must be distributed in some combination to the two parties in the lease contract. If one party receives a specific unit of benefit, the other cannot consume that same benefit. In the preceding paragraphs, an explanation of how the total basket of net benefits (benefits minus costs and risks) can be maximized through an advantageous use of tax law was given. However there are other ways either the lessor or lessee may be able to increase their present value of the lease contract, which may or may not affect the net benefit of the lease contract. While the preceding scenarios isolate the rationale for potential gains via leasing, the magnitude of the gains will depend upon several factors. To appreciate the potential for gain to either party to the lease, first consider the limits to the size of the lease payment. For the lessee, the lease payment must be consistent with a present value cost that is less than the present value cost of ownership. Likewise, the lease payment must be sufficient to yield a positive net present value of ownership to the lessor. That net present value of ownership includes the present value costs highlighted in the preceding scenarios and includes a present value of the benefits from the taxable leasing income. To illustrate the size of the gains, consider scenario 5. In that scenario, the present value cost to the public laboratory is consistent with an annual lease payment of $27,676, which covers operational use of the equipment and maintenance. Thus any lease terms must have a lease payment below $27,676 for the lease to be attractive to the public laboratory. For the lessor, an annual lease payment of $20,805 is consistent with a return that covers the after-tax cost of borrowing to purchase the equipment and maintenance contract. Any lease payment must be at least as high as $20,805.

The distribution of the gains to the lessee and the lessor will be determined by factors that affect any exchange. These factors include assessments of risk, required returns, resale markets, and the competitive nature of the markets for leasing. For example, the more specialized the equipment, the greater the risk to the lessor and the higher the lease payment. However, the ability of the public sector to tax and/or borrow will temper much of the risk and should provide negotiating power to the public laboratory.

Concluding Remarks

The increased demands for forensic laboratory services coupled with calls for austerity across the public sector have forced laboratories to explore means to better optimize service delivery. Most laboratories will continue to explore process improvements, consider the efficacy of all investigative services, defer long-term investments, reduce investment in human capital, and perhaps leave open positions unfilled. The present analysis suggests that laboratories may be able to increase the spending power of their budgets through a strategic use of leasing. Project FORESIGHT data submissions suggest that leasing is an un-derused financial vehicle for many laboratories. As shown in the example, there may be as much as 25% of the cost of capital investment that may be a shared gain between the lessee laboratory and a lessor. When considered against alternative cost-cutting measures, the extension of leasing may save and potentially expand the number of positions for bench scientists, thus expanding the services rendered by the laboratory.

Appendix—Lease Versus Purchase Savings Technical Details

Investment in capital assets involves an upfront expenditure with the expectation of future cash flows that will be received over a number of years. For inter-temporal decisions, alternatives such as lease versus purchase require a mechanism to adjust for the time period in which the laboratory spends cash. To illustrate the potential savings from leasing versus the purchase of laboratory equipment, consider first the corrections for the time value of money. Consider an allowance for the time value of money through conversion to the present value cost (PVC) in which all future cash flow is converted to the present. The general descriptions of the PVC for the public laboratory and the private laboratory are presented in equations (1) and (2), respectively.

$$PVC_{pub} = \sum_{t=1}^{N} CF_t \frac{1}{(1+r_{pub})^t} \tag{1}$$

$$PVC_{priv} = \sum_{t=1}^{N} CF_t \frac{1}{(1+r_{priv})^t} \tag{2}$$

where CF_t = cash flow at time t; N = total number of periods for which the equipment is maintained; r_{pub} = interest rate on bonds of the public sector jurisdiction; r_{priv} = interest rate after tax on bonds of the private corporation.

Consider both a public laboratory and a for-profit private laboratory that are faced with the same equipment and maintenance expenses, but a tax rate on ordinary income and a tax rate applied to capital gains income applies to the for-profit entity. The cash flows for the private laboratory will be the same as the public laboratory for equipment and maintenance with a few exceptions. The private laboratory will have the upfront equipment expense and the beginning of period maintenance expenditure, but it will also deduct depreciation in each period before consideration of their tax liability. That depreciation deduction offers a tax shield for each period t that is effectively a subsidy via a reduction in tax liability.

The present value cost to the public laboratory, PVC_{pub}, represents the amount that the jurisdiction requires in present dollars to pay for this investment.

$$\begin{aligned} PVC_{pub} &= \sum_{t=0}^{N} CF \frac{1}{(1+r_{pub})^t} = EQUIP_{pub} \\ &- MAINT_{pub} \sum_{t=0}^{N-1} \frac{1}{(1+r_{pub})^t} + PVCG_{pub} \end{aligned} \tag{3}$$

CF_t is the cash flow in each period t, and r_{pub} represents the cost of debt or discount rate for the public jurisdiction. The cash flows for the public laboratory are comprised of the initial cost of the equipment at time zero ($EQUIP_{pub}$) and the annual maintenance fee ($MAINT_{pub}$) paid at the beginning of each period of operation. The final term, $PVCG_{pub}$, represents the present value of any capital gain from the sale of used equipment at the end of the project at time N.

For the private entity, expenses are tax deductible. As a result, the present value cost to the private laboratory (PVC_{priv}) is offset by reductions in the tax bill, and the private entity ends up paying less for equipment and maintenance. The tax benefits to the maintenance charge are taken each year that maintenance is paid, while the tax benefits from the equipment is spread over time through deductions for depreciation in each period t (D_t), following applicable tax code.

$$PVC_{priv} = \sum_{t=0}^{N} CF_t \frac{1}{(1+r_{priv})^t}$$

$$= -EQUIP_{priv} - MAINT_{priv} \times (1-T_I) \sum_{t=0}^{N-1} \frac{1}{(1+r_{priv})^t} \tag{4}$$

$$+ \sum_{t=1}^{N} \frac{T_I D_t}{(1+r_{priv})^t} + PVCG_{priv}$$

The final term, $PVCG_{priv}$, is the present value of the capital gains net of taxes. In a comparison of equations (3) and (4), the potential gains from leasing emerge. As noted in the previous section, if the present value cost is less for the private entity, then by making the equipment purchase in the private sector, the reduced cost from tax deductibility of depreciation and maintenance can potentially be shared with the public sector laboratory through annual lease payments that are lower than the annualized costs for a public laboratory as represented by the present value cost in equation (3).

The CF_t for each period t for either (3) or (4) may be described as

$$CF_t = -EQUIP_{jt} - (MAINT_{jt} + D_t) \times (1 - T_I)$$
$$+SV_{jt} + (SV_{jt} - BV_t) \times (1 - T_C) + D_t \tag{5}$$

where $EQUIP_{jt}$ = Price of the laboratory equipment for entity j at time t; $MAINT_{jt}$ = Cost of a maintenance contract for entity j at time t; D_t = Depreciation taken at time t; T_I = Tax rate on ordinary income from combined local, state, and federal taxes; TC = Tax rate on capital gains from combined local, state, and federal taxes; SV_{jt} = Salvage value for entity j at time t, $SV_{jt} = 0$ for t = 1 ... N-1; and BV_t = Book value at time t, if t = N, 0 otherwise.

To illustrate the source of the gains from leasing in place of ownership, first assume that the terms of the operating lease meet the guidelines of the IRS. The five example scenarios and the corresponding present value costs are detailed below.

Scenario 1

In this scenario, the cash flows to the public entity and the private entity are identical and there would be no advantage for leasing over purchase.

- x = 0%; y = 0%, z = 0%
- $EQUIP_{pub} = EQUIP_{priv}$ = $100,000
- $MAINT_{pub\ t} = MAINT_{priv\ t}$ = $5,000 for t = 0, ... ,3
- T_I = 0%, T_C = 0%
- $r_{pub} = r_{priv}$ = 5.5%
- $SV_{pub} = SV_{priv}$ = $20,000

The cash flows that would be inserted into (3) and (4) above are:

TIME	0	1	2	3	4
CF_{pub}	−$105,000	−$5,000	−$5,000	−$5,000	$20,000
CF_{priv}	−$92,392	$4,844	$4,844	$4,844	$34,212
CF_{priv} - CF_{pub}	$0	$0	$0	$0	$0
PVC_{priv} - PVC_{pub}	$0	$0	$0	$0	$0

Note that in scenario 1, the cash flows to the public entity and the private entity are identical in each and every period. Further, from the discount rate assumption, the cash flows are discounted at the same rate. Thus, there would be no advantage for leasing over purchase since the net present cost would be identical.

Scenario 2

There are no taxes for the public entity, but an ordinary income tax rate $T_I > 0$ and a capital gains tax rate $T_G > 0$ are applied to the private entity. The ordinary income tax rates include taxes at the national (T_{IN} = 35%), state (T_{IS} = 7%), and local (T_{ILS} = 1%) level. Additionally there are capital gains taxes at each level of government (T_{CN} = 20%, T_{CS} = 7%, and T_{CL} = 1%).

- x = 0%; y = 0%, z = 0%
- $EQUIP_{pub} = EQUIP_{priv}$ = $100,000
- $MAINT_{pub\ t} = MAINT_{priv\ t}$ = $5,000 for t = 0, ... ,3
- T_I = 40.2%, T_C = 26.4%[12]
- $r_{pub} = r_{priv}$ = 5.5%
- $SV_{pub} = SV_{priv}$ = $20,000

While positive tax rates on corporate profits might seem to have a negative connotation, it is the impact on cost reduction that is emphasized here. Tax deductibility of expenses, combined with the depreciation tax shield, offers an advantage that the private entity has over the public sector laboratory.

In Scenario 2, tax deductibility begins to show effects immediately with the initial period. That is, each entity is charged the same price for the equipment and maintenance contract but the private entity is able to deduct the annual maintenance charge and its effective cost is reduced by the tax rate times that annual cost. It becomes an indirect subsidy for their operations that continues in each period (through time N-1) that it pays its maintenance fee. In subsequent periods the private entity receives another indirect subsidy as it is able to depreciate the initial investment and deduct the depreciation expense prior to the calculation of its tax bill. This reduces their effect cost further by TI * Dt for each subsequent period.

TIME	0	1	2	3	4
CF_{pub}	−$105,000	−$5,000	−$5,000	−$5,000	$20,000
CF_{priv}	−$92,392	$4,844	$4,844	$4,844	$34,212
CF_{priv} - CF_{pub}	$2,010	$10,050	$10,050	$10,050	$8,040
PVC_{priv} - PVC_{pub}	$12,608	$9,456	$9,086	$8,731	$14,620

Given this information, consider the present value cost to the public laboratory (PVC_{pub}) versus the present value cost in the private sector (PVC_{priv}) for the purchase and maintenance of the equipment for a four-year period. From (3) the present value cost to the public laboratory for the equipment and the maintenance contract is $102,345. From (4) the present value cost to the private entity is only $64,149. The difference in these present value costs of $38,196 represents the gains to the private sector over the public sector. Potential gains via leasing arise from the present value of reductions in tax collections of the federal government, state government, and local government. The gains to the individual laboratory, and its political jurisdiction, depend upon the tax jurisdiction of the leasing entity. Regardless, there are involuntary revenue transfers taking place.

Scenario 3

There are no taxes for the public entity, but an ordinary income tax rate $T_I > 0$ and a capital gains tax rate $T_G > 0$ are applied to the private entity. The ordinary income tax rates include taxes at the national ($T_{IN} = 35\%$), state ($T_{IS} = 7\%$), and local ($T_{IL} = 1\%$) level for both ordinary and capital gains taxes ($T_{CN} = 20\%$, $T_{CS} = 7\%$, and $T_{CL} = 1\%$). Additionally, the private entity is able to negotiate equipment prices that are x% lower for the private entity versus the public entity via volume discounts and maintenance contracts costs which are y% lower for the private entity versus the public entity from diversification of repair risk.

- x = 10%; y = 20%, z = 0%
- $EQUIP_{pub}$ = $100,000; $EQUIP_{priv}$ = $90,000
- $MAINT_{pub\ t}$ = $5,000 for t = 0, ... , 3, $MAINT_{priv\ t}$ =
- 4,000 for t = 0, ... ,3
- $T_I = 40.2\%$, $T_C = 26.4\%$
- $r_{pub} = r_{priv} = 5.5\%$
- $SV_{pub} = SV_{priv}$ = $20,000

While the most dramatic transfers occur from the trading of tax positions, the private sector negotiation advantages may be shared with a public sector lessee.

TIME	0	1	2	3	4
CF_{pub}	−$105,000	−$5,000	−$5,000	−$5,000	$20,000
CF_{priv}	−$92,392	$4,844	$4,844	$4,844	$26,732
CF_{priv} - CF_{pub}	$12,608	$9,844	$9,844	$9,844	$6,732
PVC_{priv} - PVC_{pub}	$12,608	$9,429	$9,033	$8,654	$7,342

Given this information, consider the present value cost to the public laboratory (PVC_{pub}) versus the present value cost in the private sector (PVC_{priv}) for the purchase and maintenance of the equipment for a four-year period. From (3) the present value cost to the public laboratory for the equipment and the maintenance contract is \$102,345. From (4) the present value cost to the private entity is only \$55,280. The difference in these present value costs of \$47,066 represents the gains to the private sector over the public sector. Potential gains via leasing arise from the present value of reductions in tax collections of the federal government, state government, and local government. The gains to the individual laboratory, and its political jurisdiction, depend upon the tax jurisdiction of the leasing entity. Regardless, there are involuntary revenue transfers taking place.

Scenario 4

There are no taxes for the public entity, but an ordinary income tax rate $T_I > 0$ and a capital gains tax rate $T_G > 0$ are applied to the private entity. The ordinary income tax rates include taxes at the national ($T_{IN} = 35\%$), state ($T_{IS} = 7\%$), and local ($T_{IL} = 1\%$) level for both ordinary and capital gains taxes ($T_{CN} = 20\%$, $T_{CS} = 7\%$, and $T_{CL} = 1\%$). The private entity is able to negotiate equipment prices that are x% lower for the private entity versus the public entity via volume discounts and maintenance contracts costs which are y% lower for the private entity versus the public entity from diversification of repair risk. The private entity also has an advantage in the resale market at the end of the lease and is able to sell the equipment at a salvage value that is z% higher than received by the public laboratory.

- x = 10%; y = 20%, z = 50%
- $EQUIP_{pub}$ = \$100,000; $EQUIP_{priv}$ = \$90,000
- $MAINT_{pub\ t}$ = \$5,000 for t = 0, … , 3, $MAINT_{priv\ t}$ =
- 4,000 for t = 0, … ,3
- $T_I = 40.2\%$, $T_C = 26.4\%$
- $r_{pub} = r_{priv} = 5.5\%$
- SV_{pub} = \$20,000; SV_{priv} = \$30,000

The advantage in the after-market offers a realistic advantage for a lessor. This is particularly true when the equipment manufacturer takes on the role of lessor.

TIME	0	1	2	3	4
CF_{pub}	–\$105,000	–\$5,000	–\$5,000	–\$5,000	\$20,000
CF_{priv}	–\$92,392	\$4,844	\$4,844	\$4,844	\$34,212
CF_{priv} - CF_{pub}	\$12,608	\$9,844	\$9,844	\$9,844	\$14,212
PVC_{priv} - PVC_{pub}	\$12,608	\$9,429	\$9,033	\$8,654	\$13,914

Given this information consider the present value cost to the public laboratory (PVC_{pub}) versus the present value cost in the private sector (PVC_{priv}) for the purchase and maintenance of the equipment for a four-year period. From (3) the present value cost to the public laboratory for the equipment and the maintenance contract is \$102,345. From (4) the present value cost to the private entity is only \$48,708. The difference in these present value costs of \$53,637 represents the gains to the private sector over the public sector. Potential gains via leasing arise from the present value of reductions in tax collections of the federal government, state government, and local government. The gains to the individual laboratory, and its political jurisdiction, depend upon the tax jurisdiction of the leasing entity. Regardless, there are involuntary revenue transfers taking place.

Scenario 5

The final example accounts for the difference between the public jurisdiction's borrowing power and that of the private entity. The after-tax cost of debt may be higher or lower for the private sector and hence can increase or decrease the present value cost. In this scenario, the discount rate reflects a higher pre-tax rate for the private entity, but an after-tax discount rate lower than the public jurisdictions discount rate. The higher the after-tax cost of debt, the lower is the present value net costs from future periods.

Begin with the conditions of scenario 4 where the laboratory is considering the purchase of the scientific instrument, but add the conditions where the private entity has a pre-tax cost of debt that is an additional 1% lower than the public jurisdiction. Since the interest payments on debt are tax deductible, the after-tax cost of debt, given the tax rates used previously for local, state, and federal levels, is lower for the private entity.

- x = 10%; y = 20%, z = 50%
- $EQUIP_{pub}$ = \$100,000; $EQUIP_{priv}$ = \$90,000
- $MAINT_{pub\ t}$ = \$5,000 for t = 0, ... , 3, $MAINT_{priv\ t}$ =
- 4,000 for t = 0, ... ,3
- T_I = 40.2%, T_C = 26.4%
- r_{pub} = 5.5%; r_{priv} (pre-tax) = 4.5%; r_{priv} (after-tax) =
- 2.69%
- SV_{pub} = \$20,000; SV_{priv} = \$30,000

The present value cost for the private entity falls to \$47,844 and the potential gains are derived partly from lower taxes and the remainder from purchasing power/resale savings to the private sector.

TIME	0	1	2	3	4
CF_{pub}	–\$105,000	–\$5,000	–\$5,000	–\$5,000	\$20,000
CF_{priv}	–\$92,392	\$4,844	\$4,844	\$4,844	\$34,212
PVC_{priv} - PVC_{pub}	\$12,608	\$9,456	\$9,086	\$8,731	\$14,620

Endnotes

1 The acronym PIGS was applied to Portugal, Ireland (and sometimes Italy), Greece, and Spain as these were the first European countries for which the cracks in the global economy appeared. Each country began to show the strain of budgets where national deficits exceeded the annual gross domestic product.

2 The origination of the term *sequestration* as it is currently known is a result of the Gramm-Rudman-Hollings Balance Budget Act of 1985, in which sequesters are defined as a form of automatic spending cuts when deficit targets are missed.

3 There are a broader set of categories for leases that include characteristics of both the operating lease and the financial lease, such as combination leases, sale-and-leaseback arrangements, and synthetic leases.

4 IRS allowances for leases provide guidelines to ensure that the lease is a true lease and not just a means to circumvent rules on the deductibility of expenses. So-called 'guideline' leases are described in the text to follow.

5 The lessee is the party that takes control of the asset or equipment for its intended use, whereas the lessor is the party that purchases the asset or equipment and for a fee gives control of the asset to the lessee.

6 In addition to the standards established by the accounting profession, legal standards also impose guidelines for the permissible depreciation affiliated with capital equipment. In the United States, for example, several laws restrict the lessor's use of investment tax credits and accelerated depreciation when the lessee is a tax-exempt organization.
7 Depreciation is a non-cash expense that is acknowledged each period for the calculation of taxes, but does not represent an actual cash outflow.
8 The depreciation deduction is a non-cash expense to the laboratory. They deduct the amount in the calculation of their tax bill, but do not pay it out. The determination of cash flow for any operating period includes the after-tax operating profit plus the depreciation added back.
9 One example of abusing leasing undertaken before DEFRA by a tax exempt entity at the detriment to the federal government's budget is presented by Blose et al. (1989), where they discuss the Navy's financing of 13 cargo ships in 1982. At this time the Navy could have purchased these cargo ships at an estimated cost of $178.2 million per ships, but instead they leased them as the present value cost to the Navy of using these same cargo ships for their useful life was only $141.2 million per ship for a cost savings of $37 million per ship or $481 million dollars total. On the surface this seems like a good deal, and for the Navy it was indeed preferable to purchase through retained earnings or debt financing given the law. The tax benefits received and shared between the private lessor and Navy were estimated to be $57.8 million dollars per ship, which is exactly the per ship loss experienced by the U.S. Treasury. Thus the societal per ship cost of the lease is equal to $141.2 million plus $57.8 million, which equals $199 million. This cost is greater than the cost of direct purchase by $20.8 million which displays well why DEFRA was enacted by Congress.
10 The sale at the end of four years is merely for simplicity of the calculations. The authors realize that laboratories generally keep equipment in operation for a much longer period of time. IRS guidelines restrict the lease to a period that is 80% or less of the useful life, not the length of the depreciation class. In this example, a 10-year engineering life would have permitted the lease to have lasted up to eight years.
11 Note that the discount rate for a local or state entity has advantages in the tax law that make them tax-exempt from federal taxes and may lead to lower pre-tax rates than the public sector. On the other hand, lessors tend to be high tax rate and low-risk providers who benefit from lower risk and lower rates. The net effect of these factors may result in rates for the private entity that are higher or lower than the public sector laboratory. This scenario is merely exemplary of a reduction.
12 Since payments to state and local taxing authorities are deductible from federal taxes, then the effective tax rate from all sources includes the allowance for these deductions at the federal level.

References

Blose, L.E., and J.D. Martin 1989. Federal Government Leasing: Costs, Incentives, and Effects. *Public Budgeting and Finance* 9(2): 66–75.

Granof, M.H. 1984. Tax Exempt Leasing: A Framework for Analysis. *Public Administration Review* 44(3): 232–240.

Heald, D., and G. Georgiou 2011. The Substance of Accounting for Public-Private Partnerships. *Financial Accountability & Management* 27(2): 217–247.

Houck, M.M., R.A. Riley, P.J. Speaker, and T.S. Witt 2009. FORESIGHT: A Business Approach to Improving Forensic Science

Services. *Forensic Sciece Policy & Management: An International Journal* 1(2): 85–95.

Lewellen, W.G., M.S. Long, and J.J. McConnell 1976. Asset Leasing in Competitive Capital Markets. *The Journal of Finance* 31(3): 787–798.

Masse, I., J.R. Hanrahan, and J. Kushner 1988. The Effect of Changes in Tax Legislation on the Purchase/Lease Decision in the Public Sector. *National Tax Journal* 41(1): 123–130.

Newman, J., D. Dawley, and P.J. Speaker 2012. Strategic Management of Forensic Laboratory Resources: From Project FORESIGHT Metrics to the Development of Action Plans. *Forensic Science Policy & Management: An International Journal.*

Shall, L.D. 1974. The Lease-Or-Buy and Asset Acquisition Decisions. *The Journal of Finance* 29(4): 1203–1214.

Speaker, P.J., and A.S. Fleming 2009. Monitoring Financial Performance: An Approach for Forensic Crime Labs. *The CPA Journal* 79(8): 60–65.

Speaker, P.J., and A.S. Fleming 2010. Benchmarking and Budgeting Techniques for Improved Forensic Laboratory Management. *Forensic Science Policy & Management: An International Journal* 1(4): 199–208.

Article: Strategic Management of Forensic Laboratory Resources

6.4

From Project FORESIGHT Metrics to the Development of Action Plans

JONATHAN NEWMAN,[1] DAVID DAWLEY,[2] AND PAUL J. SPEAKER[2]

[1]Centre of Forensic Sciences, Toronto, Ontario, Canada

[2]College of Business and Economics,West Virginia University, Morgantown, West Virginia, USA

Introduction

The first decade of the new millennium was marked by dramatic changes in the state of the world economy from boom to recession. The lingering recession has resulted in reduced revenues to governments at all levels and that, in turn, has been met with a reduction in budget allocations for many publicly funded organizations including forensic science laboratories (Speaker & Fleming 2009). Simultaneously, there have been dramatic changes in the demands on the justice system and the expectations of forensic laboratories to support the needs of the justice system. The combination of these two phenomena opened the second decade with great expectations placed upon the shoulders of the managers of forensic science services. How will these laboratories be able to provide a greater number of services with fewer resources at their disposal? Fortunately, efforts in the last decade have established the groundwork to address this question*.369*.369n. The European QUADRUPOL study (2003) and the North American FORESIGHT[1] study (Houck et al. 2009) offer a means to begin the measurement of forensic science services through the provision of a common nomenclature upon which to have a discussion about performance. Once a common language had been established, participating laboratories could begin to collect data and compare that data across laboratories to establish baseline industry information. The collected data was drawn from a variety of laboratory databases including casework, financial, and human resources records and separated by area of investigation. For each investigative area, the information regarding budgets and budget allocation, human resource data on the provision of personnel, and finally casework were united for each laboratory to evaluate performance.

In the present paper, we address the issue as to what a laboratory can and should do with the data for improved laboratory performance. We turn to specific business lessons regarding forensic laboratories with special attention to the laboratory management issues raised by the project FORESIGHT participants. The review of this literature includes discussions of strategic missions, key performance indicators, budgeting, and specific short-term and

 DOI: 10.4324/9780429318528-21

long-term problems in the management of forensic laboratories. With approximately 70% of the typical laboratory's budget devoted to compensation of laboratory personnel, personnel issues dominate the lessons highlighted in the literature (Speaker 2009b). We follow that discussion with a review of relevant literature in the field of strategic management. While forensic science laboratories represent a unique industry, they share a great number of commonalities with other service industries. The foundation of these shared lessons from strategic management is presented in the next section.

Combining a look at mission, vision, and values, we are able to connect the budget allocation process to a feedback loop through which the laboratory uses the FORESIGHT metrics to evaluate performance and reformulate strategic plans. To highlight the process, we provide an example of how one laboratory, Ontario's Centre of Forensic Sciences, has created a program inclusive of the management team across the laboratory to monitor and manage for continual improvement in laboratory performance. Concluding remarks follow with some suggestions on the implementation of similar processes in other laboratories.

The Business of Forensic Sciences

To date, there are few studies that have taken a strategic management approach to the operations of individual forensic crime laboratories or the role of these laboratories as an industry within the justice system. The forensic laboratory industry has traditionally lacked standard definitions and collection of data to assist in the measurement and improvement of performance. An early study (Dale & Becker 2003) noted that personnel expenditures account for the majority of forensic laboratory expenditures and the pressure for forensic laboratories to process more casework will require advanced planning in the preparation of scentists and analysts for the increased workload. While Dale and Becker (2003) call for better data to assess the market demand, they do provide a rough estimate of the numbers of scientists and analysts needed as a function of population served. The call for better industry data was answered with the European QUADRUPOL study (2003), which set the groundwork for such strategic management processes to be conducted by defining standard measures and quantifying performance across laboratories. The FORESIGHT project (Houck et al. 2009) continued the development of standardized measures and expanded participation to include a larger sample of laboratories with representation at the federal, provincial, county, and metropolitan levels and a broadening geographic profile. Additional cross-industry investigations (e.g., Peterson et al. 2010) place the forensic laboratory within the context of the entire justice system and suggest metrics for a strategic evaluation of the system as a whole.

A few recent studies (Speaker & Fleming 2009; Schade 2009; Becker et al. 2009) have undertaken aspects of strategic management issues within the context of singular issues faced by an individual laboratory. The management issues discussed in the literature (Dale and Becker 2003; 2004; Heames and Heames 2010) have tended to center on personnel matters. One example is an investigation into alternative staffing models as a strategic response to turnover in a metropolitan forensic laboratory (Dale & Becker 2004). This investigation offers some insights into the generality of a strategic planning process via a look at one specific problem. The case illustrates the connection between the identfication of the problem, the metrics to be monitored, and the review of results and feedback into the planning process. Heames and Heames (2010) offer some additional insights into measurement alternatives for the development of action plans to address similar laboratory staffing problems. And Houck (2009) offers yet another strategic perspective of personnel and staffing strategy through a discussion of how forensic sciences may provide the hook to attract women to scientific study and increase the talent pool for laboratories.

The FORESIGHT study has provided the base information to formulate a strategic planning process that may be followed by laboratories. While the project began with a plan to replicate the collection of data outlined by the QUADURPOL study, FORESIGHT expanded this collection to include issues raised by the participating laboratories (Houck et al. 2009). Subsequent review of the FORESIGHT data has been guided by the topics identified by the study's participating laboratories. These issues included the need to maintain high standards of quality while addressing an ever-increasing demand for services. Participants noted the need to correct for differences in facility sizes, geographic areas served, populations served, regional and cultural concerns, and the extent to which the CSI effect (Houck 2006) has been pressed upon them. Following the identification of those areas of concern, the search for potential key performance indicators (KPI) began with an examination of mission, vision, and values and the connection to the guiding objectives of a forensic science laboratory. The initial review of appropriate performance measures (Speaker 2009a) offers a host of potential ratio metrics to collect. These metrics offer insight into each of the areas of concern and include measures of return, quality, efficiency, analytical process, and market conditions. The very use of common size statements and ratios offered the ability to correct for differences in size population, geographical areas served, and for asset allocation by the sponsoring government entity. While the presentation of potential KPI marked a step forward, a subsequent article on the decomposition of ratios (Speaker 2009b) provided greater insight into how a laboratory might detect an explanation for its performance and identified potential red flags to highlight areas of concern. Armed with viable metrics and the means to analyze them, the foundation for the important planning stages of performance review and coordination with strategic goals is in place.

A review of the performance metrics helps to guide modifications in strategy, and along with the adjustments in strategy will come corresponding changes in budget allocations. Schade (2009) offers one example of a change in budget allocations to accompany a change in the division of duties in the analytical processing for fingerprint identification. Dale and Becker (2004) offer another example, where profound budget changes accompany a shift in duties from more highly paid scientists to lower paid analysts with oversight by scientists. Such reallocations are a natural part of the strategic planning process. A broader approach is taken through a look at the budget allocation process for a forensic laboratory (Speaker & Fleming 2009) in which performance is evaluated by connecting planned budgets to actual expenditures and adjusting to future use.

More recent analysis has begun to set the stage for the second stage of inquiry, namely to "preserve what works, and change what does not" (Houck et al. 2009, p. 85). Following a balanced scorecard approach (Kaplan & Norton 1992; Kaplan & Norton 1996), the metrics decomposing the return on investment can be considered as part of a larger set of metrics that address the broader mission of individual laboratories (Houck et al., forthcoming). Recognition of that broader mission suggests that comparison of the financial metrics to the rest of the forensic laboratory industry is only part of the strategic review that should be regularly undertaken by laboratory leadership, and many of the financial issues have at their heart some other aspect of laboratory management. Becker et al. (2010) provide laboratory leaders with a set of six specific recommendations to assist in the strategic management process. These include adopting common measures across laboratories, common quality standards, benchmarking standards in all areas of concern, identification and dissemination of best practices, constant monitoring of metrics, and use of a cost-benefit review of practices.

The Strategic Planning Process

Strategic management involves managerial decisions and resultant actions that help determine the performance of an organization. Such decisions and actions are derived from a careful internal and external analysis of the firm—a process referred to as environmental scanning. Once a manager undergoes a thorough environmental scan, s/he formulates strategies deemed to best fit with the organization's environment. Each strategy should be articulated to the extent that it can be implemented. The strategic loop is closed with a final step that evaluates the effectiveness of implemented strategies. Corrections can be made to any aforementioned step thought to be defective. This process is depicted in Figure 1.

Environmental Scanning

During an environmental scan, the laboratory manager should look at all relevant internal and external influences that have an impact or a potential impact on the laboratory. Internal influences are dichotomized into strengths and weaknesses. Similarly, external influences are divided into opportunities and treats.

Internal Analysis

In evaluating the laboratory from an internal perspective, the manager should visualize the organization as a set of resources—or capabilities—that act as the laboratory's foundation. Examples might include personnel, equipment, building, productivity, technology,

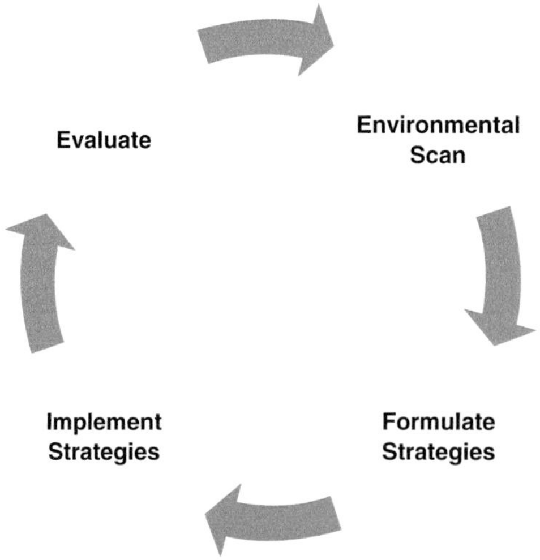

Figure 1 The strategic management process (color figure available online).

supplies, and comparison ratios. A ratio gaining current interest in the forensic science literature is that of return on investment as a cost minimization metric (Speaker 2009b). Strengths and weaknesses are only determined in a comparative sense. That is, it might not be accurate to stake a claim that a laboratory has a specific strength or weakness without comparing a capability in question to that of an industry benchmark. Benchmarks could be obtained through membership in a consortium of forensic labs, such as FORESIGHT. Only through such a comparison can a manager stake a claim that an activity in his/her laboratory is necessarily strong or weak.

Resource strengths might be considered to be core competencies of the laboratory. This is not to say that all resource strengths should be left alone. Forward-thinking organizations must consider resource strengths in the context of continuous improvement (Deming 1986). In other words, what can the laboratory do to improve on that which it already does well?

The bigger challenge often comes with the discovery of comparative weaknesses. Weaknesses need to be eliminated or improved upon. One of the most often cited strategy tools for discovering and repairing resource weakness can be observed by using value chain analysis (Porter 1985). The value chain is a sequentially linked set of activities beginning with materials entering a laboratory through test reporting and billing. The focus of the value chain analysis is to examine each activity with the question in mind, is this activity adding value? If the answer is no, the manager must find a way to fix the malfunction. Often, techniques such as brainstorming and process mapping will uncover methods for improving upon a weak link in the value chain.

While the above offers a very generic description of how value chain analysis works, in practice the process should take a fine-grained analysis of all laboratory activities. Blood alcohol, DNA, firearms and ballistics, and toxicology might be among the major value chain activities that a forensic laboratory undergoes. If so, then these four activities should be analyzed individually. For each activity, the strategist needs to address the following questions.

- Which links in that value chain can be called weaknesses? Can they be fixed?
- What are the closest upstream and downstream linkages to the activity under question?
- Are the linkages identified above partially responsible for deficiencies in the focal link? Is there a potential for synergies between the focal link and links with the other value chains? For example, can any links between the value chains of serology be combined into that of DNA analysis?

External Analysis

An environmental scan also includes the environment external to the lab. Major external forces to be scanned include economic, technological, political-legal, and socio-cultural variables. Table 1 lists some of the relevant external variables in each of these categories.

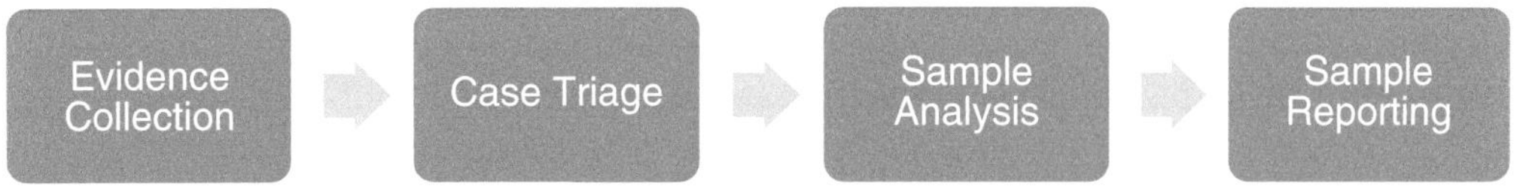

Figure 2 Typical value chain for a forensic crime laboratory (color figure available online).

Table 1 Environmental Scan External Force Variables

Economic variables	Technological variables	Political-legal variables	Socio-cultural variables
• Unemployment levels • Wage controls • Availability of laboratory scientists, analysts, and technologists	• Industry spending on R&D • New analytical products • Productivity improvements through automation	• Legislative mandates • Government funding • Environmental protection laws • Laws on hiring and promotion	• Population growth • Employee health care • Crime rates

While the above variables are used for purposes of illustration, each forensic laboratory faces its own set of external issues. External issues are opportunities or threats to the organization. Once the opportunities and threats have been identified, they need to be subjected to the following questions (Lederman 1984):

- What is the probability of occurrence?
- What is the probability of impact on the laboratory?

If the manager used a simple classification of high, medium, and low in the context of these two questions, then opportunities and threats graded as high in both probabilities should be given the utmost consideration going forward. Similarly, opportunities and threats scoring high/medium or medium/high to the two questions should also be identified as major issues and should be included in the strategic planning process.

Formulate Strategies

A commonly used tool in strategy formulation is the threat, opportunity, weakness, and strength (TOWS) matrix. Once a laboratory has identified the major issues it faces through environmental scanning, it can apply the TOWS matrix. To generate a TOWS matrix, refer to the graphic in Figure 3. From the internal portion of the environmental scan, the strategist lists the major strengths of the laboratory in the strengths (S) cell and lists the identified weaknesses in the weaknesses (W) cell. From the external environmental scan, the major opportunities are recorded in the opportunities (O) cell, with the threats (T) cell listing the major threats facing the laboratory.

The management team would then generate a host of strategies for the laboratory based on four possible strategy combinations from the TOWS matrix. SO strategies are created by brainstorming ways in which the laboratory can take advantage of opportunities using its identified strengths. WO strategies should consider exploiting opportunities while reducing weaknesses. ST strategies are generated in an effort to leverage a strength to minimize a threat. Finally, WT strategies simultaneously attempt to minimize weaknesses while avoiding threats. Once completed, all SO, WO, ST, and WT strate-gies should be listed as plausible strategies. In practice this list can become very lengthy. Therefore it is advisable to further brainstorm all plausible strategies with the intent of whittling the list down to a manageable few. This short list should contain only the most important—and viable—strategies critical to the lab's success. These selected strategies form the backbone of the lab's strategic plan.

	STRENGTHS – S List strengths (5-10)	WEAKNESSES – W List weaknesses (5-10)
OPPORTUNITIES – O List opportunities (5-10)	SO STRATEGIES Use strengths to take advantage of opportunities	WO STRATEGIES Overcome weaknesses by taking advantage of opportunities
THREATS – T List threats (5-10)	ST STRATEGIES Use strengths to avoid threats	WT STRATEGIES Minimize weaknesses and avoid threats

Figure 3 The TOWS matrix.

The Centre of Forensic Sciences Case

The Centre of Forensic Sciences (CFS) laboratory in Toronto, Ontario, Canada, was among the initial participants in the FORESIGHT study. As detailed in Houck et al. (2009), the FORESIGHT participants guided the definitions and collection of data through a series of discussions (an environmental scan) of the internal and external influences on their ability to meet their missions. From a discussion of those concerns have come the initial FORESIGHT-spawned studies outlined above, which have been concentrated in the identification and decomposition of appropriate metrics to assess the strategy of the individual laboratories.

The Auditor General of Ontario conducts value-for-money-oriented audits of selected government activities and programs. In December 2007 the Auditor published a report (Auditor General of Ontario 2007) concerning its assessment as to whether CFS had adequate systems and procedures in place to "provide efficient timely and reliable services and measure and report on the effectiveness of its services in supporting the administration of justice in Ontario."

The auditor made a recommendation that CFS improve its strategic management processes, specifically:

> In order to better monitor and report on its financial and operational performance the CFS should:
>
> - Establish measures to monitor the cost-effectiveness of its operations;
> - Benchmark its performance against that of other laboratories

In the next two years CFS implemented a strategic management cycle that integrated annual operational goals and objectives with human resources and financial performance planning cycles.

The framework for the integrated planning cycles was an overarching long-term strategic plan that was aligned with government priorities. The CFS strategic plan articulated mission, vision, and values as a direction statement with three supporting goals and associated objectives underpinned by the stated values of the organization. The strategic goals related to the service delivery, the workforce, and strategic management process.

The strategic management goal was: "To ensure the effective employment of resources in support of business objectives through the process of acquiring, allocating, controlling, and reporting on the use of fiscal and human resources." One of the supporting objectives was to establish measures to improve the monitoring and reporting of financial and operational performance.

Participation in the FORESIGHT study presented an opportunity that allowed CFS to respond to the recommendation of the auditor and as a result improve its strategic management system. In his follow up report published in December 2009 (Auditor General of Ontario 2009), the Ontario Auditor General reported, "The Centre has made a substantial effort to update its systems and procedures to address our recommendations and increase its efficiency. Of particular note is the Centre's participation in a multi-jurisdictional performance-benchmark project."

The need for laboratory leaders to seek continuous improvements in the performance of their organizations has been highlighted in the literature (Speaker 2009a). The determination of metrics to be used as indictors of progress toward this objective presents significant challenges to the managers of forensic science laboratory operations. Speaker (2009a) articulated the value of an examination of the vision, mission, and values, and the goals of an organization to be a good start point in the quest for metrics that will enable the development of objective pointers towards success.

Following the initial FORESIGHT studies, the CFS engaged in a strategic management process in which the metrics from the 2008–2009 fiscal year were used as a baseline from which to assess performance in subsequent periods.[2] Following the decomposition of average cost in Speaker (2009b), the baseline metrics came from the following relationship.

$$\frac{Cost}{Case} = \frac{Average\ Compensation \times Testing\ Intensity}{Labor\ Productivity \times Labor\ Expense\ Ratio}$$

where

- Average compensation includes the wages, salaries, and benefits to the average laboratory employee;
- Testing intensity reflects the ratio of the tests performed to the cases processed;
- Labor productivity reflects the output per laboratory employee; and
- Labor expense ratio measures the percentage of total expenses used for employee compensation.

The CFS performance in the 2008-2009 fiscal year was distributed to the managers accountable for laboratory operations. This performance included detail on casework and expenditures with breakdowns by distribution of personnel in the laboratory. Prominent among the metrics were ratio measures within each investigative area with comparisons to the FORESIGHT study averages.[3] These ratio metrics included the return on investment metric (average cost per case), an efficiency metric (tests performed relative to the employees in the investigative area), a quality metric (tests performed per case), a proxy for analytical process (percentage of expenditures for personnel), and an economic market metric (average compensation).

CFS used the concepts of the value chain and also of the TOWS matrix in their internal analysis. The first step was to identify the value chains being performed by the lab.

These chains were interpreted to be the division into investigative units as practiced by the laboratory prior to the strategic audit. Section managers were asked to review the value chain for each investigative area and contribute to the environmental scan and strategy formulation. Table 2 provides an annotated depiction of the TOWS matrix with more detailed description in the text that follows.

Identification of the relevant value chains and provision of performance metrics and benchmarks made the assessment of strengths and weaknesses a convenient process for the managers.

Weaknesses

Managers used a combination of an analysis of internal metrics and comparisons to benchmark averages in order to identify weaknesses in each investigative area. Areas in which to concentrate strategic efforts followed from the results of this analysis. A number

Table 2 CFS Threats, Opportunities, Weaknesses, and Strengths Matrix

	STRENGTHS – S	WEAKNESSES – W
	• Quality Assurance • Compensation and Benefits • Scientific staff is young and highly educated • Low employee turnover • Upward trend in productivity • Data collection and monitoring	• Limited Cross training • Some aging equipment • Capacity utilization • Staff Absences • Several analytical processes with low productivity • Unfilled positions
OPPORTUNITIES – O	SO STRATEGIES (examples)	WO STRATEGIES (examples)
• Federal Government funding initiatives • Technological advances • Funding for outsourcing Toxicology postmortem testing. • Growth in forensic management literature	• Use Federal government funding for new equipment to meet increased demands on DNA testing due to changes to the CCC • Improve data quality to ensure accurate accounting of expenditures and casework output.	• Use temporary funding for outsourcing testing of a historical backlog of Toxicology postmortem cases. • Improve cost per case and efficiency in Toxicology Ante Mortem through batching of same case type for processing and shifting case related duties.
THREATS – T	ST STRATEGIES (examples)	WT STRATEGIES (examples)
• Government budget constraints • Privatization of services • Changes to the Criminal Code of Canada (CCC) • Redistribution of the demand for services across investigative areas • Declining demand for some services • Generation Y work expectations	• Develop and implement new methods e.g. new combined drug assays in Toxicology. • Restructuring of operations in Toxicology Ante Mortem and Post Mortem and streamlining of testing processes	• Refurbish older SEM/EDX instruments to extend life without need for full replacement • Reduce Trace cost per case by using technologists rather than scientists for evidence collection

of weaknesses came to light including the following examples: Analytical processes in toxicology ante-mortem and toxicology postmortem were operating below capacity; toxicology postmortem showed room for improvement in productivity and demonstrated an excessive level of testing intensity; ageing instrumentation (SEM) that was potentially costly to replace was identified in GSR; and DNA casework suffered from a high number of staff absences due to the current staffing profile combined with employment benefits that provide for maternity leave of absence of one year. It was also recognized that performance metrics in low-volume, high-expertise areas such as explosives and trace can be impacted dramatically by slight fluctuations in demand and allocation of staff resources.

Strengths

In a similar fashion to the derivation of weaknesses, the identification of strengths came from individual investigative units' self-analysis of internal metrics and comparisons to benchmark averages. These strengths include: an attractive living and work environment in Toronto; superior benefits; a youthful and highly educated scientific staff; an upward trend in productivity; improvements in the availability and accuracy of laboratory performance metrics; a budget to replace/upgrade technology; and a high standard for quality scientific investigations.

Opportunities

Turning the attention to factors outside the laboratory, the following opportunities were noted: additional funding from the federal government; technological advances across investigative areas especially toxicology and DNA; short-term outsourcing of toxicology post-mortem analysis to reduce a historical backlog; and the growth in forensic management literature.

Threats

As with opportunities, the threats are similar to the issues faced by other forensic laboratories including: widespread government budget constraints as a response to the global economic downturn; consideration of alternative models to privatize services; the so-called "CSI effect" and the increase in demand for forensic services without corresponding increases in budgets; a redistribution of the demand for services across investigative areas; and the work expectations of Generation Y. Most notably in 2008, new legislation had come into force changing the Criminal Code of Canada. These changes created the potential for law enforcement to increase demands for scientific testing in toxicology ante-mortem (impaired driving by drugs offenses) and DNA casework (offenses investigated through the use of the National DNA Data Bank).

TOWS Matrix Application

The above listed strengths, weaknesses, opportunities, and threats become input for the TOWS matrix as shown in Table 2.

SO Strategies

The SO strategies were created by brainstorming ways in which the laboratory can take advantage of opportunities using its identified strengths. Comparing the strengths to the opportunities, CFS developed SO strategic initiatives in several investigative areas. Examples developed in the investigative area of DNA casework are shown in Table 2. Strategies included the use of federal government funding to acquire new equipment to meet increased demands on DNA testing due to changes in the Criminal Code of Canada. Another strategy was to continue to improve data accuracy and develop a strategy to address unusual productivity and costing numbers in the DNA casework investigative area. This included the need to improve staff performance management and the identification and mitigation of any delayed cases in the system.

ST Strategies

The ST strategies represent opportunities to address the threats by taking advantage of significant strengths. Examples of strategies in both of the toxicology investigative areas are shown in Table 2. One strategy was to leverage staff expertise to develop and implement new methods (e.g., new combined drug assays) that would improve return on investment through increased productivity. Toxicology operations were also restructured to improve performance management and to streamline testing processes. The objective of this strategy was to improve efficiencies by reviewing the value chain quality measure of testing intensity. The elimination of unnecessary testing in the toxicology post-mortem investigative area would mitigate the anticipated increase in testing intensity that would be required in the analysis of new drug-impaired driving cases in toxicology ante-mortem investigative area.

WO Strategies

The WO strategies make use of opportunities in order to address the weaknesses. Examples of strategies in both of the toxicology investigative areas are shown in Table 2. The opportunity for short-term funding to outsource toxicology post-mortem cases to reduce a historical backlog was used as a strategy to allow toxicology management to address weaknesses in analytical processes with low productivity. To improve the cost per case and efficiency in the toxicology ante-mortem investigative area, some caserelated duties were shifted from higher to lower paid staff and a new method of batch processing similar cases was implemented.

WT Strategies

The WT strategies address the weaknesses that have been identified in order to position the laboratory to avoid the threats. Examples of strategies identified in the investigative areas of GSR and trace evidence are shown in Table 2. To maintain the return on investment in the investigative area of GSR refurbishment of the aging SEM was preferred over replacement. This controlled capital investment costs in an area where technology is operating below capacity. The weakness of a lack of cross training and underutilized capacity in the

Table 3 Two-Year Percentage Change in Toxicology Investigative Areas Following CFS Strategic Initiatives

Measure	Metric	Toxicology ante mortem	Toxicology post mortem
ROI	Δ Cost/Case	2.02%	−11.98%
Demand	Δ Cases	−4.6%	4.6%
Demand	Δ Items	5.5%	−33.5%
Productivity	Δ Cases/FTE	6.6%	16%
Productivity	Δ Tests/FTE	87%	−21%
Testing Intensity	Δ Tests/Case	75%	−30.42%
Labor	Δ Staff FTE/area	−9%	−9%
Labor expense ratio	Δ Capital Expense/Total Expense	49.54%	40.13%
Turnaround time	Δ Median TAT	−51%	−52%

trace investigative area was addressed by training technologists rather than scientists for evidence collection in cases such as paint/glass hairs/fibers. This would have the net effect of reducing costs per case by reducing labor costs associated with the use of higher paid scientists.

CFS began to make its changes as suggested by these and other strategies and experienced some meaningful results by the end of the first year and stability in the changes by the end of the second year. Improvements were realized most quickly in the investigative areas of toxicology ante-mortem and toxicology post-mortem. Table 3 illustrates the rate of change in demand (casework inputs) and productivity outputs and the allocation of resources required such as full-time equivalent (FTE) staff and expenditures.

In the toxicology ante-mortem investigative area, the average cost per case increased (5%). However, the return on this modest increase was a 51% decrease in the median turnaround time required to complete the analysis required and to produce a report. Productivity improved by a 6.6% increase in the number of cases per person and an 87% increase in the number of tests per person. The improvement in the tests per person measure was important for the impaired driving by drugs cases where testing for a wide range of drugs is needed. The increase in cost per case was also due in part to increases in the investment in capital as a contribution to the higher cost. In the toxicology post-mortem investigative area, the average cost per case decreased by almost 12%. Changing processes had a significant impact by reducing testing intensity (tests/case) by over 30% and productivity (cases/FTE) increased by 16%. Most importantly for the customer was the 52% decrease in the median turnaround time required to complete the analysis required and to produce a report.

For many of the successful strategies, the improvement in the return on investment had a positive influence through the strategies that improved the productivity of laboratory personnel (i.e., more tests performed per person or more cases processed).

Concluding Comments

Global recession has forced governments around the world to reassess the provision of all public sector services, and forensic science laboratories are not immune from such scrutiny. Fortunately, the last decade has seen growing attention to the business of forensic science and that attention has begun to offer insight as to how a laboratory may meet its mission without sacrificing quality. Past research has offered the adaptation of base business tools and techniques for use by forensic laboratories, and this paper suggests ways to envelope a wide range of tools into a unified process. The initial adoption of a strategic management process will enable a laboratory to set the stage for continual improvement. The combination of these basic management tools with the data being generated industry wide will enable a laboratory to manage that process from production through quality assurance and budget allocation. And, as seen through the experience of the Centre of Forensic Sciences, the sharing of experiences offers a critical addition to the establishment of best practices.

Endnotes

1. As noted in its mission statement, "FORESIGHT is a business-guided self-evaluation of forensic science laboratories across North America. The participating laboratories represent local, regional, state, and federal agencies... The process involved standardizing definitions for metrics to evaluate work processes, linking financial information to work tasks and functions. Laboratory managers can use these functions to assess resource allocations, efficiencies, and value of services—the mission is to measure, preserve what works, and change what does not."
2. The metrics follow Speaker (2009b), where the objective of the laboratory is assumed to be to process as many cases in each investigative area as the budget permits. That objective translates into a minimization of the cost per case. Cost per case may be decomposed into a series of four ratios that help to explain performance. These ratios provide indications of quality, efficiency, analytical process, and local economic market conditions.
3. CFS has more investigative areas than described in the FORESIGHT study. Data across areas was grouped according to the FORESIGHT study to take advantage of the benchmarking benefits from the study. This enabled a review of investigative areas in common including: blood alcohol, DNA casework, document examination (including handwriting), explosives, fire analysis, firearms and ballistics, gun shot residue (GSR), marks and impressions, toxicology ante-mortem, toxicology post-mortem, and trace evidence (including hairs and fibers, and paint and glass). Discussion has been limited to a subset of possible areas to highlight the connection to strategic management.

References

Auditor General of Ontario. 2007. Annual report of the office of the Auditor General of Ontario: Chapter 3.02 Centre of Forensic Sciences. Accessed June 11, 2010, https://www.crto.on.ca/pdf/Reports/2009AuditorsReport.pdf.

Auditor General of Ontario. 2009. Annual report of the office of the Auditor General of Ontario: Chapter 4.02 Centre of Forensic Sciences. Accessed June 11, 2010, https://www.crto.on.ca/pdf/Reports/2009AuditorsReport.pdf.

Becker, W.S., W.M. Dale, and E.J. Pavur. 2010. Forensic science in transition: Critical leadership challenges. *Forensic Science Policy & Management* 1(4):214–223.

Dale, W.M., and W.S. Becker. 2003. Strategy for staffing forensic scientists. *Journal of Forensic Sciences* 48(2):465–466.

Dale, W.M., and W.S. Becker. 2004. A case study of forensic scientist turnover. *Forensic Science Communications*, 6(3).

Deming, W.E. 1986. *Out of the crisis.* Boston: MIT Center for Advanced Engineering Study.

Heames, J.T., and J.T. Heames 2010. Forensic science staffing: Creating a working formula. *Forensic Science Policy & Management* 2(1):5–10.

Houck, M.M. 2006. CSI: Reality. *Scientific American*, 295(1):84–89.

Houck, M.M. 2009. Is forensic science a gateway for women in science? *Forensic Science Policy & Management* 1(1):65–69.

Houck, M.M., R.A. Riley, P.J. Speaker, and T.S. Witt. 2009. FORESIGHT: A business approach to improving forensic science services. *Forensic Science Policy and Management* 1(2):85– 95.

Houck, M.M., P.J. Speaker, A.S. Fleming, and R. Riley. Forthcoming. The balanced scorecard: Sustainable performance assessment for forensic laboratories. *Science & Justice.*

Kaplan, R.S., and D.P. Norton. 1996. *The balanced scorecard.* Boston: Harvard Business Press.

Kaplan, R.S., and D.P. Norton. 1992. The balanced scorecard: Measures that drive performance. *Harvard Business Review* Jan/Feb:71–80.

Lederman, L.L. 1984. Foresight activities in the U.S.A.: Time for a reassessment? *Long Range Planning* 17(3):41–50.

Peterson, J., I. Sommers, D. Baskin, and D. Johnson. 2010. *The role and impact of forensic evidence in the criminal justice process. Document no. 231977. Award #2006-DN-BX-0094.* Washington, DC: U.S. Department of Justice, National Institute of Justice.

Porter, M.E. 1985. *Competitive advantage: Creating and sustaining superior performance.* New York: The Free Press.

QUADRUPOL. 2003. *European Network of Forensic Science Institutes.*

Schade, W. 2009. Budget crisis or management opportunity. *Forensic Science Policy & Management* 1(1):57–61.

Speaker, P.J. 2009a. Key performance indicators and managerial analysis for forensic laboratories. *Forensic Science Policy and Management* 1(1):32–42.

Speaker, P.J. 2009b. The decomposition of return on investment for forensic laboratories. *Forensic Science Policy and Management* 1(2):96–102.

Speaker, P.J., and A.S. Fleming. 2009. Benchmarking and budgeting techniques for improved forensic laboratory management. *Forensic Science Policy and Management* 1(4):199– 208.

What Next? 7

FSPs have long been compared to the for-profit sector. As far back as the 1970s, efforts to define the forensic industry were made, with little advancement. Yet, some researchers noted how FSPs operated like private businesses and how they differed:

> In one sense, the crime laboratory can be considered and compared with a business operation which has three basic operating elements: production, distribution, and marketing. The crime laboratory production capacity is more analogous to that of a business which provides services rather than one manufacturing a specific product. If the business is to prosper, it must advertise or market the availability of its services, it must have a location that is convenient to its users (or provide a pickup and delivery services), and it must do quality work on a timely basis if it is to enjoy the confidence and repeat business of its customers.
>
> *(Benson et al., 1972, page 26)*

That last point is important and the focus of this book. How do you know your FSP is doing quality work? FORESIGHT has outlined the internal metrics that matter, as discussed previously. But what about your supervisor's or parent agency's metrics? Many FSPs default or adopt metrics like "prosecutions aided," "arrests," "convictions," and other law enforcement measures of putative effectiveness.

> As part of the total police function, the laboratory is expected to justify the resources budgeted for its scientific services. This pressure has led, in some instances, to record keeping which stresses convictions, clearances, or positive findings–a practice which, to some, contradicts the theoretical goals of scientific objectivity and impartiality.
>
> *(Benson et al., 1972, page 5)*

Factors like guilt and innocence should have no bearing on the scientific results your FSP provides. Those are legal decisions, not scientific ones. Why worry about that?

Why This Matters

Using legal or policing measures like "convictions" is troublesome because the FSP could slide into a police or prosecutorial mindset and develop biases in that direction. It makes sense for police or prosecutors to think that way: That's how they are evaluated. But an FSP is different and needs to strive to be neutral. Using policing metrics or cherry-picking metrics to look best to supervisors is a recipe for trouble:

> All criminal justice researchers have observed the tendency of agencies within the system to stress performance measures which are impressive and "look" as good as possible...[t]his

DOI: 10.4324/9780429318528-22

> practice of projecting an amplified image of caseloads and competence may have an adverse effect on the laboratory's contribution to the criminal justice system.
>
> *(Peterson, 1974, p. 5)*

If the methods of assessment of laboratory efficiency are focused on policing or judicial metrics (arrests, suspects identified, guilt, etc.) rather than on the internal processes that would affect quality and turnaround times, the stated determinants in those reports of police use of forensic services, then the process is self-defeating. If the numbers used to improve the laboratory's efficiency, and, hence, its usefulness to the police, are the wrong ones, the laboratory is doomed to failure and the police have lost a valuable resource. These factors make it abundantly clear that the metrics used to evaluate a laboratory must be relevant, accurate, and free from influence or bias. Otherwise, why bother?

British economist Charles Goodhart noted in a 1975 article on the UK monetary policy that any observed statistical regularity will tend to collapse once pressure is placed upon it for control purposes. This has been more clearly restated as Goodhart's Law: "When a measure becomes a target, it ceases to be a good measure." The statement is a warning against a type of bias. If a nail manufacturer's assembly line measures productivity by the *number of nails*, they may get thousands of small nails but if the shipping department wants to measure productivity by *weight*, they may get a few very small nails. Thus, if metrics are chosen by one group for their own purposes, it likely will be a bad metric for other groups in the organization. Therefore, metrics must be chosen carefully and in context (preferably *yours*).

Metrics, however, may become enticing tools for seeking out what you think you ought to do, rather than explore what you are doing. Let's unpack that statement. Assume your turnaround time for sexual assault kits is 76 days—not bad but improvements are definitely possible. You meet with your biology team and push for a 10% decrease in turnaround time, that is, a reduction of eight days, taking it from 76 to 68. First of all, why the *need* for a reduction in turnaround time? Yes, getting answers to investigators and to victims quickly is important, without question, but is that the real goal? Second, why 10%? A nice round number? Pending or existing legislation requiring a report to be issued within 70 days of kit submission? Finally, *how will you reduce turnaround time by 10%*? Chasing what you think you *ought* to do leads you away from *what you are doing*. Instead, ask these questions:

- Why is your turnaround time currently 76 days? (Try using the "5 whys" method referenced in the next chapter.)
- How does your turnaround time compared to other FSPs?
- How long does each component of producing a report take? Where are the bottlenecks?
- Is there something else that needs to be addressed that might bring about a reduction in turnaround time?

Flatly stating a goal ("ought") without understanding where you are now and what you are doing ("is") can lead to employees "gaming the system." Where the stated goals are complex, sophisticated, or subtle, the personnel involved may seek their own goals to the detriment of the assigned tasks. They may create "workarounds" or cut corners or other ethically and procedurally questionable acts. Goals are all well and good, just be sure not to make them the purpose or mission of your FSP.

Do More Good

This book has intentionally used the phrase forensic service providers to describe those kinds of agencies or units. The phrase acknowledges that forensic science is the service being provided as a public good or service, like the fire department or clean tap water. Good science must be provided equally, effectively, and efficiently for it to fulfill its purpose. FSPs compete with other governmental entities for resources. Unless you can provide a winning argument for why monies should go to your FSP and not another agency, your FSP will remain under-resourced. Using the data Project FORESIGHT provides to laboratories and laboratory systems about their performance and that of the industry to monitor and improve your FSP operations, the resourcing for and provision of forensic services can lead to more effective science informing the criminal justice system and good stewardship of the public's investment in forensic services.

Go Deeper

Benson, W., Stacy, J. and Nicol, J., 1972. Systems analysts look at the crime laboratory. *Journal of Forensic Sciences*, 17(1), pp.25–49.

Houck, M.M., McAndrew, W.P., Porter, M. and Davies, B., 2015. A review of forensic science management literature. *Forensic Science Review*, 27(1), pp.53–68.

Houck, M.M. and Speaker, P., 2020. Project FORESIGHT: A ten-year retrospective. *Forensic Science International: Synergy*, 2, pp.275–281.

McAndrew, W.P. and Houck, M.M., 2020. Interpol review of forensic science management literature 2016–2019. *Forensic Science International: Synergy*, 2, pp.382–388.

McAndrew, W.P., Speaker, P.J. and Houck, M.M., 2023. Interpol review of forensic management, 2019–2022. *Forensic Science International: Synergy*, 6, p.100301.

Peterson, J., 1974. *The utilization of criminalistics services by the police: An Analysis of the physical evidence recovery process*. Washington, DC: Law Enforcement Assistance Administration.

Strathern, M., 1997. Improving ratings: Audit in the British university system. *European Review*, 5(3), pp.305–321.

Examples and Exercises

8

Tell, Show, Do, Apply

So far, this book has told you about benchmarking and Project FORESIGHT and shown you some examples of how they work and are used by FSPs. It is now time to do it for yourself and apply it to your own FSP.

The examples are scenarios that may seem simple at first but the choices you need to make can be quite difficult—after all, cutting a job means one thing to an organization but quite another to the employee who gets let go. Read through the scenario and take time to consider the implications and options. Once the solutions (and it will always be plural: There are more solutions than problems) are drafted, think about:

- What comes next, that is, given the changes you have suggested, consider the downstream events.
- About those downstream events…did you consider any *negative* ones? Brainstorm as if your solution failed spectacularly. How does that change your plan?
- Did you plan for a way to demonstrate that your solution worked? Can you prove it?
- What did you leave out? What did you not consider?
- Can your solution be edited? Could it be made more streamlined or efficient?

The example data set provided here is based on a real FSP, with its messy inconsistencies and accumulated history of decisions. Each exercise has one or more scenarios that present different challenges to management and benchmarking. Once you have dealt with the example data set, try the scenarios on your own FSP data.

State FSP Example Data

Jurisdiction:	State
Number of facilities:	1
Standard work week:	40 hours

The data appears at the end of this section.

Scenario 1: Budget Cuts

Your jurisdiction's tax base has taken a hit this year and therefore next year will be a lean one. This may be a one-time cut or the new baseline, you can't be sure. Also, you do not know just how deep the budget cuts will be. Your parent agency has asked for separate

DOI: 10.4324/9780429318528-23

budget plans that address a 5%, 7%, and a 12% budget reduction while maintaining the maximum service provision possible.

Scenario 2: Clean-Up

Phase 1

A nearby jurisdiction's FSP has been shut down due to improper practices. Upon review, 3,051 untested sexual assault kits (SAKs) have been found in evidence storage. Your FSP has been tasked with processing these kits. The legislature has allotted $1.6 million and 18–24 months to complete the work. The lowest cost for a private laboratory to work an SAK is $800 each.

Develop a strategic plan and proposed budget to address the untested SAKs—*while maintaining jurisdictional service levels as much as possible*—within the resources allotted. Detail what services will suffer, to what degree, and why.

Phase 2

Six months into the "clean-up" project, you discover that one of your analysts has "dry-labbed" a number of cases due to pressures associated with the increased workload. The details are:

- Cases assigned to analyst: 215
- Reports issued to date: 75
- Detected incidents of dry-labbing: 29

All of the evidence in the assigned cases is available for re-testing. Develop a strategic plan for addressing this situation, including costs for re-work; assume the analyst is on paid administrative leave pending further investigation.

Scenario 3: Legislative Bonus

Due to increased lottery sales, your jurisdiction has a surplus of money for the fiscal year. The additional funds allocated to your FSP are 4% of your annual budget to be spent over a two-year period. Develop a plan and budget for how that additional *one time* allocation will be spent. Other than the two-year spending period, there are no other restrictions on how the money can be spent.

Scenario 4: Policy Change for a Different FSP

The State Bureau of Investigation Crime Laboratory has experienced rapid growth in the number of Drug-Controlled Substances cases submitted to the laboratory (9,250 in the most recent year). Resource allocations from the State budget have not been able to keep up with this rapid growth and as a result, turnaround time has increased 4% per year,

	Current	Routine	Complex
Cases	9,250	8,800	450
Items Examined Internally	30,000	9,500	20,500
Samples Analyzed	67,500	12,500	55,000
Tests Conducted	205,000	37,500	167,500
Turnaround Time (days)	120		
Open Cases	3,000		
Cases Open 30 or more days	1,500		
FTE	27.00		
Capital Expenditures	$275,000		
Labor Expenditures	$2,750,000		
Variable Expenditures	$375,000		
Other Expenditures	$275,000		
Total Expenditures	$3,675,000		
Cost/Case	$397.30		
Cost/Item In	$122.50		
Cost/Sample	$54.44		
Cost/Test	$17.93		
Case/FTE	343		
Items/FTE	1,111		
Samples/FTE	2,500		
Tests/FTE	7,593		

open cases have increased 5% annually, and the number of cases open 30 or more days has increased 7% annually.

The laboratory director believes that some complex cases (cases with over 100 samples) are the source of the problem, given the laboratory policy of processing every sample, weighing each sample and conducting presumptive and confirmatory tests. Although these complex cases involve fewer than 5% of submitted cases, they involve the lion's share of tests conducted. One bench scientist in the unit has suggested that the laboratory policy be changed from a test all samples policy to a hypergeometric sampling[1] of the complex cases.

The table provides a breakdown of some of the key data submitted by the laboratory to Project FORESIGHT. Should the suggestion be adopted, and if so, what happens to the key FORESIGHT metrics for the laboratory?

Investigation area	Cases	FTE
Blood Alcohol	5,521	7.00
Crime Scene Investigation		
Digital evidence		
DNA Casework	2,142	16.00
DNA Database (including CODIS)		
Document Examination (including handwriting)		
Drugs - Controlled Substances	25,041	27.00
Evidence Screening & Processing		
Explosives		
Fingerprints	5,731	10.00
Fingerprints Database (including IAFIS)	80	2.00
Fire analysis	574	5.00
Firearms and Ballistics		
Firearms Database (including NIBIN)		
Forensic Pathology	93	2.00
Gun Shot Residue (GSR)	1,489	11.00
Marks and Impressions		
Serology/Biology		
Toxicology ante mortem (excluding BAC)	196	3.00
Toxicology post mortem (excluding BAC)		
Trace Evidence (includes Hairs & Fibers, Paint & Glass)		
Administration and Support		
Other Specialties (describe below)		
TOTAL		83.00

Figure 8.1 Cases/FTE by investigative area example for Scenarios 1–3.

Investigation area	Personnel Expenditures: Salary, Benefits, Overtime, and Temporary	Capital Expenditures current year; Service of instruments, Equipment Leasing	Chemicals, reagents, consumables, gases	Other Expenditures (all other charges & expenditures not included in the prior columns)	Total Expenditures
Total Laboratory Expenditures	**$9,331,250**			$9,633,750	**$18,965,000**
Blood Alcohol	$513,583				
Crime Scene Investigation					
Digital evidence					
DNA Casework	$2,173,056				
DNA Database (including CODIS)					
Document Examination (including handwriting)					
Drugs - Controlled Substances	$2,587,437				
Evidence Screening & Processing					
Explosives					
Fingerprints	$1,018,780				
Fingerprints Database (including IAFIS)	$232,246				
Fire analysis	$544,090				
Firearms and Ballistics					
Firearms Database (including NIBIN)					
Forensic Pathology	$189,928				
Gun Shot Residue (GSR)	$1,089,352				
Marks and Impressions					
Serology/Biology					
Toxicology ante mortem (excluding BAC)	$332,778				
Toxicology post mortem (excluding BAC)	$650,000				
Trace Evidence (includes Hairs & Fibers, Paint & Glass)					
Administration and Support					
Other Specialties (describe below)					

Figure 8.2 Expenditures by investigative area example for Scenarios 1–3.

Investigation area	cases	items	items outsourced	items examined internally	total samples from items examined internally	tests on samples examined internally	reports	Median turn around time, days from last submission (internal)	Median turn around time, days from first submission (internal)	open area cases at end of year	open area cases older than 30 days at end of year (backlog)	Operational Staff FTE	Digital Evidence gigabytes examined	DNA Casework Sexual Assault Kits examined	DNA Database offender entries	DNA Database arrestee entries
Blood Alcohol	5,521	5,760			5,565	12,500	5,625		28		158					
Crime Scene Investigation	0															
Digital evidence	0															
DNA Casework	2,142	10,250			15,000	55,000	1,875		320		1,360					
DNA Database (including CODIS)	0															
Document Examination (including handwriting)	0															
Drugs - Controlled Substances	25,041	52,500			82,500	200,000	24,350		145		3,950					
Evidence Screening & Processing	0															
Explosives	0															
Fingerprints	5,731	5,700			8,625	29,750	1,165		205		725					
Fingerprints Database (including IAFIS)	80															
Fire analysis	574	290			650	800	75		150		32					
Firearms and Ballistics	0	1,150			2,250	4,500	455		90		90					
Firearms Database (including NIBIN)	0															
Forensic Pathology	93															
Gun Shot Residue (GSR)	1,489															
Marks and Impressions	0	225			525	600	85		135		27					
Serology/Biology	0	6,250			30,000	35,000	1,385		85		280					
Toxicology ante mortem (excluding BAC)	196															
Toxicology post mortem (excluding BAC)	0															
Trace Evidence (includes Hairs & Fibers, Paint & Glass)	0	2,000			3,000	6,250	180		350		170					
Administration and Support	0															
Other Specialties (describe below)																
	0															
	0															
	0															
	0															
	0															
	0															
	0															
	0															
TOTAL												0.00				

Figure 8.3 Raw data by investigative area example for Scenarios 1–3.

Investigation area	Personnel Expenditures: Salary, Benefits, Overtime, and Temporary	Capital Expenditures current year; Service of instruments, Equipment Leasing	Capital Expenditures, Service of Instruments, Equipment Leasing for the prior four years	Chemicals, reagents, consumables, gases	Other Expenditures (all other charges & expenditures not included in the prior columns)	Total Expenditures (Personnel, Capital, Consumables, & Other Expenditures)
Total Laboratory Expenditures	**$9,331,250**	**$0**		**$0**	**$9,633,750**	**$18,965,000**
Blood Alcohol	**$513,583**					
Crime Scene Investigation						
Digital evidence						
DNA Casework	**$2,173,056**					
DNA Database (including CODIS)						
Document Examination (including handwriting)						
Drugs - Controlled Substances	**$2,587,437**					
Evidence Screening & Processing						
Explosives						
Fingerprints	**$1,018,780**					
Fingerprints Database (including IAFIS)	**$232,246**					
Fire analysis	**$544,090**					
Firearms and Ballistics						
Firearms Database (including NIBIN)						
Forensic Pathology	**$189,928**					
Gun Shot Residue (GSR)	**$1,089,352**					
Marks and Impressions						
Serology/Biology						
Toxicology ante mortem (excluding BAC)	**$332,778**					
Toxicology post mortem (excluding BAC)	**$650,000**					
Trace Evidence (includes Hairs & Fibers, Paint & Glass)						
Administration and Support	**$0**					
Other Specialties (describe below)						
TOTAL	$9,331,250	$0	$0	$0	$0	$0

Figure 8.4 Expenditures by investigative area example for Scenarios 1–3.

Investigation area	Expenditures Total	Personnel Expenditures	Capital Expenditures	Consumables Expenditures	Other Expenditures	FTE	Area Cases	Items	Samples	Tests	Reports
Blood Alcohol	$1,351,191	$513,583	$0	$0	$837,608	7.00	5,521	NA	5,565	12,500	5,625
Crime Scene Investigation	$0	$0	$0	$0	$0	0.00	0	NA	NA	NA	NA
Digital evidence - Audio & Video	$0	$0	$0	$0	$0	0.00	0	NA	NA	NA	NA
DNA Casework	$4,087,588	$2,173,056	$0	$0	$1,914,532	16.00	2,142	NA	15,000	55,000	1,875
DNA Database	$0	$0	$0	$0	$0	0.00	0	NA	NA	NA	NA
Document Examination	$0	$0	$0	$0	$0	0.00	0	NA	NA	NA	NA
Drugs - Controlled Substances	$5,818,210	$2,587,437	$0	$0	$3,230,773	27.00	25,041	NA	82,500	200,000	24,350
Evidence Screening & Processing	$0	$0	$0	$0	$0	0.00	0	NA	NA	NA	NA
Explosives	$0	$0	$0	$0	$0	0.00	0	NA	NA	NA	NA
Fingerprints	$2,215,363	$1,018,780	$0	$0	$1,196,583	10.00	5,731	NA	8,625	29,750	1,165
Fingerprints Database (including IAFIS)	$471,563	$232,246	$0	$0	$239,317	2.00	80	NA	NA	NA	NA
Fire analysis	$1,142,381	$544,090	$0	$0	$598,291	5.00	574	NA	650	800	75
Firearms and Ballistics	$0	$0	$0	$0	$0	0.00	0	NA	2,250	4,500	455
Firearms Database (including NIBIN)	$0	$0	$0	$0	$0	0.00	0	NA	NA	NA	NA
Forensic Pathology	$429,245	$189,928	$0	$0	$239,317	2.00	93	NA	NA	NA	NA
Gun Shot Residue (GSR)	$2,405,593	$1,089,352	$0	$0	$1,316,241	11.00	1,489	NA	NA	NA	NA
Marks and Impressions	$0	$0	$0	$0	$0	0.00	0	NA	525	600	85
Serology/Biology	$0	$0	$0	$0	$0	0.00	0	NA	30,000	35,000	1,385
Toxicology ante mortem (excluding BAC)	$691,753	$332,778	$0	$0	$358,975	3.00	196	NA	NA	NA	NA
Toxicology post mortem (excluding BAC)	$0	$0	$0	$0	$0	0.00	0	NA	NA	NA	NA
Trace Evidence	$0	$0	$0	$0	$0	0.00	0	NA	3,000	6,250	180
Other Specialties (describe below)											
	$0	$0	$0	$0	$0	0.00	0	NA	NA	NA	NA
	$0	$0	$0	$0	$0	0.00	0	NA	NA	NA	NA
	$0	$0	$0	$0	$0	0.00	0	NA	NA	NA	NA
	$0	$0	$0	$0	$0	0.00	0	NA	NA	NA	NA
	$0	$0	$0	$0	$0	0.00	0	NA	NA	NA	NA
	$0	$0	$0	$0	$0	0.00	0	NA	NA	NA	NA
	$0	$0	$0	$0	$0	0.00	0	NA	NA	NA	NA
	18,612,887.00	8,681,250.00	0.00	0.00	9,931,637.00	83.00					

Figure 8.5 Detailed expenditures by investigative area example for Scenarios 1–3.

Investigation area	Cases/100K Population	Items Examined Internally/ 100K Population	Samples Examined Internally/ 100K Population	Tests Performed Internally/ 100K Population	Reports Generated Internally/ 100K Population	Cost per Case	Cost per item	Cost per Sample	Cost per Test	Cost per Report	Average Compensation
Blood Alcohol	NA	NA	NA	NA	NA	$244.74	NA	$242.80	$108.10	$240.21	$73,369
Crime Scene Investigation	NA	NA	NA	NA	NA	NA	NA	NA	NA	NA	NA
Digital evidence - Audio & Video	NA	NA	NA	NA	NA	NA	NA	NA	NA	NA	NA
DNA Casework	NA	NA	NA	NA	NA	$1,908.30	NA	$272.51	$74.32	$2,180.05	$135,816
DNA Database	NA	NA	NA	NA	NA	NA	NA	NA	NA	NA	NA
Document Examination	NA	NA	NA	NA	NA	NA	NA	NA	NA	NA	NA
Drugs - Controlled Substances	NA	NA	NA	NA	NA	$232,35	NA	$70,52	$29.09	$238.94	$95,831
Evidence Screening & Processing	NA	NA	NA	NA	NA	NA	NA	NA	NA	NA	NA
Explosives	NA	NA	NA	NA	NA	NA	NA	NA	NA	NA	NA
Fingerprints	NA	NA	NA	NA	NA	$386,56	NA	$256,85	$74.47	$1,901,60	$101,878
Fingerprints Database (including IAFIS)	NA	NA	NA	NA	NA	$5,894.53	NA	NA	NA	NA	$116,123
Fire analysis	NA	NA	NA	NA	NA	$1,990.21	NA	$1,757.51	$1,427.98	$15,231.75	$108,818
Firearms and Ballistics	NA	NA	NA	NA	NA	NA	NA	NA	NA	NA	NA
Firearms Database (including NIBIN)	NA	NA	NA	NA	NA	NA	NA	NA	NA	NA	NA
Forensic Pathology	NA	NA	NA	NA	NA	$4,615.53	NA	NA	NA	NA	$94,964
Gun Shot Residue (GSR)	NA	NA	NA	NA	NA	$1,615.58	NA	NA	NA	NA	$99,032
Marks and Impressions	NA	NA	NA	NA	NA	NA	NA	NA	NA	NA	NA
Serology/Biology	NA	NA	NA	NA	NA	NA	NA	NA	NA	NA	NA
Toxicology ante mortem (excluding BAC)	NA	NA	NA	NA	NA	$3,529.35	NA	NA	NA	NA	$110,926
Toxicology post mortem (excluding BAC)	NA	NA	NA	NA	NA	NA	NA	NA	NA	NA	NA
Trace Evidence	NA	NA	NA	NA	NA	NA	NA	NA	NA	NA	NA
Other Specialties (describe below)											
	NA	NA	NA	NA	NA	NA	NA	NA	NA	NA	NA
	NA	NA	NA	NA	NA	NA	NA	NA	NA	NA	NA
	NA	NA	NA	NA	NA	NA	NA	NA	NA	NA	NA
	NA	NA	NA	NA	NA	NA	NA	NA	NA	NA	NA
	NA	NA	NA	NA	NA	NA	NA	NA	NA	NA	NA
	NA	NA	NA	NA	NA	NA	NA	NA	NA	NA	NA
	NA	NA	NA	NA	NA	NA	NA	NA	NA	NA	NA
	0	0	0	0	0						

Figure 8.6 Various metrics by investigative area example for Scenarios 1–3.

Investigation area	Items per Case	Samples per Case	Tests per Case	Reports per Case	Samples per Item	Tests per Item	Reports per Item	Tests per Sample	Reports per Sample	Cases per FTE	Items per FTE	Samples per FTE	Tests per FTE	Reports per FTE
Blood Alcohol	NA	1.01	2.26	1.02	NA	NA	NA	2.25	1.01	788.71	NA	795.00	1,785.71	803.57
Crime Scene Investigation	NA	NA	NA	NA	NA	NA	NA	NA	NA	NA	NA	NA	NA	NA
Digital evidence - Audio & Video	NA	NA	NA	NA	NA	NA	NA	NA	NA	NA	NA	NA	NA	NA
DNA Casework	NA	7.00	25.68	0.88	NA	NA	NA	3.67	0.13	133.88	NA	937.50	3,437.50	117.19
DNA Database	NA	NA	NA	NA	NA	NA	NA	NA	NA	NA	NA	NA	NA	NA
Document Examination	NA	NA	NA	NA	NA	NA	NA	NA	NA	NA	NA	NA	NA	NA
Drugs - Controlled Substances	NA	3.29	7.99	0.97	NA	NA	NA	2.42	0.30	927.44	NA	3,055.56	7,407.41	901.85
Evidence Screening & Processing	NA	NA	NA	NA	NA	NA	NA	NA	NA	NA	NA	NA	NA	NA
Explosives	NA	NA	NA	NA	NA	NA	NA	NA	NA	NA	NA	NA	NA	NA
Fingerprints	NA	1.50	5.19	0.20	NA	NA	NA	3.45	0.14	573.10	NA	862.50	2,975.00	116.50
Fingerprints Database (including IAFIS)	NA	NA	NA	NA	NA	NA	NA	NA	NA	40.00	NA	NA	NA	NA
Fire analysis	NA	1.13	1.39	0.13	NA	NA	NA	1.23	0.12	114.80	NA	130.00	160.00	15.00
Firearms and Ballistics	NA	NA	NA	NA	NA	NA	NA	2.00	0.20	NA	NA	NA	NA	NA
Firearms Database (including NIBIN)	NA	NA	NA	NA	NA	NA	NA	NA	NA	NA	NA	NA	NA	NA
Forensic Pathology	NA	NA	NA	NA	NA	NA	NA	NA	NA	46.50	NA	NA	NA	NA
Gun Shot Residue (GSR)	NA	NA	NA	NA	NA	NA	NA	NA	NA	135.36	NA	NA	NA	NA
Marks and Impressions	NA	NA	NA	NA	NA	NA	NA	1.14	0.16	NA	NA	NA	NA	NA
Serology/Biology	NA	NA	NA	NA	NA	NA	NA	1.17	0.05	NA	NA	NA	NA	NA
Toxicology ante mortem (excluding BAC)	NA	NA	NA	NA	NA	NA	NA	NA	NA	65.33	NA	NA	NA	NA
Toxicology post mortem (excluding BAC)	NA	NA	NA	NA	NA	NA	NA	NA	NA	NA	NA	NA	NA	NA
Trace Evidence	NA	NA	NA	NA	NA	NA	NA	2.08	0.06	NA	NA	NA	NA	NA

Figure 8.7 Additional metrics by investigative area example for Scenarios 1–3.

Investigation area	Personnel Budget / Total Budget	CapitalBudget / Total Budget	Consumables Budget / Total Budget	Other Expenditures / Total Expenditures	TAT last	TAT first	Backlog / CASE
Blood Alcohol	38.01%	0.00%	0.00%	61.99%	NA	28	NA
Crime Scene Investigation	NA	NA	NA	NA	NA	NA	NA
Digital evidence - Audio & Video	NA	NA	NA	NA	NA	NA	NA
DNA Casework	53.16%	0.00%	0.00%	46.84%	NA	320	NA
DNA Database	NA	NA	NA	NA	NA	NA	NA
Document Examination	NA	NA	NA	NA	NA	NA	NA
Drugs - Controlled Substances	44.47%	0.00%	0.00%	55.53%	NA	145	NA
Evidence Screening & Processing	NA	NA	NA	NA	NA	NA	NA
Explosives	NA	NA	NA	NA	NA	NA	NA
Fingerprints	45.99%	0.00%	0.00%	54.01%	NA	205	NA
Fingerprints Database (including IAFIS)	49.25%	0.00%	0.00%	50.75%	NA	NA	NA
Fire analysis	47.63%	0.00%	0.00%	52.37%	NA	150	NA
Firearms and Ballistics	NA	NA	NA	NA	NA	90	NA
Firearms Database (including NIBIN)	NA	NA	NA	NA	NA	NA	NA
Forensic Pathology	44.25%	0.00%	0.00%	55.75%	NA	NA	NA
Gun Shot Residue (GSR)	45.28%	0.00%	0.00%	54.72%	NA	NA	NA
Marks and Impressions	NA	NA	NA	NA	NA	135	NA
Serology/Biology	NA	NA	NA	NA	NA	85	NA
Toxicology ante mortem (excluding BAC)	48.11%	0.00%	0.00%	51.89%	NA	NA	NA
Toxicology post mortem (excluding BAC)	NA	NA	NA	NA	NA	NA	NA
Trace Evidence	NA	NA	NA	NA	NA	350	NA

Figure 8.8 Additional metrics by investigative area example for Scenarios 1–3.

Go Deeper

Heames, J.T. and Heames, J.T., 2011. Forensic science staffing: Creating a working formula. *Forensic Science Policy & Management: An International Journal*, 2(1), pp.5–10.

Houck, M.M., 2016. Risk, reward, and redemption: Root cause analysis in forensic organizations. *Forensic Science Policy & Management: An International Journal*, 7(3–4), pp.106–112.

Note

1 See UNODC/ENFSI report *Guidelines on Representative Drug Sampling*, https://www.unodc.org/documents/scientific/Drug_Sampling.pdf

Article: Forensic Science Staffing
Creating a Working Formula

8.1

JOYCE THOMPSON HEAMES
AND JON TIMOTHY HEAMES
College of Business & Economics, West Virginia University, Morgantown, West Virginia, USA

Introduction

In 2003, Dale and Becker stated, "Our nation needs a strategy for staffing forensic scientists in public forensic laboratories" (pg. 1). Seven years later, there is still a dearth on the subject in the forensic science literature. As a profession, forensic science was formally given a legal code in the early 1900s, but only in the past 20 years has it come to be recognized by the public (Houck 2006); this demonstrates that the field is still in its infancy in developing policy and procedures. Like many other professions whose primary human capital is based in scientific understandings, those promoted to management positions in forensic labs typically have been individuals who previously excelled in forensic science operations. Few, however, have had any formal management training. Thus, most forensic labs have had difficulty in creating a business case for the staffing issues they face.

To make a compelling business case, a lab director must be able to identify the costs/benefits and risks/returns on proposed investments in the recruiting, hiring, training, and conservation of personnel. He or she needs to be able to answer the questions of who to hire, how to hire, how and who to train, and, most importantly, why hire now?

Tough economic times with severe salary crunches add to the difficulty of building an airtight business case for a forensic science lab to hire additional staff or be innovative and hire different types of employees. Four years ago, a team of researchers at West Virginia University (WVU) sponsored by the National Institute of Justice (NIJ) implemented the FORESIGHT project with the primary objective of collecting data in a longitudinal format to enhance the understanding of creating a business case for forensic science laboratories (Houck et al. 2009). The FORESIGHT project, modeled after the European Quadrupol study, started the arduous task of creating a centralized data collection platform with standardized terminology that would give lab directors access to historical data that might lead to better business practices. That data pool has helped labs identify problems, develop metrics to measure said problems, and strategize about solutions in many different management areas (Houck et al. 2009).

In a recent conversation with the senior executive vice president of HR for one of the world's largest companies, the VP commented that the single most important objective for any company should be to hire the right people for the right jobs and then retain them. He went on to say that, unfortunately, this is the one thing that most organizations do with the least success. Our work with the FORESIGHT project and the training we have conducted

 DOI: 10.4324/9780429318528-24

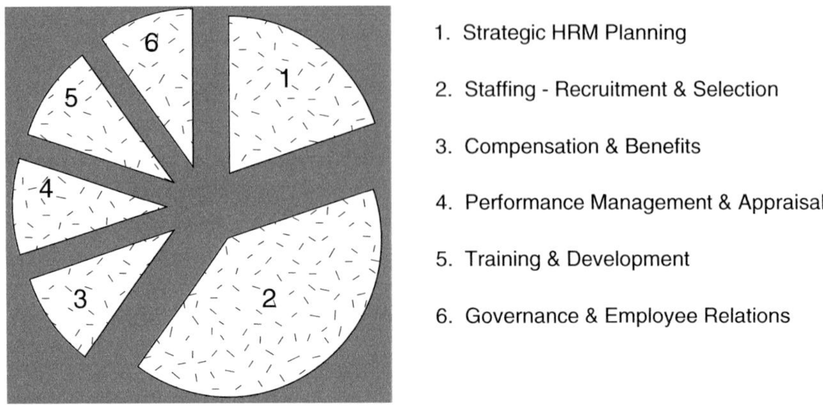

Figure 1 Human Resource Management Functions.

with forensic lab personnel across the country suggest that this is, indeed, a major problem in forensic labs as well.

Figure 1 illustrates the six different functional areas of human resource management (HRM) and their respective percentage absorption of administrative resources. These HR functions are central to any organization and drive the business model. Forensic labs are no different. In fact, since the essence of what forensic labs do is imbedded in the talents of the lab personnel, it may be even more important in this field. As Figure 1 illustrates, HRM is so much more than just hiring and firing, but the two key areas and the focus of this article are 1) strategic HRM planning and 2) staffing—recruitment and selection. These two functions should collectively consume at least 50% of the time, effort, and administrative costs of the HR department. If these two are done effectively and efficiently, all the other functions will take less time and administrative resources. The purpose of this article is to identify staffing measurements from the FORESIGHT data that would create a working formula to help lab directors better manage facilities.

Strategic Human Resource Management (S-HRM) Planning

"The most important labor demand forecasts are those for the positions and skills that will be central to the organizations' intended strategic direction" (Phillips & Gulley 2009, p. 127). Strategic direction flows from an organization mission statement, the organization's short- and long-term objectives, and from its strategic plan. For a forensic lab, the first step in this process would be to identify the lab's mission statement if it has been imposed externally, or to develop one if the lab has that freedom. Following is the development of its short- and long-term business objectives.

From those two foundations, the lab needs to develop its long-term strategic plan and determine its primary customer groups and the range of services to be offered. Raw data for these decisions flows from historical data and from demand forecasts developed for the service areas to be targeted. The absence of a deliberate effort here often leads to a situation where the lab tries to serve every customer in its geographic domain and offer a wider range of services than is economically feasible. Business firms have learned that they are more likely to be successful if they narrow their focus and not try to be all things to all people. Well-known strategy expert Michael Porter (1996) argues, "Competitive strategy is about being different . . . It means deliberately choosing a different set of activities to deliver a unique mix of value." Of course, business firms have greater freedom to choose which customers to serve and how to serve them, while those organizations in the public sector often have these requirements imposed upon them externally. In cases where this is not true, however, the lab would be well served by making similar focused choices.

Part and parcel with the development of the strategic plan should be an accompanying strategic HR plan (S– HRP), which identifies the human resource implications of the strategic service choices outlined in the strategic plan. After all, a forensic lab is essentially a service organization, and its competitive and operating capacity rests in the human resources (lab technicians) that it employs. Unfortunately, most labs do not spend adequate time at this stage and pay for it later by spending excess capital in the recruiting and training, and then later replacing inappropriate hires.

The lab's HR officer/director should be involved in both the strategic and the S–HRP processes from the beginning, so that those plans can be integrated appropriately at the staffing level. This level of analysis should be focused on how the lab's HR policies or programs are strategically interfaced with the organization primary resource (a.k.a. quality, motivated, talented employees, a lab's primary resource) to increase lab performance and services while reducing labor costs (Barney 1991). If done holistically and with consideration for the overall business goals of the lab, the outcome will be more efficiency and sustainability (Kearns 2003).

This type of alignment allows organizations to match critical success factors designed to support the organization's strategic business objectives with employees who not only possess the desired knowledge, skills, and abilities but also have the right organizational fit (Torraco & Swanson 2001). This step in the process lays the ground rules for all of the other HR functions to follow, especially the staffing (e.g., recruitment and selection). It creates the web that laces all of the other functions together in a way that provides the best service to the lab. Ultimately, the common goal of S–HRP is to maximize the value derived from human capital (Kearns 2003; Torraco & Swanson 2001). Once the lab can identify its purpose and strategic focus, the process begins with hiring the right personnel and putting them in the right places. Metrics and measurements supply the needed analysis to substantiate that human capital efforts (e.g., staffing efforts) have been successful in meeting the organization's strategic goals and maximizing the return on investment of people employed (Holton & Naquin 2001).

Staffing: Recruitment & Selection

How do you quantify how many and of what kind of scientist, law enforcement officer, or crime scene specialist is needed? Without knowledge from historical data and/or legal requirements mandated by regulatory boards, the task is sure to be laced with difficulties and even more failures. Equally important, however, is that these staffing requirements must flow from the demand forecasts for lab services that accompanied the development of the strategic plan. These forecasts should be developed from historical data maintained at the individual lab level. One of the primary objectives of the FORESIGHT project was to create a database of historical demand patterns for various lab services across a population of multiple labs. Note: For those labs not a member of the FORESIGHT project, please go the official FORESIGHT website for more information (www.be.wvu/forensic).

The DNA of a forensic lab is laced with a variety of skills and educational backgrounds that create an interesting and fascinating mosaic of people. It is one of the few environments where you have such an assortment of trained specialists working for, in many cases, the same boss. One lab director joked he wore so many hats trying to deal with the issues associated with all the different types of personnel that he should own a millinery shop. Lay on another layer of complexity with the advanced technology and lack of financial resources and it is easy to understand how the personnel fabric of most labs is worn thin. A colorful picture of the chemical structure of DNA is representative of the complex personnel structure of a forensic lab. It is so descriptive of the myriad of relationships, talents, skills, and needs of the various people that make up the personnel of a forensic lab and represents a wonderfully complicated mixture of sworn vs. unsworn, private vs. public, scientist vs. law enforcement, researchers vs. administrators, and formally educated vs. on-the-job-trained workers. The mosaic nature of forensic labs complicates the handling of personnel issues.

To further confound issues, for the first time in history, there is a cross section of four generations working side by side in the labs: traditionalists (born before 1945), baby boomers (1946–1964), Gen Xers (1965–1980), and "millennials" or Gen Ys (born after 1980) now comprise the working population of many labs. The population of the traditionalists is larger than one might have expected because of the downturn in the economy and its impact on intended retirements, but a mass exit is expected in the next five years. The growing number of Gen Xers and Millennials in the field is attributed somewhat to the CSI factor, whose glamorous gory depiction of crime labs and their capabilities has enticed many young people to enter the field. Unfortunately, there is a gap between the reality of what labs can actually do and what TV magic can make happen (Houck 2006). In any event, the baby boomers and Gen Xers are being sandwiched in between the young, formally trained millennials and the older, on-the-job-trained traditionalists. They find themselves in managerial roles and are unsure of how to handle a constant change in personnel, an ever-changing environment, and a growing demand for services (in part because the TV-educated population has become more aware of these services).

Beyond the numbers, replacing one scientist with another scientist of the same training and educational background is not necessarily the best course of action. An area that has not received enough attention can be tied to organizational, job, and person fit. Information on these is harder to collect and is subject to some degree of personal interpretation, but one still should not throw out the practice simply because it is difficult. It is a matter of understanding the DNA of a lab or department to truly understand which personalities will best fit, but there are mathematical calculations that will generate sets of numbers that lab directors can use to improve the operations of the lab. Learning how to generate the numbers and interpret them is a critical step in creating a more efficient work environment. The next section provides some examples.

Developing Key Staffing Performance Indicators

A review of both labor demand and labor supply are critical to the overall success of a lab in developing key staffing performance indicators. An internal analysis requires determining how many new employees are needed, but just as critical is what type of new personnel (e.g., knowledge, skills, and abilities). Identifying which recruiting source provides the best and brightest staff is a study of internal and external labor markets. However, to be thoroug it is necessary to also identify market trends that will show the growth or decline in availability of certain talent pools. A gap analysis then highlights the areas of concern that require action on behalf of the lab manager.

Labor Demand Forecasting

The three most common ways to forecast labor demand are judgmental forecasting, staffing ratio analysis, and trend analysis (Phillips & Gulley 2009). Considered a viable tool in most situations, *judgmental forecasting* relies on the knowledge and experience of people inside the organization to envisage future employment needs. Unfortunately, the infancy of forensic science and the lack of historical data, prior to FORESIGHT, belies the ability of most lab directors to truly understand the trends and make accurate predictions of future personnel needs. Thus, most labs have taken a "well, that is what we have done in the past" approach. This leads to the old adage, "getting what we have always got because we do what we have always done," which can cripple growth and innovation.

Staffing ratio analysis allows for the mathematical interpretation of the number of employees needed to conduct a particular service and then make a comparison of this ratio with industry or other internal process data. For example, the measure *Test/FTE* offers a quick gauge of the output (number of tests performed, i.e., DNA, hair & fiber, entomology, etc.) per person (full-time-equivalent (FTE) employees) in the laboratory. In a 2009 article, Speaker, one of the original architects of FORESIGHT, gave the following sets of personnel ratios for consideration: Items processed/FTE; Samples processed/FTE; Reports completed/FTE; Actual FTE employees/budgeted FTE employees. Another round of analysis may concern the use of facilities to number of employees (e.g., FTE/laboratory square feet area; FTE/laboratory square meter area). Other ratios that may be of importance to lab performance are revenue per employee, managers to employees, inventory levels to employees, labor costs to all operational expense costs.

Trend analysis is using past patterns to predict future needs. The FORESIGHT project is in its infancy too and needs to have time to collect more data to truly have longitudinal information with which to develop recommendations, but it is a good start. Calculations on turnover by position type, location, salary, employee tenure, and education background are just examples of some of the analyses that can be derived from historical data. As good practice, lab directors should dig into past data to develop some understanding of the number of past services performed, both by type of service and by type of customer groups requiring the service. It is useful to do this over a multiple year period to get some sense of year-to-year variations in these demands. Moreover, these should then be adjusted for any changes in service and/or customer emphasis developed in the strategic plan for the future.

Other aids to forecasting labor needs include a *transition analysis* (also referred to as a Markov analysis), which is a technique to quantifiably track the movement of existing employees across positions and thus show the number of potential openings by position or area (Heneman and Judge 2009). For example, this type of graph would allow tracking of how many "scientists in training" actually became "reporting scientists/analysts" over the past x years and thus how many scientists in training would need to be replaced in future years if past stats are predictive of future trends. As a sidebar, a Markov chart could also illustrate the success of development or training programs. The drawback to this type of analysis is that it is reporting in aggregate numbers, i.e., if an assistant/analyst split their time 50/50 in different investigative areas of the lab (e.g., fire analysis vs. computer analysis), the recording would show.5 for each position. Thus, responsibilities would have to be recorded based on the proportion of time spent on each type of crime scene analysis, which could be very helpful but very time consuming. Also this type of analysis gives you the "how many" but not necessarily the type of skill needed nor which specific employee might be available for transfer.

Talent inventories, which summarize each individual employee's specific set of skills, and *replacement charts,* which outline the specific skills needed for a given position, are also ways to track the potential replacements for any given position. These techniques are more detailed, but most can be combined manually or in some form of automation so that the two sets of data can serve one another. The drawback to these systems is that they tell you who and what, but not how many.

Finally, a lab should always be poised to calculate the *return on investment* (ROI) for new positions or a new hire, which is based on costs versus outcomes and provides very important information. The difficulty is assigning dollar amounts to all of the outcomes desired (e.g., processing a DNA test is valued at x) and that ROI can be based on any temporal time frame (monthly, quarterly, or annually). Detailing the cost of recruiting, interviewing, testing, and selecting is a critical step in the calculation of the cost per new hire.

Labor Supply Forecasting

Knowing what is forecasted for the *external labor market* for forensics is a necessity not just a nicety. Turnover will eventually drive every lab to turn to the marketplace for scientists, technicians, and clerical personnel. Generating data on the external market comes from two sources, *personal observation and government/industry reports.* The following is just an example of some sources from which to draw information on the labor market, such as projected growth in each job area, list of tasks, KSAs, work values, and wages/employment trends.

- Labor market information by state: www. workforcesecurity.doleta.gov/map.asp
- Manpower Employment Outlook Survey: www.us. manpower.com
- SHRM/Rutgers Leading Indictor of National Employment Index(LINE): www.shrm.org/line
- Local employment dynamics from the U.S. Census Bureau: http://lehd.did.census.gov/led
- O*Net Online, created by the U.S. Department of Labor: online.onetcenter.org

The U.S. Bureaus of Labor Statistics (www.bls.gov) is the most comprehensive and widely used government source. Table 1 is an extract from the BLS website that gives a projected growth of 20% in forensic science technicians (FST) needed by 2018. Table 2 is excerpted from the O*Net Online website, created by the U.S. Department of Labor, that gives "detailed descriptions of the world of work for use by job seekers, workforce development, researchers, and HR professionals." The table illustrates that for FSTs, the median wage for 2009 is $24.75 hourly and $51,480 annually. Additional important information is that there

Table 1 Projections Data from the National Employment Matrix (Excerpted from www.bls.gov)

Occupational Title	SOC Code	Employment, 2008	Projected Employment, 2018	Change, 2008–18	
				Number	Percent
Science technicians	—	270,800	302,600	31,800	12
Agricultural and food science technicians	19–4011	21,900	23,800	1,900	9
Biological technicians	19–4021	79,500	93,500	14,000	18
Chemical technicians	19–4031	66,100	65,500	-500	-1
Geological and petroleum technicians	19–4041	15,200	15,400	200	2
Nuclear technicians	19–4051	6,400	7,000	600	9
Environmental science and protection technicians, including health	19–4091	35,000	45,200	10,100	29
Forensic science technicians	19–4092	12,800	15,300	2,500	20
Forest and conservation technicians	19–4093	34,000	36,900	2,900	9

NOTE: Data in this table are rounded. See the discussion of the employment projections table in the *Handbook* introductory chapter on *Occupational Information Included in the Handbook.*

Table 2 Wages & Employment Trends Summary Report for *Forensic Science Technicians*—National (O*Net Online, http://online.onetcenter.org)

Median wages (2009)	$24.75 hourly, $51,480 annual
Employment (2008)	13,000 employees
Projected growth (2008–2018)	Much faster than average (20% or higher)
Projected job openings (2008–2018)	8,000
Top industries (2008)	Government

were approximately 13,000 FSTs in the United States in 2008 with an anticipated growth of 8,000 additional jobs by 2018, a 60% increase. Note that the O*Net forecast is considerably higher than the 20% issued by the BLS.

Gap Analysis

Running a *gap analysis* between labor demand and labor supply is the next critical stage in this phase of the staffing process. This step provides the information needed to proactively resolve the gaps by creating an action plan that deals with any projected shortages or surpluses and that demonstrates the number of people that should be recruited, what resources are needed to hire, and how much time will it take to hire the necessary employees. These action plans are unique to each lab and should be aligned with the original S–HRP and talent philosophy. Calculating *staffing yields* (proportion of applicants moving from one stage of the hiring process to the next) and *hiring yields* (percent of applicants ultimately hired) (Phillips & Gulley 2009) provides additional quantitative information that guides the decisions on the quality of recruiting sources and hiring processes.

Discussion

The bottom line is that any reported piece of data can be easily obtained and compared to a second piece of data for interpretation. The real question labs must ask is: what is important to our lab that will help us meet overall goals and objectives? The task now falls to lab directors to exercise as much due diligence in the analysis of staffing data as they do to the analysis of forensic data. Understanding the DNA of the lab will assist in better determining the appropriate actions to take in efficiently and effectively managing the staffing process. Moreover, that knowledge will assist in creating a working formula that better serves the lab's long-term plans to create a business case that allows for sustainability. The primary goal is to hire the right people, the right number of people, and put them in the right places.

References

Barney, J. 1991. Firm resources and sustained competitive advantage. *J. Manage.* 17:99–120.

Dale, W.M., and W.S. Becker. 2003. Strategy for staffing forensic scientists. *J. Forensic Sci.* 48:465–466.

Heneman, H.G., and T. Judge. 2009. Staffing Organizations, 6th Ed. Middletown, WI: McGraw Hill.

Houck, M.M. 2006. CSI: Reality. *Sci. Am.* 295:84–89.

Houck, M.M., R.A. Riley, P.J. Speaker, and T.S. Witt. 2009. FORESIGHT: A business approach to improving forensic science services. *Forensic Sci. Pol. Manage.* 1:85–95.

Kearns, P.L. 2003. *HR strategy: Business focused, individually centered.* Burlington, MA: Elsevier Butterworth-Heinemann.

Phillips, J.M., and S.M. Gully. 2009. *Strategic staffing.* Upper Saddle River, NJ: Pearson/Prentice Hall.

Porter, M.E. 1996. What is strategy? *Harvard Bus. Rev.* November- December, 61–78.

Speaker, P.J. 2009. Key performance indicators and managerial analysis for forensic laboratories. *Forensic Sci. Pol. Manage.* 1:32- 42.

Torraco, R.J., and R.A. Swanson. 2001. The strategic roles of human resource development. In *Foundations of human resource development,* Eds. R.A. Swanson and E.F. Holton, 339–357. San Francisco, CA: Berrett-Kohler.

Article: Risk, Reward, and Redemption
Root Cause Analysis in Forensic Organizations

8.2

MAX M. HOUCK
Forensic & Intelligence Services, LLC, St. Petersburg, FL

Introduction

Forensic science service providers are high reliability organizations, that is, they avoid catastrophic outcomes in an environment where those outcomes can be expected (Weick and Sutcliffe 2007). They operate in unforgiving social and political environments with technologies that potentially have life-altering outcomes if an error is made; the complexities of what they work with reduces the ability for experimentation to mitigate catastrophic outcomes. Thus, risk management is—or should be—a core responsibility of forensic organizations. Managing risk should be comprehensive and systematic, scaled appropriately to the potential outcomes based on likelihood, consequence, and impact (Robertson 2012). Although in a forensic context risk management can sometimes be viewed transactionally as an audit and compliance sequence, this narrow view comes at a high cost for the organization: A lack of coordinated strategy ignores the benefits of a holistic systems approach.

When an event has an unexpected outcome, an investigation should follow, even if the outcome is positive. The "why" is more important than the "what" for an organization to learn and grow (Behn 2014). Beyond good practice for a learning organization, such investigations are recommended by the National Commission on Forensic Science (NCFS 2015) and are mandated by applicable accreditation standards (ISO/IEC 17025).

Risk management

Consequences of global events are difficult to predict both globally and locally (Meadows 2008; Robertson 2012). Therefore, the effect of adverse or uncertain outcomes to an organization's goals (the definition of risk) can only be managed to a greater or lesser degree. Risks with negative outcomes are *hazard risks* and those with positive outcomes are *opportunity risks*. All organizations should develop risk management plans to help with their operational resilience (Weick and Sutcliffe 2007) but forensic organizations, in particular, because they are high-reliability organizations (HRO). A high-reliability organization is one that avoids catastrophes in a risky setting or complex environment where accidents are expected. Hospitals, airlines, and the military are examples of high-reliability organizations. With the potential for error and the possible severity of that error (lifelong imprisonment or capital cases, for example), forensic science surely counts as a HRO. Forensic

 DOI: 10.4324/9780429318528-25

		Consequences				
		Insignificant	Minor	Moderate	Major	Severe
	Almost Certain	Medium	Medium	High	High	Extreme
	Likely	Medium	Medium	Medium	High	
Likelihood	Possible	Low	Medium	Medium	High	High
	Unlikely	Low	Low	Medium	Medium	High
	Rare	Low	Low	Low	Medium	High

Figure 1 A risk consequence likelihood table (redrawn from Robertson 2012).

service providers need to be almost-constantly operational, have failure rates close to zero, and work with limited resources (Houck 2015) and these pose severe challenges to anticipating and mitigating risk. Forensic organizations, therefore, should have a very low tolerance of risk.

A risk management plan should be proportionate, aligned, comprehensive, embedded, and dynamic (Robertson 2012). Given these requirements, a risk management plan needs to be appropriate and cost-effective; an agency may not be able to afford or tolerate zero risk. Thus, risk management needs to be appropriate to the level of risk or, that is, proportionate and scaled to the potential adverse or unwanted outcomes (Figure 1). Risk management plans do not exist in a vacuum or are they singular events (like so-called "annual strategic planning") but must be comprehensive to the whole of the organization's commitment to its purpose, mission, and values. Risk management must be an on-going process, dynamic in response to changing circumstances, and supportive of an organization's resilience.

Generally, there are four responses to risk: Tolerate, treat, transfer, or terminate (Robertson 2012) (Figure 2).

Knowing which action to take to ward off a hazard risk or make the most of an opportunity risk requires knowing the root cause of the risk-producing activity. Without a grounded, evidence-based understanding of what is causing the problem, the organization may spend its limited resources on treating symptoms rather than the disease.

Why root cause analysis?

Quality is an acceptable range of variation in an outcome or product; that is, it is an explicit statement of the level of acceptable risk manifest in a product. Root cause analysis (RCA) is a method of problem solving (ultimately, risk mitigation) that tries to identify the necessary and sufficient reasons for a quality event, typically a fault, problem, or a nonconformity in a quality system that exceeds the acceptable range. Something is necessary for an event if the event cannot occur without it being present; something is sufficient if it guarantees the occurrence of the event. For example, while air is necessary for human life, it is not sufficient because other things are also needed (food, water, and so on). A root cause is both necessary and sufficient to have initiated the nonconformity; no prior causes can be used to explain the nonconformity as the root cause is the basis for the chain of actions that

Tolerate	The risk may be tolerable; even so, the ability to do anything about some risks may be limited or the cost of taking any action may be disproportionate to the potential benefit gained.
Treat	While the organization continues the activity giving rise to the risk, action is taken to constrain the risk to an acceptable level.
Transfer	Shifting the responsibility for the risk may be the best option for the organization; examples include insurance or liability waivers.
Terminate	Some risks can be remediated only by terminating the activity giving rise to the risk. Governmental organizations may have limited options to terminate risky activities, especially as compared with the private sector.

Figure 2 The four responses to a risk.

occurred. Once a root cause is removed from the sequence, the nonconformity should be prevented from happening again and the risk to the organization is mitigated.

RCA assumes that systems and events are related. An action in one area can create other actions or results in others, continuing on in what is known as the "causal chain." RCA traces the remnants of the chain of events back to its source, the root cause.

Overview of RCA

The method in this article is based on the work of Shewhart (1980), Deming (2006), Snowden and Boone (2007), and Ohno (2007). RCA is an iterative process and has the Plan-Do-Act-Check cycle of Shewart and Deming as its backbone (Figure 1). The process of RCA starts with the problem statement: What happened? Why is it a nonconformity? How is it outside the quality system requirements? This is done using a technique of continually asking "Why?" until the root cause is identified (Ohno 2007). Once identified, the chain of events leading to the root cause of the nonconformity is characterized as simple, complicated, or complex (Snowden and Boone 2007); a root cause may be characterized as chaotic, but rarely. The characterization will indicate the kinds of changes that need to be made to test if the identified root cause is the actual source of the nonconformity; additional observations may be needed to confirm this decision. The confirmed changes are then made to the root cause, taking the rest of the quality system into account, to see if the nonconformity is prevented from recurring. Periodic checks are made to provide feedback on the continual prevention of the nonconformity, that is, are the changes still preventing the nonconformity.

The RCA process

The RCA process is intended to be followed step by step; it may be necessary, however, to revisit or repeat a step if something does not produce the expected results. An RCA should be a team effort. It is recommended that, at a minimum, a quality specialist, a subject matter expert not involved directly in the nonconformity, and a manager be members of the team, with others joining as necessary.

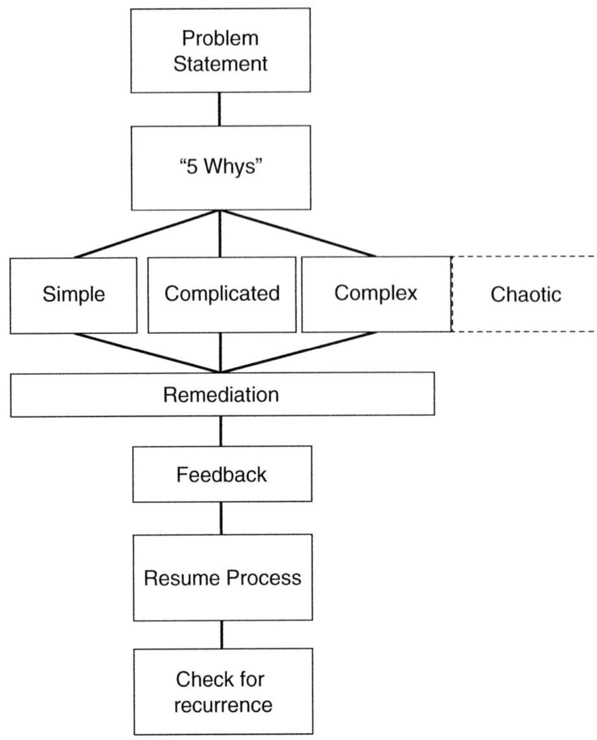

Figure 3 Overview of RCA process. The method in this article is based on the work of Shewhart (1980), Deming (1986), Snowden and Boone (2007), and Ohno (2007).

Stating the problem

The problem statement is a concise, complete, and accurate single sentence describing what the problem (nonconformity) is. Examples include:

- the proficiency test was not passed;
- the case jacket is missing documentation;
- the wrong individual was identified;
- the specimen was contaminated; and
- the results from a different case were reported.

A timeline of the nonconformity should be re-created to help determine the sequence of events leading up to the nonconformity. The problem statement should be kept general, as in the examples, so that the focus of the RCA does not miss potential contributing, or even root, causal factors. Being too specific can suggest avenues of investigation based on misleading assumptions. It should also be specified in a way that the chance of looking for assignable causes when they are not present does not exceed some prescribed value; that is, it shouldn't cost more to look for problems that aren't there than it does to make the mistake in the first place. In new processes or in the early stages of implementation, variation is to be expected but this does not mean that an RCA should not be conducted.

The problem statement is central to the RCA: it acts as the reference point and compass for the analysis. It should be understood and agreed upon by all members of the RCA team.

It may be necessary to have participants sign a declaration of the problem statement to document the RCA's starting point. Solutions should not be offered at this stage of the process.

Getting to the root cause

The key to getting to the root cause of a non-conformity is to avoid assumptions and mental traps (both a type of bias). The chain of causality should be traced backwards in direct increments from the event through each link in the chain to a root cause. Typically, the next-to-last link in the causal chain is a process itself and this is central to a RCA: finding a valid root cause that illuminates a process that is not working well or does not exist. Superficial and obvious answers, such as "Not enough [people, time, money, etc.]", point to processes out of (immediate) control and usually do not solve the issue, regardless. If a process is not working well, putting more people in place to use it will only increase problems, not resolve them. A standard phrase in RCA thought is, "People do not fail; processes do."

A simple but effective approach in getting to a root cause is the "5 Whys" approach (Ohno 2007). Starting with the problem statement, ask "Why (did this happen)?" repeatedly until the root cause is identified. For example:

Problem Statement. The analysis of a sample/specimen was not completed by the deadline.

1 Why? *The instrument failed to complete the run.*
2 Why? *The instrument ran out of carrier gas.*
3 Why? *The tank of carrier gas emptied mid-run.*
4 Why? *More gas was not ordered.*
5 Why? *An employee forgot to order more gas.*

The temptation at this point is to blame the employee but the root cause lies with the process, not the person, prompting one more "Why?": *Why did the employee forget?* Note that the last question points to a failed process or one that did not exist: A reminder system for ordering gas.

Problem Statement. The vehicle will not start.

1 Why? - *The battery is dead.*
2 Why? - *The alternator is not functioning.*
3 Why? - *The alternator belt has broken.*
4 Why? - *The alternator belt was well beyond its useful service life and was not replaced.*
5 Why? - *The vehicle was not maintained according to the recommended service schedule.*

It may take more or less than 5 "whys" to get to a root cause; there is nothing magical about 5, other than experience has shown that it typically takes at least that many to get to most root causes. For example, it may take multiple "whys" to come to the conclusion *the battery is dead.*

The "whys" provide a problem *analysis* (the problem has already been *stated*), a breakdown of effects and causes leading from the problem to the root cause. A timeline can be helpful to understand the relationships between the elements involved; visual

	Characteristics	Response	Danger Signs	Response to Danger
Simple	Repeating patterns and consistent events Clear cause-and-effect relationships evident to every one; right answer exists	Sense, categorize, respond Ensure that proper processes are in place Delegate Use best practices Communicate in clear, direct ways	Complacency and comfort Desire to make complex problems simple Entrained thinking. "the way we've always done it" No challenge of received wisdom Overreliance on best practice if context shifts	Create communication channels to challenge history Stay connected without micromanaging Don't assume things are simple Recognize both the value and the limitations of best practice
Complicated	Expert diagnosis required Cause-and-effect relationships discoverable but not immediately apparent 10 everyone More than one right answer possible	Sense, analyze, respond Seek expert advice Listen to conflicting advice	Experts overconfident in their own solutions or in the efficacy of past solutions Analysis paralysis Expert panels Viewpoints of non-experts excluded	Encourage external and internal stakeholders to challenge expert opinions to combat entrained thinking Use experiments and games to force people to think outside the familiar
Complex	Flux and unpredictability No right answers; emergent instructive patterns Unknown unknowns Many competing ideas A need for creative and innovative approaches	Probe, sense, respond Create environments and experiments that allow patterns to emerge Increase levels of Interaction and communication Use methods that can help generate ideas Manage starting conditions and monitor for emergenice	Temptation to fall back into habitual, command-and-control mode Temptation to look for facts rather than allowing patterns to emerge Desire for accelerated resolution of problems or exploitation of opportunities	Be patient and allow time for reflection Use approaches that encourage interaction so patterns can emerge
Chaotic	High turbulence No clear cause-and-effect relationships, so no point in looking for right answers Unknowables Many decisions to make and no time to think High tension Pattern-based leadership	Act, sense, respond Look for what works instead of seeking right answers Take immediate action to reestablish order (command and control) Provide clear, direct communication	Applying a command-and-control approach longer than needed Missed opportunity for innovation Chaos unabated	Set up mechanisms to take advantage of opportunities afforded by a chaotic environment Encourage advisers to challange your point of view once the crisis has abated Work to shift the context from chaotic 10 complex

Figure 4 The categories of situation for decision-making and appropriate responses. Redrawn from Snowden and Boone (2007).

representations are also a good way to look at the processes and share the analysis with the team (Straker 2009).

Things to avoid in a "5 why" approach are mostly mental blocks or biases on the part of the team. The team should not stop at superficial symptoms rather than continuing on to deeper causal events (but stopping short of global causes, like not enough staff or world hunger). The team should have the resources needed to ask the questions and identify the root cause; under-resourcing a root cause can create further non-conformities. For many RCAs, materials are cheap (markers, notes, paper) and the main cost is personnel time. In an RCA, different individuals should come to the same conclusions in the process; if different team members come to different conclusions, then the causal chain must be re-evaluated, even back to the problem statement which may need to be revised.

The decision framework

The decision framework, designed as a construct for executive decision making (Snowden and Boone 2007), sorts a root cause into one of five contexts defined by the nature of the cause and effect. Four of these lead to a diagnosis of the non-conformity and suggest contextually appropriate ways of acting on it; these four are Simple, Complicated, Complex, and Chaotic, the latter being rare in RCA. The fifth, disorder, applies when it is not clear which of the other four the root cause fits into. A description of each is below.

- **Simple.** Stability, clear cause-and-effect relationships; the "right" answer is often obvious.
- **Complicated.** Clear cause-and-effect relationship but not immediately obvious; at least one "right" answer exists; requires expertise; testing may be required to obtain necessary information; better practices, rather than best, may be appropriate.
- **Complex.** Unpredictable; no one "right" answer may exist; "adequate" may be the best answer; answers reveal themselves upon exploration.
- **Chaotic.** Searching for the right answer is pointless; turbulence predominates; cause-and-effect relationships are impossible to determine due to constant change and flux.

Each context calls for a different set of responses, as explored below.

A good RCA will explore the causal chain thoroughly; other instances of the event may have occurred and the root cause may be more wide-spread and disrupting than initially assumed. Thinking that only one cause exists is dangerous: *A cause never stands alone.* Going back to the car example, if four people were asked what the most important part of the car is, one might say the engine, another the breaks, a third the key, and so on. There is *no one right answer* because the car is a system and without significant elements, it will not operate (Meadows 2008). A root cause is not one cause but a system of causes working together, understanding the interaction between multiple factors.

Remediation

Also known as a corrective or preventative action, remediation sets in place actions and processes intended to stop and prevent the root cause that lead to the problem. Corrective actions are responsive to an event; preventative actions are predictive and are meant to stop a problem before it occurs (possibly again). Remediations can take many forms; in fact, there are more remediations than nonconformities, that is, there are more solutions than problems. This should be a source of optimism for those involved in RCA.

A remediation is essentially a hypothesis to be tested, a proposed solution that can be checked against how well it solved the problem, if it did at all. A remediation may be as simple as signage or as complicated as a fully developed project to be managed. Timeframes, responsibilities, and milestone for remediation must be established. Objective criteria must be established by which to evaluate the planned remediation, both immediately and at later intervals; these are not necessarily milestones, which are meaningful achievements, but values or goals against which the remediation is judged. The team must state the relevant criteria by which the planned remediation will be evaluated.

In the proficiency test example, there might be sev-eral remediations suggested, such as:

- request gas with each order placed;
- create a calendar reminder to send a message to an employee to order more gas; and
- work with the vendor to schedule deliveries.

The remediation changes are made to the root cause process, taking the rest of the quality system into account, to see if the nonconformity is prevented from recurring. The remediation may need to be repeated, with changes, to properly prevent the non-conformity from recurring. If so, how much, how often, and by what criteria need to be specified so the team or responsible parties know what to do.

Successful remediation need not be a binary result, that is, all or nothing; a range of success is possible, a partial remediation, because the problem may be too big, the context may have been misjudged, or the planned remediation may have been insufficient. In these cases, the problem statement and its context needs to be re-evaluated, then the planned remediation should be revised as necessary.

Feedback

Remember, RCA assumes that systems and events are related. Feedback is necessary to inform the system about the results of the solution attempted, the remediation. The results of a remediation attempt will inform whether it was successful or not and, if so, to what degree. This information needs to be returned to the team as feedback for evaluation, discussion, and further action.

A formal plan for review of the remediation and response with feedback must be delineated and executed by the team. An action in one area can create other actions or results in others. During the remediation process, look for feedback or other nonconformities occurring in other areas of the organization.

Resume process

If the remediation is appropriately successful and the team is satisfied with the results, the original process can resume with the remediated changes in place. Clear criteria for resuming the process must be established and agreed upon by the team prior to the remediation.

Check for recurrence

Periodic checks should be made to provide feedback on the continual prevention of the nonconformity to ensure that the remediation continues to prevent the nonconformity. The periodicity should be relevant and meaningful to the remediation and the timing of the new process.

Conclusions

Using RCA as a risk management tool has significant positive potential for high-reliability organizations. These benefits include more normal operations with fewer disruptions, leveraging opportunities by being stable and prepared to engage in acceptable risk, identifying risky events prior to their happening, and being ready to convert hazard risks to opportunity risks to improve service, processes, or quality (Robertson 2012). Laboratories accredited under ISO/IEC 17025 are required to establish a policy and a procedure for implementing corrective actions when nonconforming work or departures from policy are identified. The document also states that the corrective action begins with an investigation into the root cause(s) of the problem. Incorporating RCA into a forensic organization's quality process not only helps to ensure continuity of operations, a key component of their purpose, and to reduce or remove risks before they occurs, but is required if the laboratory is accredited under 17025. Moreover, the National Commission on Forensic Science has released a final draft, *Root Cause Analysis (RCA) in Forensic Science*, that recommends the adoption of root cause analysis protocols for all federal forensic agencies or those that receive federal funds (NCFS 2015). Given the success of RCA in other HROs (Patankar 2012; Tamuz and Harrison 2006), the forensic industry has a wealth of resources to help it adopt RCA into its policies and procedures. The current article offers a hybridized, streamlined approach for conducting an RCA in a forensic setting.

References

Behn, R. 2014. *The PerformanceStat Potential: A Leadership Strategy for Producing Results*. Washington, DC: Brookings Institution Press.

Ohno, T. 2007. *Workplace Management*. Translated by Jon Miller, Tokyo: Gemba Press.

Deming, W. 2006. *Out of the Crisis*. Cambridge, MA: MIT Press.

Houck, M. 2015. *Professional Issues in Forensic Science*. Amsterdam: Elsevier.

ISO/IEC 17025:2005(E). General requirements for the competence of testing and calibration laboratories.

Meadows, D. 2008. *Thinking in Systems: A primer*. White River Junction, VT: Chelsea Green Publishing.

National Commission on Forensic Science. 2015. *Directive Recommendation: Root Cause Analysis (RCA) in Forensic Science*. https://www.justice.gov/ncfs/file/786581/download

Patankar, M.S. 2012. *Safety Culture: Building and Sustaining a Cultural Change in Aviation and Healthcare*. Surrey, UK: Ashgate Publishing, Ltd.

Robertson, J. 2012. "238: Risk Management." In *Encyclopedia of Forensic Sciences*, 2nd ed., eds. J. Siegel and P. Saukko. Amsterdam: Elsevier.

Shewhart, W. 2008. *Economic Control of Quality of Manufactured Products*. Milwaukee, WI: ASQ Press.

Snowden, D. and Boone, M. 2007. A leader's framework for decision making. *Harvard Business Review* 85(11): 88–92.

Straker, D. 2009. *Rapid Problem Solving with Post-It Notes*. Boston, MA: Da Capo Press.

Tamuz, M. and Harrison, M. I. 2006. Improving patient safety in hospitals: Contributions of high-reliability theory and normal accident theory. *Health Services Research* 41: 1654–1676.

Weick, K. E. and Sutcliffe, K. M. 2007. *Managing the Unexpected: Resilient Performance in and Age of Uncertainty*, 2nd ed. San Fransisco, CA: Jossey-Bass.

Index

Note: **Bold** page numbers refer to tables; *italic* page numbers refer to figures and page numbers followed by "n" denote endnotes.

For Product Safety Concerns and Information please contact our EU representative GPSR@taylorandfrancis.com Taylor & Francis Verlag GmbH, Kaufingerstraße 24, 80331 München, Germany